Blue Economy in Indian Sundarbans

Abhijit Mitra · Sufia Zaman · Prosenjit Pramanick

Blue Economy in Indian Sundarbans

Exploring Livelihood Opportunities

Springer

Abhijit Mitra
Department of Marine Science
University of Calcutta
Kolkata, West Bengal, India

Sufia Zaman
Department of Oceanography
Techno India University
Kolkata, West Bengal, India

Prosenjit Pramanick
Department of Oceanography
Techno India University
Kolkata, West Bengal, India

ISBN 978-3-031-07907-8 ISBN 978-3-031-07908-5 (eBook)
https://doi.org/10.1007/978-3-031-07908-5

This Springer imprint is published by the registered company Springer Nature Switzerland AG
The registered company address is: Gewerbestrasse 11, 6330 Cham, Switzerland

Acknowledgements

The authors from the innermost core of mind and heart acknowledge the following personalities for their direct and indirect contributions.

- **Late Professor Amalesh Choudhury**: All the authors developed the base on mangrove ecology from Professor Choudhury not only in the capacity of students but also as sincere disciples. He taught the authors to work in the rigorous environment of Sundarbans with sincerity, dedication and tireless effort to reveal the ground reality of the ecosystem.
- **Dr. Goutam Roychowdhury**: The authors greatly acknowledge the infrastructural facilities offered by Techno India University, West Bengal at Salt Lake Campus while preparing the manuscript. The authors also received inspiration for touching the sky from Sri Roychowdhury. The statistical software used to reflect the spatio-temporal variation in explaining the biotic community structure in the mangrove ecosystem was offered under the guidance of Sri Roychowdhury.

Dr. Abhijit Mitra expresses his gratefulness to his wife Shampa, daughter Ankita and late mother Manjulika whose inspirations and encouragements act as boosters to complete the manuscript. The statement of late Dhanesh Chandra Mitra, father of Dr. Abhijit Mitra, to create a strong footprint in life still boosts the author with a heavenly energy. The sudden demise of mother of Dr. Mitra during the COVID phase retarded the pace significantly, but with the active support of his well-wisher and Chancellor of Techno India University, West Bengal Dr. Goutam Roy Chowdhury, the author could finally serve the ground-zero data of Sundarban on the reader's plate.

Dr. Sufia Zaman expresses her deepest gratitude to her mother Mrs. Ayesha Zaman for her unconditional love, and practical day-to-day support, father Mr. Salim-uz-Zaman who gave her immense moral support. Dr. Zaman also acknowledges the support of her beloved husband Dr. Sahid Imam Mallick and her baby girl Shanza Mallick. She wishes to accord her deep sense of gratitude to her family members including her uncle (Mr. Pradip Kumar Mitra) and aunt (Late Mrs. Kanika Mitra), younger sister (Ms. Sharmilee Zaman), brother-in-law Kazi Dr. Sazzad Manir, her nephew baby Diyan Kazi, her in-laws and her beloved grandmother (Late Mrs. Shibani Dhar) for their encouragement and inspiration throughout the strenuous period of manuscript preparation. Dr. Zaman still recalls her granny late Shibani Dhar, who shaped her life and always wished to see her as an author of a subject that can create a mark on the society.

Dr. Prosenjit Pramanick expresses his gratefulness to his father Mr. Amar Sankar Pramanick, mother Mrs. Rekha Pramanick and aunt Ms. Benodini Samui who gave him unconditional love, inspiration, support and enormous encouragement. Dr. Pramanick also acknowledges his sister Mrs. Monalisha Ghosh, brother-in-law Mr. Nilkanta Ghosh and nephew Mr. Nilax Ghosh for their inspiration to complete the manuscript.

About This Book

It is not an exaggeration to say that the present-day world population demands food, employment and a healthy environment around to survive, but to achieve these targets, several dark chapters like unplanned industrialization, urbanization and deforestation will also open simultaneously. This will result in the deterioration of land resources and ethical values of life and already the countdowns have started. At this critical juncture, it is extremely important to cater information on various causes of environmental degradation and their proper conservation to people of all ranks of the society so that we can develop a sustainable blue economy for the betterment of the society and nation at large. We believe that such steps can trigger the sustainability of blue economy by integrating all the industrial sectors through a green belt with zero emission of GHGs.

The roots of different prospective sectors of blue economy in the lower Gangetic Delta is not restricted to India, but scattered and inter-linked with countries like Bangladesh, Sri Lanka, Myanmar and Thailand with combined population of more than 1.5 billion, which is 22% of the World population. It has been documented that about 200 million people in the Bay of Bengal region live in the coastal areas with a substantial proportion either partially or fully dependent on fisheries. The inshore and offshore regions of Bay of Bengal in the southern sector of the lower Gangetic delta are rich in living and non-living resources and sustain a large chunk of coastal population in terms of livelihood. The region with rich genetic diversity is the home of large number of flora and fauna including Royal Bengal tiger (*Panthera tigris tigris*), which has a high potential to initiate eco-tourism in the region. However, such ecotone of the World is today facing threats of various categories and magnitudes, which have been highlighted in several chapters through respondent analysis that serves as the ground-zero voice of the relevant stakeholders. The book critically analyzes every sector of blue economy relevant to the lower Gangetic delta through an in-depth SWOT analysis. The multiple activities in the coastal zone often pose adverse impacts on the fragile ecotone for which CRZ NOTIFICATION 2018 is one of the most important safeguarding policies followed in India. We present the notification here as stated by the concerned ministry along with other technological steps to sustain the blue economy in this deltaic complex.

We have tried to the best of our ability to make the concepts of the book comprehensive, up-to-date, backed up with field level data so that the book may serve as a ready source of information to the readers who want to pursue their careers in blue economy, environmental science, ecology, oceanography, marine science, fishery, economics, geography, coastal zone management and related subjects.

Contents

About the Authors

Dr. Abhijit Mitra Associate Professor and former Head, Dept. of Marine Science, University of Calcutta (INDIA) has been active in the sphere of Oceanography since 1985. He obtained his Ph.D. as NET qualified scholar in 1994. Since then he joined Calcutta Port Trust and WWF (World Wide Fund for Nature-India), in various capacities to carry out research programmes on environmental science, biodiversity conservation, climate change and carbon sequestration. Presently Dr. Mitra is serving as the Director, Research of Techno India University, West Bengal (Hony.). He has to his credit about 660 scientific publications in various National and International journals, and 58 books of postgraduate standards. Dr. Mitra is presently the member of several committees like PACON International, IUCN, SIOS, Mangrove Society of India, The West Bengal National University of Juridical Sciences etc. and has successfully completed about 17 projects on biodiversity loss in fishery sector, coastal pollution, aquaculture, alternative livelihood, climate change and carbon sequestration. Dr. Mitra also visited as a faculty member and invited speaker in several Universities of Singapore, Kenya, Oman and USA. In 2008, Dr. Mitra was invited as visiting fellow at the University of Massachusetts at Dartmouth, USA to deliver a series of lecture on climate change. Dr. Mitra also successfully guided 46 Ph.D. students. Presently his research areas include environmental science, mangrove ecology, sustainable aquaculture, alternative livelihood, climate change, carbon sequestration and blue economy.

Dr. Sufia Zaman presently serving as Head, Department of Oceanography in Techno India University, West Bengal started her career in the field of Marine Science since 2001. She worked in the rigorous region of Indian Sundarbans and has a wide range of experience in exploring the floral and faunal diversity of Sundarbans. She has published 5 books on carbon sequestration, 296 scientific papers and contributed chapters in several books on biodiversity, environmental science, aquaculture and livelihood development. Dr. Zaman is presently a member of Fisheries Society of India. She is also running projects on carbon sequestration by mangroves of Indian Sundarbans. She is the recipient of DST Women Scientist and Jawaharlal Memorial Doctoral fellowship awards. Her areas of research include aquaculture, fish nutrition, phytoplankton diversity, climate change mangrove ecology and alternative livelihood. Dr. Zaman is also the first researcher in the maritime state of West Bengal (India), who initiated trial experiments on iron fertilization and subsequent enhancement of primary (phytoplankton) and secondary (fish) productions in the brackish water ponds of Indian Sundarbans with the financial assistance of Department of Science and Technology, Government of India. Dr. Zaman is also providing consultancy on green technology to several industries, NGOs and corporate sectors.

Dr. Prosenjit Pramanick is presently holding the position of Post-Doctoral Researcher, Department of Oceanography, Techno India University, West Bengal. He had passed Bachelor of Science in Biochemistry in 2010 from the University of Calcutta, West Bengal and Master of Science in Biochemistry (in 2012) from Vidyasagar University, West Bengal and then obtained his Ph.D. degree in 2017 from Techno India University, West Bengal. He has to his credit about 106 scientific publications, 20 book chapters published by Springer International, Emerald Group, CSIR-NISCAIR, 1 Book (A, B, C of Environment) and 18 publications in conference proceedings in the sphere of Food technology, Aquaculture, Agribiotechnology, Alternative livelihood and Environmental Science. He has participated in various National and International conferences, seminars, workshops and webinars. He also served as an Associate Editor in an Inter-continental Webinar Proceedings on "Ecosystem Services and United Nations Sustainable Development Goals (UNSDG)" entitled "Natural Resources and their Ecosystem Services". He has expertise in water quality assessment (*in situ* and *ex situ*), food product preparation from mangrove floral species and carbon sequestration. He has served in several consultancy services on the carbon sequestration and Biodiversity assessment conducted by Konnagar Municipality, Government of West Bengal, Kingston Educational Institute, Forest Department, Government of West Bengal, TATA STEEL Ltd. Kalinganagar, Odisha, India and TATA STEEL Ltd. Noamundi Mines, Jharkhand, India. He also rendered his technological expertise as Research Associate in a project on innovative fish feed conducted by the Directorate of Fishery, Government of West Bengal.

Chapter 1
Blue Economy: An Overview

Abstract Blue Economy is a marine based economy that strengthens the economic backbone of the country by rational utilization of marine resources. It is an integration of the resource base of oceans, seas, bays and estuaries associated with societal development and environmental sustainability, coupled with innovative business models. The major sectors of blue economy are (i) marine living resources, (ii) marine non-living resources (like deep sea minerals, oil and natural gas), (iii) renewable natural forces encompassing wind, wave and tidal energy, (iv) shipping, (v) marine and coastal biodiversity, (vi) marine tourism etc. In addition, the endemic mangrove medicinal plants are often used to cure several diseases and thus provide immediate relief to the patients/sufferers in the coastal areas. Each of these sectors has unique opportunities to strengthen the GDP of the nations, which are addressed in this chapter with regional case studies. The present chapter highlights the possibility of India with a coastline of 7516.60 km to take the leadership role in the region based on the strong wings of innovation and low carbon emission in various sectors of blue economy.

1.1 Background of Blue Economy

The twenty-first century is experiencing an unprecedented growth in population throughout the World (Table 1.1) due to which land resources are depleting at a rapid pace. The exponential rise of population has also resulted in the degradation of environment and a big question has been raised on the issue of sustainability of resources. According to Richards (1986) population and environment are intricately related and this relationship is mediated by a number of socio-economic, cultural, political and developmental aspects.

The increase in population almost in all the countries of the World (with few exceptions denoted by '–' sign in Table 1.1) has posed an adverse impact on the forest resources of the World, which provide several ecosystem services that can be categorized into provisioning services (like timber, fuel wood, honey, fodder, wax, lac, medicinal plants etc.), regulating services (like flood control, control of erosion, carbon sequestration, oxygen generation, air purification, temperature, humidity,

A. Mitra et al., *Blue Economy in Indian Sundarbans*,
https://doi.org/10.1007/978-3-031-07908-5_1

Table 1.1 Country wise World's population during 2013 and 2021

Sl. No	Country	2013	2021	% of change
1	China	1,38,55,66,537	1,44,42,16,107	4.23
2	India	1,25,21,39,596	1,39,34,09,038	11.28
3	United States	32,00,50,716	33,29,15,073	4.02
4	Indonesia	24,98,65,631	27,63,61,783	10.60
5	Pakistan	18,21,42,594	22,51,99,937	23.64
6	Brazil	20,03,61,925	21,39,93,437	6.80
7	Nigeria	17,36,15,345	21,14,00,708	21.76
8	Bangladesh	15,65,94,962	16,63,03,498	6.20
9	Russia	14,28,33,689	14,59,12,025	2.16
10	Mexico	12,23,32,399	13,02,62,216	6.48
11	Japan	12,71,43,577	12,60,50,804	−0.86
12	Philippines	9,83,93,574	11,10,46,913	12.86
13	Egypt	8,20,56,378	10,42,58,327	27.06
14	Vietnam	9,16,79,733	9,81,68,833	7.08
15	Turkey	7,49,32,641	8,50,42,738	13.49
16	Iran	7,74,47,168	8,50,28,759	9.79
17	Germany	8,27,26,626	8,39,00,473	1.42
18	Thailand	6,70,10,502	6,99,50,850	4.39
19	United Kingdom	6,31,36,265	6,82,07,116	8.03
20	France	6,42,91,280	6,54,26,179	1.77
21	Tanzania	4,92,53,126	6,14,98,437	24.86
22	Italy	6,09,90,277	6,03,67,477	−1.02
23	South Africa	5,27,76,130	6,00,41,994	13.77
24	Kenya	4,43,53,691	5,49,85,698	23.97
25	Myanmar	5,32,59,018	5,48,06,012	2.90
26	Colombia	4,83,21,405	5,12,65,844	6.09
27	Uganda	3,75,78,876	4,71,23,531	25.40
28	Spain	4,69,26,963	4,67,45,216	−0.39
29	Argentina	4,14,46,246	4,56,05,826	10.04
30	Sudan	3,79,64,306	4,49,09,353	18.29
31	Algeria	3,92,08,194	4,46,16,624	13.79
32	Ukraine	4,52,38,805	4,34,66,819	−3.92
33	Iraq	3,37,65,232	4,11,79,350	21.96
34	Afghanistan	3,05,51,674	3,98,35,428	30.39
35	Canada	3,51,81,704	3,80,67,903	8.20
36	Poland	3,82,16,635	3,77,97,005	−1.10

(continued)

Table 1.1 (continued)

Sl. No	Country	2013	2021	% of change
37	Morocco	3,30,08,150	3,73,44,795	13.14
38	Saudi Arabia	2,88,28,870	3,53,40,683	22.59
39	Uzbekistan	2,89,34,102	3,39,35,763	17.29
40	Angola	2,14,71,618	3,39,33,610	58.04
41	Peru	3,03,75,603	3,33,59,418	9.82
42	Malaysia	2,97,16,965	3,27,76,194	10.29
43	Mozambique	2,58,33,752	3,21,63,047	24.50
44	Ghana	2,59,04,598	3,17,32,129	22.50
45	Yemen	2,44,07,381	3,04,90,640	24.92
46	Nepal	2,77,97,457	2,96,74,920	6.75
47	Venezuela	3,04,05,207	2,87,04,954	−5.59
48	Madagascar	2,29,24,851	2,84,27,328	24.00
49	Cameroon	2,22,53,959	2,72,24,265	22.33
50	Australia	2,33,42,553	2,57,88,215	10.48
51	Niger	1,78,31,270	2,51,30,817	40.94
52	Sri Lanka	2,12,73,228	2,14,97,310	1.05
53	Burkina Faso	1,69,34,839	2,14,97,096	26.94
54	Mali	1,53,01,650	2,08,55,735	36.30
55	Malawi	1,63,62,567	1,96,47,684	20.08
56	Chile	1,76,19,708	1,92,12,361	9.04
57	Romania	2,16,98,585	1,91,27,774	−11.85
58	Kazakhstan	1,64,40,586	1,89,94,962	15.54
59	Zambia	1,45,38,640	1,89,20,651	30.14
60	Guatemala	1,54,68,203	1,82,49,860	17.98
61	Ecuador	1,57,37,878	1,78,88,475	13.67
62	Senegal	1,41,33,280	1,71,96,301	21.67
63	Netherlands	1,67,59,229	1,71,73,099	2.47
64	Cambodia	1,51,35,169	1,69,46,438	11.97
65	Chad	1,28,25,314	1,69,14,985	31.89
66	Somalia	1,04,95,583	1,63,59,504	55.87
67	Zimbabwe	1,41,49,648	1,50,92,171	6.66
68	Guinea	1,17,45,189	1,34,97,244	14.92
69	Rwanda	1,17,76,522	1,32,76,513	12.74
70	Benin	1,03,23,474	1,24,51,040	20.61
71	Burundi	1,01,62,532	1,22,55,433	20.59
72	Tunisia	1,09,96,515	1,19,35,766	8.54

(continued)

Table 1.1 (continued)

Sl. No	Country	2013	2021	% of change
73	Bolivia	1,06,71,200	1,18,32,940	10.89
74	Belgium	1,11,04,476	1,16,32,326	4.75
75	Haiti	1,03,17,461	1,15,41,685	11.87
76	South Sudan	1,12,96,173	1,13,81,378	0.75
77	Cuba	1,12,65,629	1,13,17,505	0.46
78	Dominican Republic	1,04,03,761	1,09,53,703	5.29
79	Greece	1,11,27,990	1,03,70,744	−6.80
80	Jordan	72,73,799	1,02,69,021	41.18
81	Azerbaijan	94,13,420	1,02,23,342	8.60
82	Portugal	1,06,08,156	1,01,67,925	−4.15
83	Sweden	95,71,105	1,01,60,169	6.15
84	Honduras	80,97,688	1,00,62,991	24.27
85	United Arab Emirates	93,46,129	99,91,089	6.90
86	Tajikistan	82,07,834	97,49,627	18.78
87	Hungary	99,54,941	96,34,164	−3.22
88	Belarus	93,56,678	94,42,862	0.92
89	Papua New Guinea	73,21,262	91,19,010	24.56
90	Austria	84,95,145	90,43,070	6.45
91	Israel	77,33,144	87,89,774	13.66
92	Switzerland	80,77,833	87,15,494	7.89
93	Serbia	95,10,506	86,97,550	−8.55
94	Togo	68,16,982	84,78,250	24.37
95	Sierra Leone	60,92,075	81,41,343	33.64
96	Paraguay	68,02,295	72,19,638	6.14
97	Libya	62,01,521	69,58,532	12.21
98	Bulgaria	72,22,943	68,96,663	−4.52
99	Lebanon	48,21,971	67,69,146	40.38
100	Nicaragua	60,80,478	67,02,385	10.23
101	Kyrgyzstan	55,47,548	66,28,356	19.48
102	El Salvador	63,40,454	65,18,499	2.81
103	Turkmenistan	52,40,072	61,17,924	16.75
104	Singapore	54,11,737	58,96,686	8.96
105	Denmark	56,19,096	58,13,298	3.46
106	Finland	54,26,323	55,48,360	2.25
107	Norway	50,42,671	54,65,630	8.39
108	Slovakia	54,50,223	54,60,721	0.19

(continued)

Table 1.1 (continued)

Sl. No	Country	2013	2021	% of change
109	Oman	36,32,444	52,23,375	43.80
110	Palestine	43,26,295	52,22,748	20.72
111	Liberia	42,94,077	51,80,203	20.64
112	Costa Rica	48,72,166	51,39,052	5.48
113	Ireland	46,27,173	49,82,907	7.69
114	Central African Republic	46,16,417	49,19,981	6.58
115	New Zealand	45,05,761	48,60,643	7.88
116	Mauritania	38,89,880	47,75,119	22.76
117	Panama	38,64,170	43,81,579	13.39
118	Kuwait	33,68,572	43,28,550	28.50
119	Croatia	42,89,714	40,81,651	−4.85
120	Moldova	34,87,204	40,24,019	15.39
121	Georgia	43,40,895	39,79,765	−8.32
122	Eritrea	63,33,135	36,01,467	−43.13
123	Uruguay	34,07,062	34,85,151	2.29
124	Mongolia	28,39,073	33,29,289	17.27
125	Bosnia And Herzegovina	38,29,307	32,63,466	−14.78
126	Jamaica	27,83,888	29,73,463	6.81
127	Armenia	29,76,566	29,68,127	−0.28
128	Qatar	21,68,673	29,30,528	35.13
129	Albania	31,73,271	28,72,933	−9.46
130	Puerto Rico	36,88,318	28,28,255	−23.32
131	Lithuania	30,16,933	26,89,862	−10.84
132	Namibia	23,03,315	25,87,344	12.33
133	Gambia	18,49,285	24,86,945	34.48
134	Botswana	20,21,144	23,97,241	18.61
135	Gabon	16,71,711	22,78,825	36.32
136	Lesotho	20,74,465	21,59,079	4.08
137	Slovenia	20,71,997	20,78,724	0.32
138	Guinea Bissau	17,04,255	20,15,494	18.26
139	Latvia	20,50,317	18,66,942	−8.94
140	Bahrain	13,32,171	17,48,296	31.24
141	Equatorial Guinea	7,57,014	14,49,896	91.53
142	Trinidad And Tobago	13,41,151	14,03,375	4.64
143	Timor Leste	11,32,879	13,43,873	18.62
144	Estonia	12,87,251	13,25,185	2.95

(continued)

Table 1.1 (continued)

Sl. No	Country	2013	2021	% of change
145	Mauritius	12,44,403	12,73,433	2.33
146	Cyprus	11,41,166	12,15,584	6.52
147	Djibouti	8,72,932	10,02,187	14.81
148	Fiji	8,81,065	9,02,906	2.48
149	Reunion	8,75,375	9,01,686	3.01
150	Comoros	7,34,917	8,88,451	20.89
151	Guyana	7,99,613	7,90,326	−1.16
152	Bhutan	7,53,947	7,79,898	3.44
153	Luxembourg	5,30,380	6,34,814	19.69
154	Montenegro	6,21,383	6,28,053	1.07
155	Western Sahara	5,67,315	6,11,875	7.85
156	Suriname	5,39,276	5,91,800	9.74
157	Maldives	3,45,023	5,43,617	57.56
158	Malta	4,29,004	4,42,784	3.21
159	Brunei	4,17,784	4,41,532	5.68
160	Belize	3,31,900	4,04,914	22.00
161	Guadeloupe	4,65,800	4,00,020	−14.12
162	Bahamas	3,77,374	3,96,913	5.18
163	Martinique	4,03,682	3,74,745	−7.17
164	Iceland	3,29,535	3,43,353	4.19
165	Vanuatu	2,52,763	3,14,464	24.41
166	French Guiana	2,49,227	3,06,448	22.96
167	New Caledonia	2,56,496	2,88,218	12.37
168	Barbados	2,84,644	2,87,711	1.08
169	French Polynesia	2,76,831	2,82,530	2.06
170	Mayotte	2,22,152	2,79,515	25.82
171	Samoa	1,90,372	2,00,149	5.14
172	Guam	1,65,124	1,70,179	3.06
173	Curacao	1,58,760	1,64,798	3.80
174	Kiribati	1,02,351	1,21,392	18.60
175	Micronesia	1,03,549	1,16,254	12.27
176	Grenada	1,05,897	1,13,021	6.73
177	Aruba	1,02,911	1,07,204	4.17
178	Tonga	1,05,323	1,06,760	1.36
179	Seychelles	92,838	98,908	6.54
180	Antigua and Barbuda	89,985	98,731	9.72

(continued)

Table 1.1 (continued)

Sl. No	Country	2013	2021	% of change
181	Isle of Man	85,888	85,410	−0.56
182	Andorra	79,218	77,355	−2.35
183	Dominica	72,003	72,167	0.23
184	Bermuda	65,341	62,090	−4.98
185	Marshall Islands	52,634	59,610	13.25
186	Greenland	56,987	56,877	−0.19
187	American Samoa	55,165	55,100	−0.12
188	Sint Maarten	45,233	43,412	−4.03
189	Monaco	37,831	39,511	4.44
190	Liechtenstein	36,925	38,250	3.59
191	San Marino	31,448	34,017	8.17
192	Gibraltar	29,310	33,698	14.97
193	Palau	20,918	18,169	−13.14
194	Anguilla	14,300	15,117	5.71
195	Tuvalu	9,876	11,931	20.81
196	Wallis And Futuna	13,272	11,094	−16.41
197	Nauru	10,051	10,876	8.21
198	Montserrat	5,091	4,977	−2.24
199	Falkland Islands	3,044	3,533	16.06
200	Tokelau	1,195	1,373	14.90

Source https://www.worldometers.info/world-population/world-population-countries.php https://worldpopulationreview.com/countries; '–' indicates a decrease in population

nutrient cycling, bioremediation etc.), supporting services (like floral and faunal biodiversity, seed dispersal, life cycle of organisms etc.) and cultural services (like recreation, use of traditional resources and spirituality). In spite of providing a wide array of ecosystem services for the benefit of mankind, the forest area is depleting fast throughout the World (Fig. 1.1 and Table 1.2) to switch over to a new *era* of industrial development, urban expansion, aquaculture, tourism and several miscellaneous activities to fulfill the needs of human beings.

Table 1.3 reflects the country-wise forest area (in hectares), where India occupies the 10th position with a per capita forest area of 531 m^2.

The over exploitation of forest/land resources have pushed the human civilization to touch the boundary wall of the resource base of the planet Earth. Hence it is perhaps the need of the hour to think of the blue economy, which is exclusively marine based economy encompassing the living and non-living resources of the oceans, seas, bays and estuaries. Covering 71% of the Earth's surface, the saline water of the marine and estuarine ecosystems embedded in different compartments exhibits substantial area and volume (Table 1.4) with rich reservoir of resources.

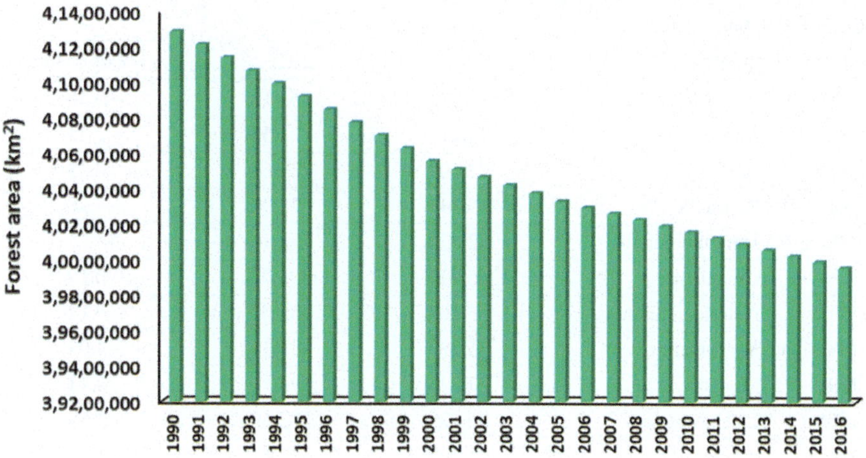

Fig. 1.1 Depletion of world forest area (in km^2) with the passage of time

Considering the resource base of the marine and estuarine ecosystems it has been observed that the wheels of modern civilization can be run by selective productive sectors that can not only generate job opportunities, but can also provide food security, health security and economic stability. In addition, some of these sectors play crucial role in the process of upgradation of environment as many of the marine based livelihoods are intricately related to low carbon emission. Details of such productive sectors of blue economy are presented in Table 1.5.

The different sectors of blue economy provide both direct and indirect job opportunities and have huge prospects in strengthening the economic backbone of the nations.

1.2 Blue Economy: Indian Scenario

India has a coastline of 8,129 km encompassing 3,600 fishing villages. The EEZ covers an area of 2.03 million km^2. The marine and estuarine ecosystems of the country support rich taxonomic diversity of the flora and fauna, as revealed from the number of species under different groups (Table 1.6).

The major non-living sectors in blue economy that have the potential to strengthen the economic backbone of the nation are (i) Transportation in the Oceans, Seas, Bays and Estuaries, (ii) Ship Breaking and Ship Waste Management, (iii) Marine and Coastal constructions, (v) Marine Based Energy (vi) Oil and Natural gas extraction and (vii) Deep Sea Minerals.

Table 1.2 Country-wise change in forest area in (km^2)

Country	2011	2020	% of change
Afghanistan	12084.4	12084.4	0.0000
Angola	716029.38	666073.8	−6.9768
Albania	7834.935	7889	0.6901
Arab World	389730.66	367980.7	−5.5808
United Arab Emirates	3173	3173	0.0000
Argentina	299,906	285,730	−4.7268
Armenia	3303.52	3284.7	−0.5697
Antigua and Barbuda	87.14	81.2	−6.8166
Australia	1302557.8	1,340,051	2.8784
Austria	38667.98	38991.5	0.8367
Azerbaijan	10415.731	11317.7	8.6597
Burundi	2110.8	2796.4	32.4806
Belgium	6897.56	6893	−0.0661
Benin	35851.5	31351.5	−12.5518
Burkina Faso	66664.8	62164	−6.7514
Bangladesh	18873.536	18,834	−0.2095
Bulgaria	37562	38,930	3.6420
Bahrain	5.36	7	30.5970
Bahamas	5098.6	5098.6	0.0000
Bosnia and Herzegovina	21142.28	21879.1	3.4851
Belarus	86,307	87,676	1.5862
Belize	13796.78	12770.5	−7.4385
Bermuda	10	10	0.0000
Bolivia	528742.52	508337.6	−3.8591
Brazil	5100415.2	4966196	−2.6315
Barbados	63	63	0.0000
Brunei Darussalam	3800	3800	0.0000
Bhutan	27072.68	27250.8	0.6579
Botswana	163,194	152,547	−6.5241
Central African Republic	225,730	223,030	−1.1961
Canada	3472809.1	3,469,281	−0.1016
Central Europe and the Baltics	373931.3	381881.8	2.1262
Switzerland	12381.58	12691.1	2.4998
Channel Islands	10.2		−100.0000
Chile	168994.24	182,107	7.7593
China	2025471.54	2199781.8	8.6059

(continued)

Table 1.2 (continued)

Country	2011	2020	% of change
Cote d'Ivoire	38526.94	28367.1	−26.3707
Cameroon	208444.8	203404.8	−2.4179
Congo, Dem. Rep	1360676.24	1261552.4	−7.2849
Congo, Rep	220,630	219,460	−0.5303
Colombia	606731.64	591419.1	−2.5238
Comoros	368.62	329.2	−10.6939
Cabo Verde	430.2	457.2	6.2762
Costa Rica	28875.66	30348.7	5.1013
Caribbean small states	366230.12	363639.2	−0.7075
Cuba	29,824	32,420	8.7044
Curacao	0.7	0.7	0.0000
Cayman Islands	127.2	127.2	0.0000
Cyprus	1728.14	1725.3	−0.1643
Czech Republic	26595.82	26770.9	0.6583
Germany	114,110	114,190	0.0701
Djibouti	56	58	3.5714
Dominica	478.7	478.7	0.0000
Denmark	5941.28	6284.4	5.7752
Dominican Republic	20792.26	21,441	3.1201
Algeria	19,256	19,490	1.2152
Egypt, Arab Rep	621.96	449.8	−27.6802
Euro area	1019686.45	1036486.8	1.6476
Eritrea	10,837	10552.6	−2.6243
Spain	185465.08	185721.7	0.1384
Estonia	23530.18	24,384	3.6286
Ethiopia	177,255	170,685	−3.7065
European Union	1569420.23	1592313.8	1.4587
Finland	222,754	224,090	0.5998
Fiji	10799.16	11400.2	5.5656
France	165,024	172,530	4.5484
Gabon	236375.28	235,306	−0.4524
United Kingdom	30,782	31,900	3.6320
Georgia	28,224	28,224	0.0000
Ghana	79304.32	79857.1	0.6970
Guinea	65,330	61,890	−5.2656
Gambia	2942.68	2426.7	−17.5344
Guinea–Bissau	20559.86	19800.1	−3.6954

(continued)

Table 1.2 (continued)

Country	2011	2020	% of change
Equatorial Guinea	25236.52	24484.2	−2.9811
Greece	39018.03	39,018	−0.0001
Grenada	177	177	0.0000
Greenland	2.2	2.2	0.0000
Guatemala	36952.4	35,278	−4.5312
Guam	242	280	15.7025
Guyana	185081.2	184153.4	−0.5013
Honduras	65531.42	63592.6	−2.9586
Croatia	19,204	19391.1	0.9743
Haiti	3752.74	3473	−7.4543
Hungary	20492.76	20530.1	0.1822
Indonesia	987329.4	921,332	−6.6844
India	697624	721,600	3.4368
Ireland	7272.38	7820.2	7.5329
Iran, Islamic Rep	106919.8	107518.7	0.5601
Iraq	8250	8250	0.0000
Iceland	453.68	513.5	13.1855
Israel	1562	1400	−10.3713
Italy	90818.48	95661.3	5.3324
Jamaica	5622.52	5968.9	6.1606
Jordan	975	975	0.0000
Japan	249,616	249,350	−0.1066
Kazakhstan	31274.349	34546.8	10.4637
Kenya	35975.54	36110.9	0.3763
Kyrgyz Republic	12341.06	13153.8	6.5857
Cambodia	102407.48	80683.7	−21.2131
Kiribati	11.8	11.8	0.0000
Korea, Rep	63,770	62,870	−1.4113
Kuwait	62.5	62.5	0.0000
Lebanon	1379.46	1433.3	3.9030
Liberia	78897.8	76174.4	−3.4518
Libya	2170	2170	0.0000
St. Lucia	207.7	207.7	0.0000
Liechtenstein	66.994	67	0.0090
Sri Lanka	21086.6	21130.2	0.2068
Luxembourg	887	887	0.0000
Latvia	33759.84	34107.9	1.0310

(continued)

Table 1.2 (continued)

Country	2011	2020	% of change
Morocco	56765.94	57424.9	1.1608
Moldova	3768.8	3865	2.5525
Madagascar	125487.62	124298.1	−0.9479
Mexico	668208.3	656920.8	−1.6892
Middle income	27076764.5	26768858.3	−1.1372
North Macedonia	9672.24	10014.9	3.5427
Mali	132,960	132,960	0.0000
Myanmar	311512.88	285438.9	−8.3701
Mongolia	141827.86	141727.8	−0.0706
Northern Mariana Islands	301.54	243.6	−19.2147
Mozambique	387657.12	367437.6	−5.2158
Mauritania	3618	3128	−13.5434
Mauritius	383.72	387.7	1.0372
Malawi	26,197	22,417	−14.4291
Malaysia	190509.64	191140.4	0.3311
North America	6562769.1	6,567,241	0.0681
Namibia	72779.82	66,389	−8.7810
New Caledonia	8389.2	8380.2	−0.1073
Niger	11914.8	10,797	−9.3816
Nigeria	230966.74	216269.5	−6.3634
Nicaragua	41320.26	34075.3	−17.5337
Netherlands	3717.5	3695	−0.6052
Norway	121,098	121,800	0.5797
Nepal	59620.3	59620.3	0.0000
New Zealand	98478.18	98925.9	0.4546
OECD members	11573194.09	11,632,704	0.5142
Oman	30	25	−16.6667
Other small states	602668.88	583546.3	−3.1730
Pakistan	40615.04	37,259	−8.2630
Panama	43165.74	42138.4	−2.3800
Peru	738787.46	723303.7	−2.0958
Philippines	68746.06	71885.9	4.5673
Palau	406.46	414.1	1.8796
Papua New Guinea	361479.924	358557.6	−0.8084
Poland	93,472	94,830	1.4528
Puerto Rico	4919.36	4963.3	0.8932

(continued)

Table 1.2 (continued)

Country	2011	2020	% of change
Korea, Dem. People's Rep	62211.6	60300.9	−3.0713
Portugal	32,640	33,120	1.4706
Paraguay	191559.56	161022.6	−15.9412
Pacific island small states	43431.04	43933.2	1.1562
Romania	65921.92	69290.5	5.1100
Russian Federation	8150945.72	8,153,116	0.0266
Rwanda	2660	2760	3.7594
South Asia	876984.756	897786.9	2.3720
Sudan	199069.22	183595.5	−7.7730
Senegal	84281.6	80681.6	−4.2714
Singapore	174.866	155.7	−10.9604
Solomon Islands	25295.48	25229.7	−0.2600
Sierra Leone	27124.34	25348.8	−6.5459
El Salvador	6243.8	5838.8	−6.4864
Somalia	66707.5	59,800	−10.3549
Serbia	27143.06	27226.5	0.3074
Sub-Saharan Africa (excluding high income)	6627233.26	6271638.7	−5.3657
Sub-Saharan Africa	6627570.26	6271975.7	−5.3654
Small states	1012330.04	991118.7	−2.0953
Sao Tome and Principe	574.8	519	−9.7077
Suriname	152902.22	151962.9	−0.6143
Slovak Republic	19186.78	19,259	0.3764
Slovenia	12,472	12378.3	−0.7513
Sweden	280,544	279,800	−0.2652
Eswatini	4866.44	4975.6	2.2431
Syrian Arab Republic	4980.8	5220.8	4.8185
Chad	54,020	43,130	−20.1592
East Asia & Pacific (IDA & IBRD countries)	4747047.024	4814808.2	1.4274
Europe & Central Asia (IDA & IBRD countries)	9003715.055	9034413.3	0.3410
Togo	12359.1	12092.7	−2.1555
Thailand	200,706	198,730	−0.9845
Tajikistan	4123.6	4238	2.7743
Turkmenistan	41,270	41,270	0.0000
Latin America & the Caribbean (IDA & IBRD countries)	9,532,008	9277297.5	−2.6722
Timor-Leste	9337	9211	−1.3495

(continued)

Table 1.2 (continued)

Country	2011	2020	% of change
Middle East & North Africa (IDA & IBRD countries)	213754.56	215507.8	0.8202
Tonga	89.5	89.5	0.0000
South Asia (IDA & IBRD)	876984.756	897786.9	2.3720
Sub-Saharan Africa (IDA & IBRD countries)	6627570.26	6271975.7	−5.3654
Trinidad and Tobago	2319.86	2281.9	−1.6363
Tunisia	6889.6	7027.3	1.9987
Turkey	211925.24	222203.6	4.8500
Tuvalu	10	10	0.0000
Tanzania	495780.08	457,450	−7.7313
Uganda	27091.68	23,379	−13.7041
Ukraine	95,698	96,900	1.2560
Uruguay	17690.4	20,310	14.8080
United States	3,089,950	3,097,950	0.2589
Uzbekistan	33895.6	36896.6	8.8537
Venezuela, RB	473,406	462,309	−2.3441
British Virgin Islands	36.36	36.2	−0.4400
Virgin Islands (U.S.)	185.76	199.1	7.1813
Vietnam	135228.2	146430.9	8.2843
Vanuatu	4423	4423	0.0000
Samoa	1660.06	1616.7	−2.6120
South Africa	173776.9	170500.9	−1.8852
Zambia	465078.2	448140.3	−3.6419
Zimbabwe	178592.1	174445.8	−2.3217

1.2.1 Marine Living Resources in Context to Blue Economy

Marine and estuarine organisms act as the foundation stone of blue economy. However, we have restricted ourselves only to the fisheries (both fin fish and shell fish) and aquaculture, coral reef and coastal vegetation (like mangroves, seagrasses, and saltmarsh grasses) that have great importance in the blue economic vertical of India.

Marine Fisheries and Aquaculture

The marine and estuarine ecosystems are the survival ground of a large number of fishes. Approximately 18,000 known species of fish with economic value have been so far documented from the ocean water. Fishermen throughout the World

Table 1.3 List of countries with forest area as per 2017 record

Sl. No	Country	Forest area (ha)	Population	m^2 per capita
1	Russia	81,48,48,460	14,55,30,082	55,992
2	Brazil	49,15,70,000	20,78,33,823	23,652
3	Canada	34,69,75,800	3,67,32,095	94,461
4	United States	31,06,45,000	32,50,84,756	9,556
5	China	21,14,05,700	1,42,10,21,791	1,488
6	DR Congo	15,19,55,200	8,13,98,764	18,668
7	Australia	12,53,67,000	2,45,84,620	50,994
8	Indonesia	8,96,41,200	26,46,50,963	3,387
9	Peru	7,36,37,800	3,14,44,298	23,418
10	India	7,10,38,800	1,33,86,76,785	531
11	Mexico	6,58,56,800	12,47,77,324	5,278
12	Colombia	5,84,48,300	4,89,09,839	11,950
13	Angola	5,76,06,400	2,98,16,766	19,320
14	Bolivia	5,41,86,000	1,11,92,855	48,411
15	Zambia	4,83,01,800	1,68,53,599	28,660
16	Venezuela	4,63,54,200	2,94,02,484	15,765
17	Tanzania	4,53,16,000	5,46,60,339	8,290
18	Mozambique	3,75,27,200	2,86,49,018	13,099
19	Papua New Guinea	3,35,53,400	84,38,036	39,764
20	Sweden	2,80,73,000	99,04,896	28,343
21	Myanmar	2,79,48,200	5,33,82,523	5,235
22	Argentina	2,65,18,400	4,39,37,140	6,036
23	Japan	2,49,54,800	12,75,02,725	1,957
24	Gabon	2,34,00,000	20,64,823	1,13,327
25	Congo	2,23,03,200	51,10,695	43,640
26	Malaysia	2,22,23,400	3,11,04,646	7,145
27	Finland	2,22,18,000	55,11,371	40,313
28	Central African Republic	2,21,38,800	45,96,023	48,169
29	Laos	1,91,39,750	69,53,035	27,527
30	Sudan	1,88,61,120	4,08,13,397	4,621
31	Spain	1,84,86,130	4,66,47,428	3,963
32	Cameroon	1,83,76,000	2,45,66,073	7,480
33	Chile	1,83,36,600	1,84,70,439	9,928
34	France	1,72,15,000	6,48,42,509	2,655
35	Guyana	1,65,06,000	7,75,222	2,12,920
36	Thailand	1,64,59,000	6,92,09,810	2,378

(continued)

Table 1.3 (continued)

Sl. No	Country	Forest area (ha)	Population	m^2 per capita
37	Suriname	1,53,24,400	5,70,496	2,68,615
38	Vietnam	1,50,31,000	9,46,00,648	1,589
39	Paraguay	1,46,72,200	68,67,061	21,366
40	Zimbabwe	1,34,37,200	1,42,36,595	9,438
41	Ethiopia	1,25,80,200	10,63,99,924	1,182
42	Madagascar	1,24,41,000	2,55,70,512	4,865
43	Ecuador	1,23,90,420	1,67,85,361	7,382
44	Mongolia	1,23,58,240	31,13,786	39,689
45	Norway	1,21,16,000	52,96,326	22,876
46	Turkey	1,19,19,800	8,11,16,450	1,469
47	Germany	1,14,23,000	8,26,58,409	1,382
48	Iran	1,06,91,980	8,06,73,883	1,325
49	Botswana	1,06,35,600	22,05,080	48,232
50	Côte d'Ivoire	1,04,00,200	2,44,37,470	4,256
51	New Zealand	1,01,52,400	47,02,034	21,592
52	Ukraine	97,00,600	4,44,87,709	2,181
53	Poland	94,77,400	3,79,53,180	2,497
54	Italy	94,04,600	6,06,73,701	1,550
55	Ghana	93,93,800	2,91,21,465	3,226
56	South Africa	92,41,000	5,70,09,756	1,621
57	Cambodia	92,02,200	1,60,09,409	5,748
58	Belarus	86,73,300	94,50,231	9,178
59	Philippines	85,20,000	10,51,72,925	810
60	Senegal	81,93,000	1,54,19,355	5,313
61	French Guiana	81,26,800	2,75,191	2,95,315
62	South Sudan	71,57,000	1,09,10,763	6,560
63	Romania	69,99,400	1,96,53,969	3,561
64	Namibia	67,70,600	24,02,633	28,180
65	Guinea	62,92,000	1,20,67,519	5,214
66	Somalia	62,09,400	1,45,89,179	4,256
67	Nigeria	61,73,800	19,08,73,244	323
68	South Korea	61,68,800	5,10,96,415	1,207
69	Morocco	56,16,000	3,55,81,255	1,578
70	Burkina Faso	52,30,400	1,91,93,234	2,725
71	North Korea	47,77,000	2,54,29,825	1,879
72	Chad	46,21,800	1,50,16,753	3,078

(continued)

Table 1.3 (continued)

Sl. No	Country	Forest area (ha)	Population	m^2 per capita
73	Panama	45,84,200	41,06,769	11,163
74	Mali	45,57,000	1,85,12,430	2,462
75	Kenya	44,86,200	5,02,21,142	893
76	Honduras	43,52,000	94,29,013	4,616
77	Benin	42,11,000	1,11,75,198	3,768
78	Turkmenistan	41,27,000	57,57,667	7,168
79	Liberia	41,19,000	47,02,226	8,760
80	Greece	41,14,400	1,05,69,450	3,893
81	Austria	38,72,600	88,19,901	4,391
82	Bulgaria	38,57,400	71,02,444	5,431
83	Nepal	36,36,000	2,76,32,681	1,316
84	Guatemala	34,67,200	1,69,14,970	2,050
85	Latvia	33,56,800	19,51,097	17,205
86	Kazakhstan	33,09,000	1,80,80,019	1,830
87	Cuba	33,07,200	1,13,39,254	2,917
88	Uzbekistan	31,97,660	3,19,59,785	1,001
89	United Kingdom	31,78,000	6,67,27,461	476
90	Sierra Leone	31,71,200	74,88,423	4,235
91	Portugal	31,59,200	1,02,88,527	3,071
92	Nicaragua	31,14,000	63,84,846	4,877
93	Malawi	31,11,000	1,76,70,196	1,761
94	Georgia	28,22,400	40,08,716	7,041
95	Costa Rica	28,16,400	49,49,954	5,690
96	Bhutan	27,74,770	7,45,563	37,217
97	Serbia	27,22,800	88,29,628	3,084
98	Czech Republic (Czechia)	26,71,000	1,06,41,034	2,510
99	Estonia	22,31,200	13,19,390	16,911
100	Bosnia and Herzegovina	21,85,000	33,51,525	6,519
100	Lithuania	21,84,000	28,45,414	7,676
101	Solomon Islands	21,73,800	6,36,039	34,177
102	Hungary	20,78,200	97,29,823	2,136
103	Sri Lanka	20,56,800	2,11,28,032	973
104	Dominican Republic	20,49,400	1,05,13,104	1,949
105	Algeria	19,71,200	4,13,89,189	476
106	Guinea-Bissau	19,52,000	18,28,145	10,677
107	Slovakia	19,40,400	54,47,900	3,562

(continued)

Table 1.3 (continued)

Sl. No	Country	Forest area (ha)	Population	m^2 per capita
108	Croatia	19,22,800	41,82,857	4,597
109	Uruguay	18,90,480	34,36,641	5,501
110	Uganda	18,06,600	4,11,66,588	439
111	Equatorial Guinea	15,44,800	12,62,002	12,241
112	Eritrea	15,01,200	34,12,893	4,399
113	Bangladesh	14,23,800	15,96,85,424	89
114	Pakistan	13,86,000	20,79,06,209	67
115	Belize	13,56,260	3,75,769	36,093
116	Afghanistan	13,50,000	3,62,96,113	372
117	Switzerland	12,61,600	84,55,804	1,492
118	Slovenia	12,48,400	20,76,394	6,012
119	Azerbaijan	11,91,840	98,45,320	1,211
120	Niger	11,17,200	2,16,02,382	517
121	Tunisia	10,61,400	1,14,33,443	928
122	Fiji	10,26,920	8,77,459	11,703
123	North Macedonia	9,98,000	20,81,996	4,793
124	Saudi Arabia	9,77,000	3,31,01,179	295
125	New Caledonia	8,39,000	2,77,150	30,272
126	Montenegro	8,27,000	6,27,563	13,178
127	Iraq	8,25,000	3,75,52,781	220
128	Albania	7,69,580	28,84,169	2,668
129	Ireland	7,65,380	47,53,279	1,610
130	Western Sahara	7,07,000	5,52,615	12,794
131	Belgium	6,84,280	1,14,19,748	599
132	Timor-Leste	6,63,600	12,43,258	5,338
133	Denmark	6,22,240	57,32,274	1,086
134	Kyrgyzstan	6,21,000	61,89,733	1,003
135	Eswatini	5,95,200	11,24,805	5,292
136	Yemen	5,49,000	2,78,34,819	197
137	Bahamas	5,15,000	3,81,755	13,490
138	Puerto Rico	5,02,570	31,63,667	1,589
139	Rwanda	4,93,600	1,19,80,961	412
140	Gambia	4,91,200	22,13,889	2,219
141	Syria	4,91,000	1,70,95,678	287
142	Vanuatu	4,40,000	2,85,510	15,411
143	Moldova	4,18,200	40,59,684	1,030

(continued)

Table 1.3 (continued)

Sl. No	Country	Forest area (ha)	Population	m^2 per capita
144	Tajikistan	4,12,800	88,80,268	465
145	Brunei	3,80,000	4,24,473	8,952
146	Netherlands	3,77,200	1,70,21,347	222
147	Jamaica	3,34,440	29,20,848	1,145
148	Armenia	3,32,400	29,44,791	1,129
149	United Arab Emirates	3,24,720	94,87,203	342
150	Burundi	2,85,200	1,08,27,019	263
151	El Salvador	2,56,200	63,88,126	401
152	Trinidad and Tobago	2,37,720	13,84,059	1,718
153	Mauritania	2,17,500	42,82,570	508
154	Libya	2,17,000	65,80,724	330
155	Cyprus	1,72,640	11,79,678	1,463
156	Samoa	1,71,000	1,95,352	8,753
157	Israel	1,69,400	82,43,848	205
158	French Polynesia	1,55,000	2,76,102	5,614
159	Togo	1,48,400	76,98,474	193
160	Lebanon	1,37,460	68,19,373	202
161	Jordan	97,500	97,85,843	100
162	Haiti	95,400	1,09,82,366	87
163	Cabo Verde	91,820	5,37,498	1,708
164	Réunion	88,000	8,76,134	1,004
165	Luxembourg	86,700	5,91,910	1,465
166	Egypt	74,200	9,64,42,591	8
167	Guadeloupe	72,560	3,99,672	1,815
168	Micronesia	64,330	5,32,899	1,207
169	Sao Tome & Principe	53,600	2,07,089	2,588
170	Iceland	51,800	3,34,393	1,549
171	Lesotho	51,000	20,91,534	244
172	Martinique	48,500	3,75,948	1,290
173	Dominica	42,790	71,458	5,988
174	Seychelles	40,670	96,418	4,218
175	Palau	40,300	17,808	22,630
176	Mauritius	38,680	12,64,499	306
177	Comoros	36,200	8,13,892	445
178	Turks and Caicos	34,400	37,115	9,268
179	Northern Mariana Islands	29,180	56,562	5,159

(continued)

Table 1.3 (continued)

Sl. No	Country	Forest area (ha)	Population	m^2 per capita
180	St. Vincent & Grenadines	27,000	1,09,827	2,458
181	Guam	25,000	1,64,281	1,522
182	Saint Lucia	20,180	1,80,954	1,115
183	Niue	17,900	1,609	1,11,249
184	American Samoa	17,460	55,620	3,139
185	U.S. Virgin Islands	17,380	1,04,751	1,659
186	Grenada	16,990	1,10,874	1,532
187	Singapore	16,350	57,08,041	29
188	Andorra	16,000	77,001	2,078
189	Cook Islands	15,100	17,507	8,625
190	Cayman Islands	12,700	63,382	2,004
191	Marshall Islands	12,640	58,058	2,177
192	Kiribati	12,150	1,14,158	1,064
193	Saint Kitts & Nevis	11,000	52,045	2,114
194	Antigua and Barbuda	9,800	95,426	1,027
195	State of Palestine	9,170	47,47,227	19
196	Tonga	9,000	1,01,998	882
197	Liechtenstein	6,900	37,800	1,825
198	Barbados	6,300	2,86,232	220
199	Kuwait	6,250	40,56,099	15
200	Wallis & Futuna	5,830	11,900	4,899
201	Djibouti	5,600	9,44,099	59
202	Anguilla	5,500	14,584	3,771
203	Mayotte	5,440	2,52,957	215
204	British Virgin Islands	3,620	29,577	1,224
205	Isle of Man	3,460	83,598	414
206	Saint Pierre & Miquelon	2,760	5,887	4,688
207	Montserrat	2,500	4,984	5,016
208	Oman	2,000	46,65,928	4
208	Tuvalu	1,000	11,370	880
209	Bermuda	1,000	63,049	159
210	Maldives	1,000	4,96,402	20
211	Channel Islands	800	1,68,665	47
212	Bahrain	620	14,94,076	4
213	Aruba	420	1,05,366	40
214	Malta	350	4,37,933	8
215	Greenland	220	56,481	39

(continued)

Table 1.3 (continued)

Sl. No	Country	Forest area (ha)	Population	m² per capita
216	Faeroe Islands	80	48,331	17

Table 1.4 Mean area and volume of oceans and seas

Name	Area (km^2)	Volume (km^3)
Atlantic Ocean	82,441,500	323,613,000
Indian Ocean	73,442,700	291,030,000
Pacific Ocean	165,246,200	707,555,000
Oceans	321,130,400 (88.94%)	1322,198,000 (96.50%)
Arctic Sea	14,090,100	16,980,000
Malay Sea	8,143,100	9,873,000
Central American Sea	4,319,500	9,573,000
Mediterranean Sea	2,965,900	4,238,000
Inter-continental Seas	29,518,600 (8.20%)	40,664,000 (2.96%)
Baltic Sea	422,300	23,000
Hudson Bay	1,232,300	158,000
Red Sea	437,900	215,000
Persian Gulf	238,800	6,000
Smaller enclosed Seas	2,331,300 (0.64%)	402,000 (0.03%)
Bering Sea	2,268,200	3,259,000
Okhotsk Sea	1,527,600	1,279,000
Japan Sea	1,007,700	1,361,000
East China Sea	1,249,200	235,000
Andaman Sea	797,600	694,000
California Sea	162,200	132,000
North Sea	575,300	54,000
English Channel andIrish Seas	178,500	10,000
Laurentian Sea	237,800	30,000
Bass Sea	74,800	5,000
Fringing Seas	8,078,900 (2.22%)	7,059,000 (0.51%)
Hydrosphere	361,059,200	137,032,500

Table 1.5 Productive sectors in the Blue Economy vertical with few associated spin-off units

No	Sector	Primary industries	Spin-off industries/Units
1	Living resources of the marine and estuarine ecosystem	Fisheries	• Processing • Net and gear making • Ice production and supply • Boat construction and maintenance • Manufacturing of fish processing equipment • Packaging • Marketing and distribution
		Aquaculture	• Water testing kits (like DO meter, refractometer, pH meter, turbidity meter, NH_3 detector etc.) • Soil testing kits (organic load, H_2S etc.) • Aerator • Pump • Generator • Civil construction for bio-treatment pond, sluice gates, inlets, outlets, feeder canal, drainage canal etc • Antibiotics • Algicides and fungicides • Processing units • Packaging
		Marine pharmaceutical products and chemicals	• Laboratory with instruments like mass spectrophotometer, chromatographs, particle size analyzer, rheometer, reciprocating cylinders, lyophilizer etc. • Packaging • Marketing • Patenting
2	Non-living resources (non- renewable)	Seabed mining for mineral extraction (preferably manganese nodules, seafloor sulphides and cobalt crust)	• Thrusters • Pickup devices • Storage bunker • Cable winch • Mother vessels

(continued)

Table 1.5 (continued)

No	Sector	Primary industries	Spin-off industries/Units
		Extraction of oil and natural gas	• Drilling rig • Strong steel pipe • Christmas tree (collection of valves mainly used to regulate the fluid flow) • Injectors • CO_2 and inert gas
		Desalination for Freshwater generation	• Membrane manufacturing units for reverse osmosis and electrodialysis • Thermal energy generation units for Multi-stage flash distillation (MSF), Multi-effect distillation (MED) and Vapor compression distillation (VCD)
3	Renewable natural forces (wind, wave, and tidal energy)	Generation of (off-shore) electricity	• Heat exchanger • Turbine • Electrical gadget
		Power plants based on renewable energy like solar energy, wind energy, tidal actions, ocean thermal energy gradient etc.	• Turbine • Drive-train (gear box, generator and power converter) • Power towers • Balance of system (BOS) • Solar panel/photo voltaic cell • Absorbing materials like silicon, thin films of light absorbing inorganic materials (like cadmium telluride (CdTe), gallium arsenide (GaAs) • Hydraulic system • Pump • Drilling injectors • Wave dragon device • Water testing kit particularly for salinity and suspended materials

(continued)

Table 1.5 (continued)

No	Sector	Primary industries	Spin-off industries/Units
4	Shipping	• Shipping and ship building	• Windlass/Mooring Winch • Steering gear • Marine air conditioner • Propellers • Side thrusters • Direction finder • Pump • Marine crane • Generator • Towing hook • Bilge water ejector • Boiler plant • Sewage plant and incinerator
		• Maritime transport	• GPS • Harbour • Canal lock • Anchor • Life saving equipment • Maritime signals like lighthouse, pillar buoy
		• Tourism and recreation	• Beach hotels • Restaurants • Marine Aquaria • Boating facilities • Recreation facilities like snorkeling, skiing, kayaking, diving, hiking and paragliding • Sunbathing facilities • Marine life watching
		• Ship wrecking and ship waste recycling	• Metal cutting gadgets • Shears • Heating gadgets • Oxygen fuel torch with flame velocity of 290–425 feet/second • Fuels like propene, butane, acetylene etc • Scalp metal from deck, frames, foundation, doors, hull attachment etc. that have high reselling value

(continued)

Table 1.5 (continued)

No	Sector	Primary industries	Spin-off industries/Units
5	Marine and coastal biodiversity	• Carbon sequestration through blue carbon (coastal vegetation like mangroves, sea grass, saltmarsh grass, mangrove associate floral community etc.)	• Nursery development • Marine Biologist, Ecologist, Environmentalist and related disciplines (Job opportunities) • Instrument like CHN analyzer • Carbon credit
		• Waste disposal from land-based industries, tourism and aquaculture farms	• Water testing kits • Laboratory development with sophisticated water analyzer, ICP, mass spectrophotometer, AAS etc • Treatment plant for treating, effluent and solid waste • Environmental monitoring related instrument development units
		• Coastal habitat protection and restoration	• Afforestation through nursery development • Water testing facilities • Biotreatment pond using seaweeds, marsh grass, oysters etc

are engaged in catching and selling these fishes and thus fetch a good economic return. However, several indirect job opportunities and activities are linked with this sector e.g., preservation and processing of fish, net making, gear making, ice production, construction of fishing vessels and trawlers, manufacturing of fish processing equipment, packaging of processed fish etc.

Aquaculture is also a vital sector under the living resources of the marine and estuarine ecosystems. Apart from making profit from several cultured species like shrimps, prawns, lobster, oysters, mussels, seaweeds, sea cucumber etc., several indirect micro, medium and small scale industries are also related and intricately associated with aquaculture like fish feed manufacturing units, water testing kits, aerator, pump, harvesting related equipments, antibiotics and several pharmaceutical products/chemicals for water quality management and prevention of fish diseases. Even in the remote villages of coastal zone in India, people manufacture shrimp feed as a part of their livelihood, a sort of local level cottage industry (Fig. 1.2).

In addition to this, the aquaculture sector needs the involvement of civil engineers and marine biologists to construct bio-treatment pond in which seaweeds, marsh grass and oysters are used to control the overloading of nutrients.

The production from the fishery and aquaculture sectors also demand a total quality management, where the involvements of biochemists, environmentalists and

Table 1.6 Marine and estuarine species diversity in India

Sl. No	Marine species	No. of species
1	Algae	724
2	Protista	750
3	Mesozoa	12
4	Prorifera	486
5	Cnidaria	812
6	Ctenophore	12
7	Gastroticha	98
8	Kinorhyncha	10
9	Plathyhelminthes	550
10	Annelida	440
11	Mollusca	3370
12	Bryozoa	200
13	Crustacean	2934
14	Merostomata	2
15	Pycnogonia	16
16	Sipuncula	35
17	Echiura	33
18	Tardigrada	10
19	Chaetognatha	30
20	Echinodermata	765
21	Hemichordate	12
22	Protochordata	119
23	Pisces	1800
24	Amphibia	6
25	Reptiles	26
26	Aves	145
27	Mammalia	29

microbiologists are needed to assure the quality of the products as per the FAO guidelines.

Marine and estuarine fishery (here we mean the finfish species) and molluscan sectors contribute significantly to the blue economy of Indian sub-continent, although the contribution of arthropods cannot be ignored because shrimps, prawns and crabs constitute a bulk quantity in the export basket of the country.

Marine fish diversity constitutes about 76% of the total fish diversity from India, comprising of about 2445 species belonging to 930 genera under 230 families and 40 orders. The Andaman and Nicobar Archipelago have the maximum marine fish diversity around 1500 species. The East Coast of India sustains about 1120 marine fish species and the West Coast has 1070 species recorded till date.

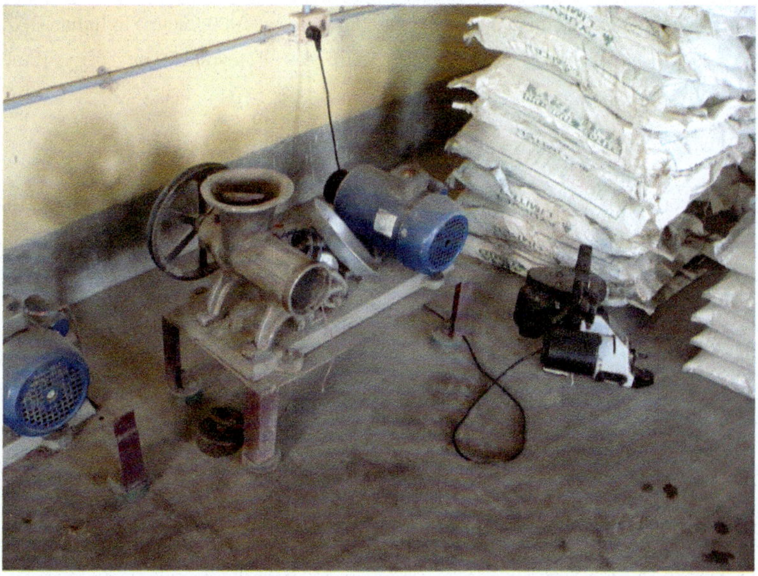

Fig. 1.2 Crude fish feed manufacturing machines used by the coastal population to sustain aquaculture

The harvestable potential of marine fishery at different depths of the ocean within the EEZ of India is presented in Table 1.7. It is clear from the table that the harvestable potential of marine fishery is 3.921 million tonnes. However, the present level of harvesting is 2.9 million tonnes, which means that 1.021 tonnes is still left for harvesting which, needs to be improved through proper conservation oriented technology to make the entire domain sustainable.

The different categories of commercially important fish available at 50 m and beyond depths in the marine and estuarine ecosystems of India are listed in Table 1.8.

These commercially important fishes not only have high demands in the internal market of the country, but also fetch foreign exchange which is an important vertical in the sector of Indian blue economy.

The molluscan species with their soft body and hard calcareous shell constitutes an important part in the tourism sector of blue economy. The decorative shells of

Table 1.7 Harvestable potential of marine fishery in the EEZ of Indian sub-continent

Depth (m)	0–50	50–200	200–500	Oceanic	Total
Demersal	1.28	0.625	0.028	–	1.993
Neretic	1.00	0.742	–	–	1.742
Oceanic pelagic	–	–	–	–	0.246
Total	2.28 (58%)	1.367 (35%)	0.028 (0.7%)	0.24 (6.3%)	3.921

Table 1.8 Category-wise estimated fishery potential yield (in 1000 tonnes) in Indian EEZ

Sl. No	Fish category	0–50 m	Beyond 50 m	Total
1	Elasmobranchs	65	103	168
2	Eels	7	–	7
3	Catfish	60	63	123
4	Oil sardine	191	–	191
5	Other sardine	96	–	96
6	White bait	91	–	91
7	Other clupeids	157	14	171
8	Bombay duck	104	–	104
9	Lizard fish	27	21	48
10	Perches	114	145	239
11	Scianids	120	22	142
12	Ribbon fish	95	216	311
13	Carangids	143	304	447
14	Silver bellies	82	4	86
15	Pomfrets	42	12	54
16	Mackeral	162	62	224
17	Seer fish	42	–	42
18	Runnies	37	242	279
19	Flat fish	38	–	38
20	Penaeid prawns	178	–	178
21	Non-penaeid prawns	54	–	54
22	Cephalopods	50	21	71
23	Priacanthus spp.	–	55	55
24	Black ruff	–	9	9
25	Indian drift fish	–	7	7
26	Deep sea prawns	–	3	3
27	Deep sea lobster	–	–	5
28	Oceanic tuna	–	209	209
29	Bill fishes	–	4	4
30	Others	254	189	443

molluscs attract the tourists to purchase home decors, key rings, fancy chandeliers, door and window curtains etc. Many molluscan organisms like mussels, oysters and clams are used as popular seafood worldwide (Fig. 1.3), although in India their demand as seafood is not very appreciable.

In India around 3,250 molluscan species belonging to 220 families and 591 genera have been reported, which comprise of bivalves (1,150–1,200 species), cephalopods (200–215 species), gastropods (180–200 species), polyplacophora (40–45 species)

Fig. 1.3 Bivalves getting sold in the market along with fishes

and scaphopods (15–25 species). The range is given to highlight the fact that the number of species is continuously getting added due to progress in bar-coding technology compared to earlier taxonomic diversity based on morphology. The importance of molluscs in the vertical of blue economy is often ignored although US$250 million, US$1.1 million and US$1.8 million/annum are fetched from cephalopod fishery, bivalve and gastropod respectively. Few commercially important clams and cockles that are available in the Indian marine and coastal waters are *Sunetta scripta* (marine clam), *Tridacna* sp. (giant clam), *Donax* sp. (surf clam), *Placenta placenta* (window pane oyster), *Geloina bengalensis* (big black clam) and *Andara granosa* (cockle). Mussels like *Perna indica* (brown mussel) and *Perna viridis* (green mussel) contribute considerable proportion to shell fish sector. Some oysters like *Pinctada margaritifera* (blacktip pearl oyster) and *Pinctada fucata* (Indian pearl oyster) produce quality pearls that have high demand in the internal and foreign markets. The tourisms sectors preferably the restaurants and seafood outlets/stores greatly depend on *Saccostrea cucullata* (rock oyster) and *Crassostrea madrasensis* (Indian backwater oyster). These oyster species are also abundantly available in the estuarine waters and mangrove patches. They are found on various hard substrata like sluice gates, jetties etc. (Fig. 1.4).

Cephalopod fishery is also gaining momentum in the marine fishery sector because of its high value in the export basket of the country. About eight species of squids like *Thysanoteuthis rhombus* (diamond squid), *Uroteuthis duvaucelii* (Indian squid), *U. edulis* (swordtip squid), *U. singhalensis* (long barrel squid), *Loliolus uyii* (little

Fig. 1.4 Oysters on the brick embankments in the estuarine region of Sundarbans

squid), *Loliolus hardwickei* (little Indian squid), *Sepioteuthis lessoniana* (Palkbay squid), *Sthenoteuthis oualaniensis* (oceanic squid) are available in the Indian coasts.

Organisms under cuttle fish group have the potential to contribute in the sector of blue economy not only of their edible value but also for their medicinal properties.

Cuttle fishes like *Sepia pharaonis* (Pharaoh cuttle fish), *S. brevimana* (short club cuttlefish), *S. prashadi* (hooded cuttlefish), *S. elliptica* (golden cuttlefish), *S aculeate* (needle cuttlefish) and *Sepiella inermis* (spineless cuttlefish) have high market value.

Octopuses like *Cistopus indicus* (old woman octopus), *Octopus vulgaris* (common octopus), *O. lobensis* (lobed octopus), *Amphioctopus neglectus* (webfoot octopus), *A. aegina* (marbled octopus), *A. marginatus* (veined octopus) are dominant species in the commercial catches from marine waters. The nutritional value of octopus is also remarkable. It has been reported that 80 gm of cooked octopus has approximately 50 gm of moisture, 140 cal of energy, 25 gm of protein, 1.60 gm of total lipid and 3.75 gm of carbohydrate. In addition to this Vitamin B12, selenium, iron, copper, zinc, magnesium, potassium, sodium and phosphorous are also present. It has high content of Omega-3 fatty acids that help to reduce the probability of heart attack and stroke in individuals. It has low content of saturated fat which helps in reducing weight. Thus octopus may be an important species in the sector of marine pharmacy.

Gastropods, available in the marine and estuarine ecosystems are mainly used for their decorative shells and their flesh is consumed as food. The dominant gastropod species in the Indian sub-continent are *Murex trapa* (rock shells), *M. pectin* (venus comb), *Murex* sp. (rock shells), *Murex badius* (rock shells), *Murex virgineus* (rock shells), *Turbinella pyrum* (scared chank), *Rapana bulbosa* (purples), *Ficus ficus* (fig shells), *Turitella attenuate* (screw shells), *Polystirasp* sp. (screw shells), *Crassisporasp* sp. (screw shells), *Phalium glaucum* (ton shells), *Naticalineata* sp. (moon snail), *Tibia curta* (wing shells), *Naticaalbula* sp. (moon snails), *Architectonica perspective* (staircase shells), *Epitonium scalaris* (ladder shells), *Xenophorasp* sp. (carrier shells), *Dentalium* sp. (tusk shell), *Nassa* sp. (button shells), *Nerita* sp. (nerite shells), *Trochus niloticus* (top shell), *Turbo marmoratus* (turban shell), *Charonia tritonis* (trumpet triton), *Lambis truncate* (spider shell), *Cyprea tigris* (cowry), *Cypraea arabica* (cowry), *Cypraea moneta* (money cowry), *Melo indica (beggars bowl), Lambis lambis* (spider conch), *Pleuroplaca trapezium* (elephant shell), *Chicoreus ramasus* (ramose murex), *Thias* sp. (dog whelk), *Strombus sp* (conch), *Turbo* sp. (turban shell), *Turbo intercostalis* (turban shell), *Cellana radiate* (limpet shell), *Umbonium vestiarium* (button shells), *Conus* sp. (cone shells), *Conus glans* (cone shells), *Harpa conoidalis* (harp shells), *Olivia* sp. (olive shells), *Olivia gibbosa* (olive shells), *Fusinus toreuma* (spindle shells), *Hemifuses pugilinus* (spindle shells), *Babylonia zeylanica* (whelk) *and B. spirata* (whelk).

We made a survey during January, 2019 to December, 2020 in the east and west coasts of India on the distribution of common gastropods (Table 1.9).

Chanks (*Turbinella pyrum*) are fished mainly for the shell and an organized fishery of considerable magnitude exists along the southeast coast of India.

Whelk fishery is also a potential component of molluscan based blue economic sector. In India, two species namely, *Babylonia spirata*, and *B. zeylanica* are landed as by-catch, mostly in the bottom trawls. The meat has high edible value and the calcareous shell is used in the shell craft industries.

The prawns and shrimps of India primarily belong to three major families, namely Pedaeidae, Palaemonidae and Sergestidae under decapoda. The penaeid prawns include three genus *Penaeus*, *Metapenaeus* and *Parapenaeopsis*. The genus *Penaeus* has a worldwide distribution and the various species belonging to it are found both in tropical and temperate latitudes. Of the 28 valid species of the genus, only 8 are usually caught in the Indian waters, which are *Penaeus japonicus, P. latisulcatus, P. canaliculature, P. monodon, P. semisulcatus, P. inducus, P. merguiensis and P. penicillatus.*

The prawns/shrimps fetch significant foreign currencies as there is a high demand of the species in various countries of the world like USA, EU, Japan, China etc. The export quantities of frozen shrimps from India were 4,34,486 tonnes and 5,65,980 tonnes during 2016–17 and 2017–18 respectively, which is a growth of 30.26% (Fig. 1.5).

This export earned considerable foreign exchange of INR 24,711.32 crores and INR 30,868.17 crores during two consecutive years 2016–17 and 2017–18 (Fig. 1.6). This is again a growth by 24.92%, although the export value reduced slightly per kg in 2017–18 compared to 2016–17. The USD/kg of 8.58 reduced to 8.87, which is a decrease by 0.12% (Fig. 1.7).

Mud crab (*Scylla serrata*) is an important exportable item, which belongs to phylum arthropoda (Fig. 1.8). The species is exported to about 45 countries. In the year 2020–2021 (Apr–Nov), India earned 54.34 USD million through crab export.

Table 1.9 Length weight analysis and occurrence of gastropods in the east and west coasts of India

Sl. No	Species	Length (cm)	Weight (gm)	Diameter (cm)	Distribution	Occurrence
1	*Architectonica laevigata*	4.3–4.5	9.15–11.15	1.5	West Coast and Rameswaram to Tuticorin	+++
2	*Babylonia spirata*	3.1–6.3	10.15–24.15	2.1–4.0	Entire West Coast and EC	++++
3	*Babylonia zeylanica*	3.2–6.4	9.15–25.15	1.5–3.0	Entire West Coast and Eat Coast	++++
4	*Bursa spinosa*	5.5–8.8	13.15–22.15	3.2–4.6	Gulf of Mannar region and West coast	+++
5	*Conusmiled edwardsii*	11.2–15.7	29.15–73.15	3.8–4.9	West Coast	+
6	*Cymatium pileare*	7.4–9.3	26.15–33.15	4.0–5.4	West Coast and Rameswaram to Kanykumari	++
7	*Ficus gracilis*	4.5–13.6	8.15–64.15	3.0–8.6	West Coast and East coast	++++
8	*Fulgoraria* sp.	11.2	29.15	3.8	West Coast	+
9	*Fusinus longicauda*	4.1–16.3	4.15–33.15	2.6–4.7	West Coast	++
10	*Harpaconoidalis*	7.6–8.5	6.35–26.15	6.0–24.3	West Coast	++
11	*Murex tribulus*	6.9	9.15	3.6	West Coast and East Coast	+++
12	*Olivia gibbosa*	5.1	4.35	14.3	West Coast and Gulf of Manar region	+++
13	*Phalium glaucum*	4.2–7.7	6.15–36.15	3.4–6.0	West Coast and Gulf of Manar region	+++
14	*Polinicies didyma*	3.6–4.7	7.15–13.15	4.5–5.9	West Coast and Gulf of Manar region	+++
15	*Strombus marginatus*	3.9–7.1	18.15–34.15	2.9–4.1	West Coast and Gulf of Manar region	++
16	*Tibia curta*	4.0–10.2	15.15–45.15	2.5–4.5	West coast and East coast	+++
17	*Tonna dolium*	4.5–5.0	10.15–18.15	3.5–4.0	West Coast and East Coast	+++
18	*Turitella duplicata*	3.8–13.6	18.15–45.15	2.4–3.7	West Coast and East Coast	++++

(continued)

Table 1.9 (continued)

Sl. No	Species	Length (cm)	Weight (gm)	Diameter (cm)	Distribution	Occurrence
19	*Xancus pyrum*	12.3	94.15	7.2	West coast and East coast entire stretch	++++
20	*Xenophora corrugate*	3.3–3.7	8.15–9.15	2.4–2.6	West Coast and East Coast	+++
21	*Umbonium vestiarium*	1.7	3.15	1.5	West Coast and East Coast	++++
22	*Chichoreus virgineus*	4.4–8.7	7.15–28.15	4.8–8.8	West Coast and East Coast	+++
23	*Turitella terebra*	3.3–15.6	15.15–55.15	2.6–4.9	West coast and East coast	+++
24	*Thais bufo*	3.8–7.0	6.15–29.15	3.8–5.6	West coast and East coast	+++
25	*Hemifusus pugilinus*	3.7–28.2	5.65–7.95	2.8–5.1	West coast and East coast	+++
26	*Melo melo*	8.5–15.7	15.15–21.15	5.7–13.8	West coast and East coast	+++
27	*Cypreae tigris*	4.7–16.5	15.15–26.15	6.6–9.8	West coast and East coast	+++

'++++' = Abundant, '+++' = Common, '++' = Moderately common and '+' = Rare

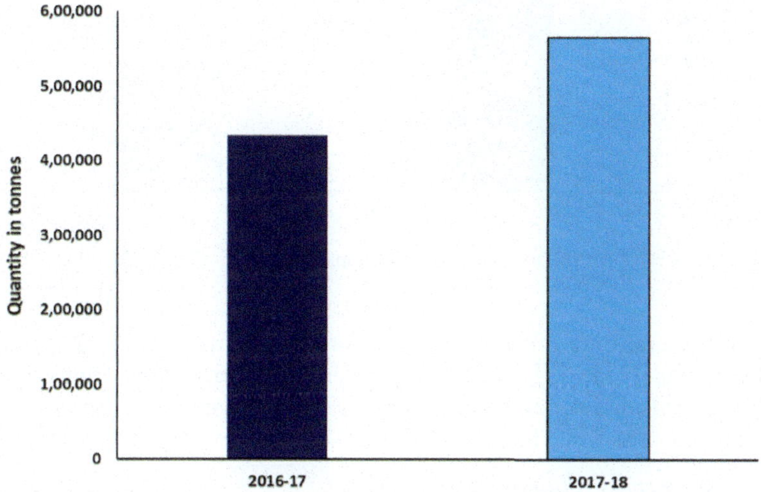

Fig. 1.5 Export quantity of Frozen Shrimp (in tonnes) during 2016–17 and 2017–18

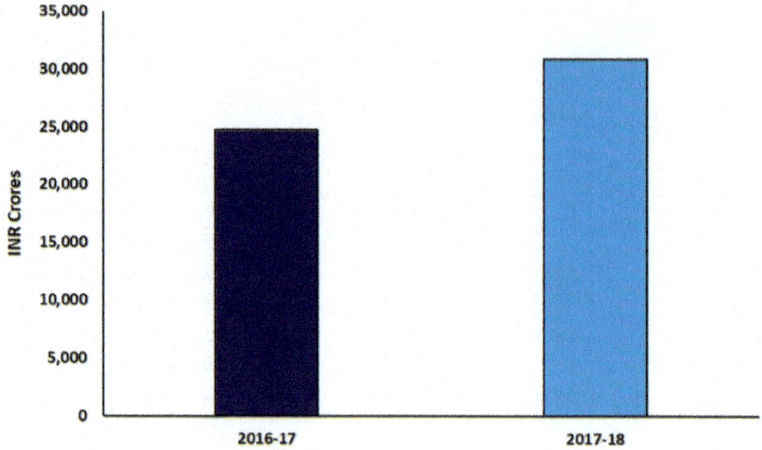

Fig. 1.6 Export value (in INR Crores) of Frozen Shrimp during 2016–17 and 2017–18

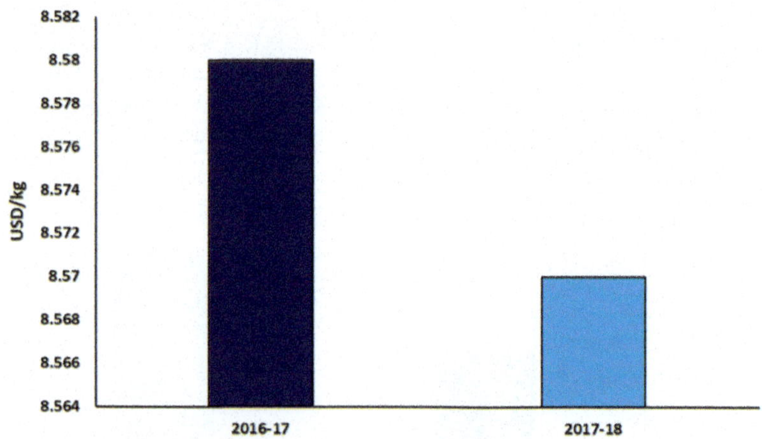

Fig. 1.7 Export value (in USD/kg) of Frozen Shrimp during 2016–17 and 2017–18

Horseshoe crabs are organisms known for their pharmaceutical properties and have the potential of taking a lead role in the marine pharmaceutical units (Fig. 1.9).

The Family Limulidae is a family of horseshoe crabs and is currently composed of several genera as listed here.

1. ***Carcinoscorpius rotundicauda***

   Phylum Arthropoda.
   Order Xiphosurida.
   Family Limulidae.
   Genus *Carcinoscorpius.*
   Species *rotundicauda.*

Fig. 1.8 Mud crabs, *Scylla serrata*—An important arthropod in the export basket of India

Fig. 1.9 Horseshoe crab on the Sundarban mangrove inter-tidal zone

2. ***Tachypleus gigas***

   Phylum Arthropoda.
   Order Xiphosurida.
   Family Limulidae.
   Genus *Tachypleus.*
   Species *gigas.*

3. ***Tachypleus syriacus***

   Phylum Arthropoda.
   Order Xiphosurida.
   Family Limulidae.
   Genus *Tachypleus.*
   Species *syriacus.*

4. ***Tachypleus decheni***

   Phylum Arthropoda.
   Order Xiphosurida.
   Family Limulidae.
   Genus *Tachypleus.*
   Species *decheni.*

5. ***Tachypleus tridentatus***

   Phylum Arthropoda.
   Order Xiphosurida.
   Family Limulidae.
   Genus *Tachypleus.*
   Species *tridentatus.*

6. ***Heterolimulus gadeai***

   Phylum Arthropoda.
   Order Xiphosurida.
   Family Limulidae.
   Genus *Heterolimulus.*
   Species *gadeai.*

7. ***Volanalimulus madagascarensis***

   Phylum Arthropoda.
   Order Xiphosurida.
   Family Limulidae.
   Genus *Volanalimulus.*
   Species *madagascarensis.*

In India, two species of horseshoe crab namely *Carcioscopius rotundicauda* and *Tachypleus gigas* are mostly seen in the North-East coast of India (Table 1.10).

In case of these two species, there is no official documentation of commercial exploitation for the production of TAL (Tachypleus Amebocyte Lysate) and CAL

Table 1.10 Distribution and availability of *Carcioscopius rotundicauda* and *Tachypleus gigas* in the maritime states of West Bengal, Odisha and Andhra Pradesh

Place/Site	Species	Location	Availability
Chemaguri (WB)	*C. rotundicauda*	Intertidal mud flat	+++
Bakkhali (WB)	*C. rotundicauda*	Intertidal mud flat	++
Digha (WB)	*C. rotundicauda*	Coastal water	+
Chandipur (Odisha)	*T. gigas*	Sandy beaches	+++
Haskura Estuary (Odisha)	*T. gigas*	Estuarine Sandy beaches	+
Panchkura Estuary (Odisha)	*T. gigas*	Estuarine Sandy beaches	+
Balaramgari (Odisha)	*T. gigas*	Estuarine beaches	+++
Dhamara (Odisha)	*T. gigas*	Creeks and beaches	+++
Puri (Orissa)	*T. gigas*	Beaches	–
Hukitola (Paradeep, Odisha)	*T. gigas*	Sandy beaches	++
Chilika (Orissa)	*T. gigas*	~25 m depth	+++
Gopalpur (Odisha)	*T. gigas*	~25 m depth	++
Visakhapatnam (Andhra Pradesh)	*T. gigas*	~25 m depth	+

(Carcinoscopius Amebocyte Lysate) as documented for *Limulous* sp. However, there is a great opportunity of linking these two species with the marine pharmaceutical units as TAL and CAL can be widely used to detect and quantify bacterial endotoxin. Basically, these lysates react with LPS of gram negative bacteria that act as the foundation of bacterial endotoxin detection.

Few researchers in Indian Sundarbans under Techno India University, West Bengal are monitoring the blue blood of *Carcinoscopius rotundicauda* to evaluate the efficacy of COVID vaccine (Fig. 1.10).

Coral Reefs

Coral reefs occupy a pivotal position in the domain of Blue Economy. The total spectrum of ecosystem services offered by coral reef may be broadly divided into four major categories as presented in Fig. 1.11.

About 850 million people throughout the World dwell within 100 km of the reefs who are greatly benefited from several ecosystem services provided by the coral reefs. Coral reef tourism is common at least in 94 countries of the World and nearly 15–20% of the GDP of 23 countries are directly related to reef tourism. Coral reef tourism not only includes scuba diving, snorkeling and boat trips, but also includes the aesthetic beauty associated with transparent water and sandy beaches. The entire reef economy spins around coral triangle and Mesoamerican reef. The Coral Triangle exist in countries like Indonesia, Philippines, Malaysia, Papua, New Gunia, Salomon

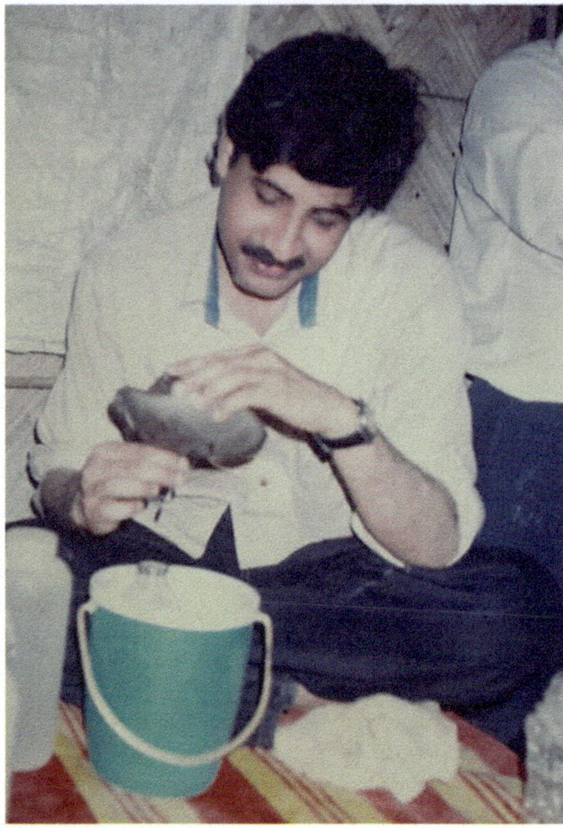

Fig. 1.10 Blood extraction (10% of the body mass) from *C. rotundicauda*

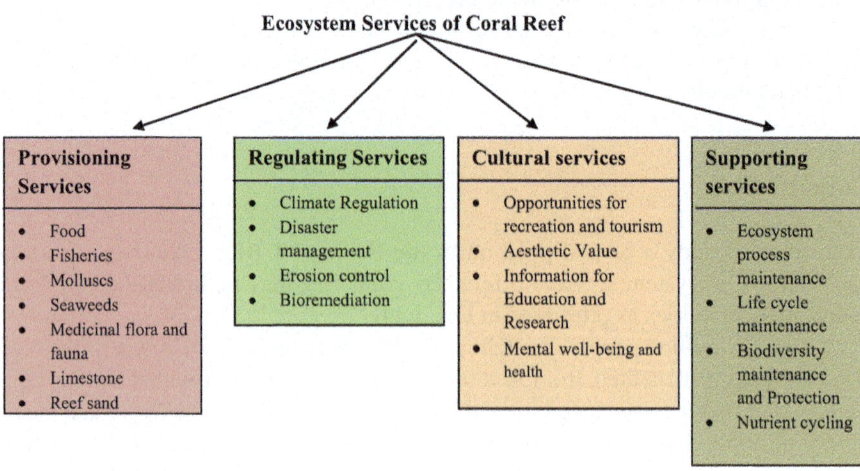

Fig. 1.11 Classification of ecosystem services of coral reef

Island and East Timor, and consist of about 75% common coral species and nearly 40% of reef fish species. The Mesoamerican reef is found in Belize, Honduras, Guatemala and Yucatan Province in Mexico. The direct and indirect economic returns from Coral Triangle and Mesoamerican reef during 2017 are presented in Figs. 1.12, 1.13 and 1.14.

In India, the coral reefs are distributed in the Gulf of Mannar, Gulf of Kutch, Andaman and Nicober and Lakshadweep Islands (Table 1.11).

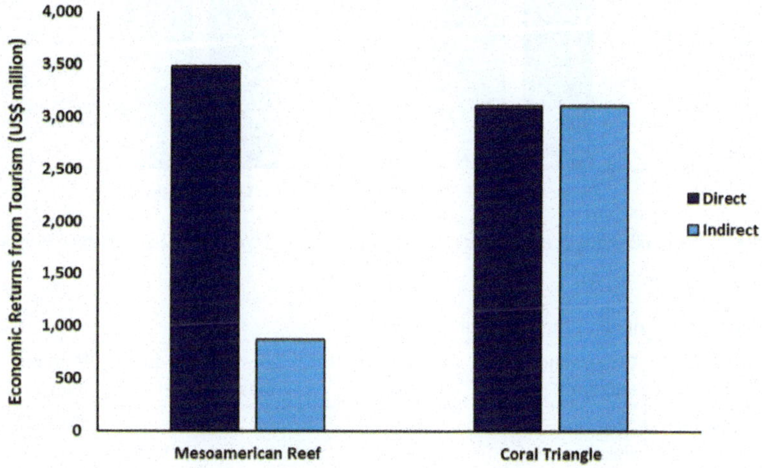

Fig. 1.12 Direct and indirect economic returns from coral reef tourism as per the record of 2017

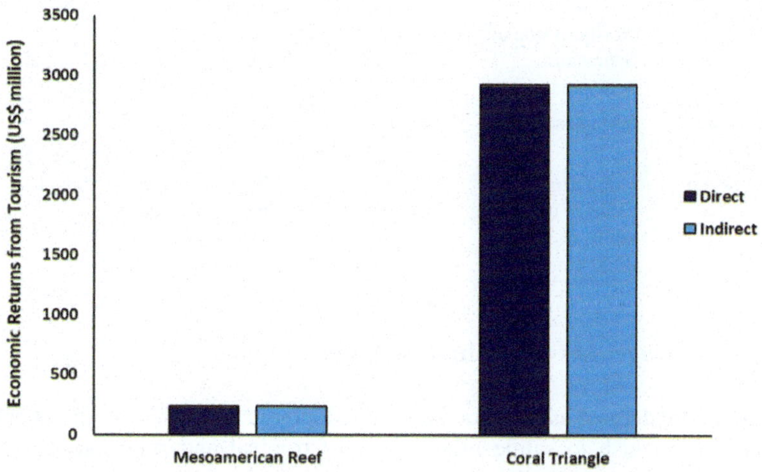

Fig. 1.13 Direct and indirect economic returns from commercial fisheries as per the record of 2017

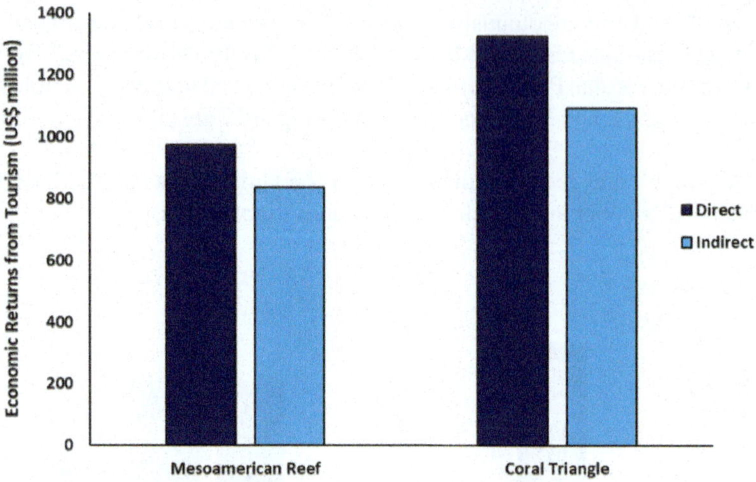

Fig. 1.14 Direct and indirect economic returns from coastal development as per the record of 2017

Table 1.11 Status of coral reef in India

Particulars	Geographical location	Area of the reef (in km^2)	Destruction during bleaching—1998	Potential for recovery
Gulf of Mannar	21 islands, South East Coast of India 140 km, between Tuticorin and Rameswaram	94.3	60–80% loss of live coral. Only 25% live corals remaining	Medium–Low
Gulf of Kutch	40 islands, Northern side of Saurashtra peninsula	325.5	50–70%	Medium–Low
Andaman and Nicobar Islands	530 islands	1021.46	15–20%	Good
Lakshadweep Islands	Uninterrupted chain of coral atolls-stretch of 2000 km	933.7	70–90% loss of live coral	Good

Mangroves, Seagrasses and Saltmarsh Grasses

Mangroves are defined as an association of halophytic trees, shrubs and other plants growing in brackish to saline tidal waters of tropical and subtropical coastlines (Mitsch and Gosselink 2007). Mangroves are generally restricted to the tidal zone. As such, mangroves in fringe areas are inundated by practically all high tides, while those at the higher topographic boundaries may be flooded only during the highest of tides (spring tides) or during storm surges. Mangroves are typically found along tropical and subtropical coastlines between about 25° N and 25° S.

The entire mangrove spectrum of the world encompasses 31 families under which 48 genera and 96 species are included as documented in the site http://www.bsienvis.nic.in/Database/IndianMangroves_3941.aspx#WORLD *DISTRIBUTION OF MANGROVES* (Table 1.12).

Mangrove forests in India are found along the coastline of 9 States and 4 Union Territories (Fig. 1.15). The mangrove habitat of India is broadly classified into three namely, Deltaic (Eastern Coast Mangroves), Estuarine & Backwater (Western Coast Mangroves) and Insular mangroves (Andaman & Nicobar Islands) (Mandal and Naskar 2008). Their overall cover is estimated to be 4740 km^2, of which about 58% is along the east coast (Bay of Bengal); 29% along the west coast (Arabian Sea) and the remaining 13% in the Andaman and Nicobar Islands (FSI 2015). The mangrove cover is larger and more widespread on the east coast compared to the west coast because of its distinctive geo-morphological setting. These differences in mangrove cover can be attributed to two reasons: (i) the east coast has large estuaries with deltas formed due to runoff and deposition of sediments, whereas the west coast has funnel-shaped estuaries with an absence of deltas; and (ii) the east coast has gentle slopes with extensive flats for colonization by mangroves, whereas the west coast has steep slopes (Kathiresan 2010). The mangroves of Andaman and Nicobar Islands (ANI) are probably the best developed in India in terms of their density and growth (Dagar et al. 1991; Mandal and Naskar 2008). Their irregular and deeply indented coastline results in innumerable creeks, bays and estuaries which facilitate the development of rich, extensive and luxuriant growth of mangrove forests with a high degree of biodiversity.

Despite its ecological and economical values, in the last two decades global mangroves have witnessed annual loss of between 0.16 and 0.39% due to rapid coastal development (Hamilton and Casey 2016). However, little is known about the effects of either widespread or localized mangrove area loss on individual mangrove species or populations due to a lack of information about the distribution of individual species (Polidoro et al. 2010). The species richness of mangroves in many geographical regions is decreasing with time as a result of the destruction of mangrove forests and exposure to various anthropogenic stresses (Hamilton and Snedaker 1984). The past and present distribution of mangrove species on a global scale has been reviewed by several authors (Tomlinson 1986; Duke 1992; Ricklefs and Latham 1993; Field 1995; Duke et al. 1998; Ellison et al. 1999; Saenger 2002; Wang et al. 2003; Spalding et al. 2010), but in India the actual number of species of mangroves that exists in different regions is not fully known due to scattered data, and lack of both comprehensive compilation and extensive field surveys (Kathiresan 2010). Without the adequate knowledge of the exact species composition it will be difficult to identify and implement conservation priorities for the mangroves of India.

Table 1.12 World distributions of common mangrove and associate species

Sl. No	Species	
1	*Acanthus ebracteatus*	
2	*Acanthus ilicifolius*	
3	*Acanthus volubilis*	
4	*Cerbera manghas*	
5	*Nypa fruticans*	

(continued)

Table 1.12 (continued)

Sl. No	Species	
6	*Phoenix paludosa*	
7	*Finlaysonia obovata*	
8	*Sarcolobus globosus*	
9	*Sarcolobus carinatus*	
10	*Tylophora tenuis*	

(continued)

Table 1.12 (continued)

Sl. No	Species	
11	*Avicennia alba*	
12	*Avicennia bicolour*	
13	*Avicennia germinans*	
14	*Avicennia integra*	
15	*Avicennia marina*	

(continued)

Table 1.12 (continued)

Sl. No	Species	
16	*Avicennia var. australasica*	
17	*Avicennia var. eucalyptifolia*	
18	*Avicennia officinalis*	
19	*Avicennia rumphiana*	
20	*Avicennia schaueriana*	

(continued)

Table 1.12 (continued)

Sl. No	Species	
21	*Dolichandrone spathacea*	
22	*Camptostemon philippinensis*	
23	*Camptostemon schultzii*	
24	*Caesalpinia bonduc*	
25	*Caesalpinia cristata*	

(continued)

Table 1.12 (continued)

Sl. No	Species	
26	*Cynometra iripa*	
27	*Cynometra ramiflora*	
28	*Intsia bijuga*	
29	*Conocarpus erectus*	
30	*Laguncularia racemosa*	

(continued)

Table 1.12 (continued)

Sl. No	Species	
31	*Lumnitzera littorea*	
32	*Lumnitzera racemosa*	
33	*Ipomoea tuba*	
34	*Fimbristylis ferruginea*	
35	*Scirpus littoralis*	

(continued)

Table 1.12 (continued)

Sl. No	Species	
36	*Excoecaria agallocha*	
37	*Excoecaria indica*	
38	*Excoecaria ovate*	
39	*Flagellaria indica*	
40	*Pemphis acidula*	

(continued)

Table 1.12 (continued)

Sl. No	Species	
41	*Hibiscus tiliaceous*	
42	*Thespesia populneoide*	
43	*Thespesia populnea*	
44	*Aglaia cucullata*	
45	*Xylocarpus granatum*	

(continued)

Table 1.12 (continued)

Sl. No	Species	
46	*Xylocarpus mekongensis*	
47	*Xylocarpus moluccensis*	
48	*Aegiceras corniculatum*	
49	*Aegiceras floridum*	
50	*Osbornia octodonta*	

(continued)

Table 1.12 (continued)

Sl. No	Species	
51	*Dalbergia spinosa*	
52	*Derris scandens*	
53	*Derris heterophylla*	
54	*Mucuna gigantea*	
55	*Pelliciera rhizophorae*	

(continued)

Table 1.12 (continued)

Sl. No	Species	
56	*Aegialitis rotundifolia*	
57	*Aegialitis annulata*	
58	*Myriostachya wightiana*	
59	*Porteresia coarctata*	
60	*Urochondra setulosa*	

(continued)

Table 1.12 (continued)

Sl. No	Species	
61	*Acrostichum aureum*	
62	*Acrostichum danaeifolium*	
63	*Acrostichum speciosum*	
64	*Bruguiera cylindrica*	
65	*Bruguiera exaristata*	

(continued)

Table 1.12 (continued)

Sl. No	Species	
66	*Bruguiera gymnorhiza*	
67	*Bruguiera hainesii*	
68	*Bruguiera parviflora*	
69	*Bruguiera sexangula*	
70	*Ceriops australis*	

(continued)

Table 1.12 (continued)

Sl. No	Species	
71	*Ceriops decandra*	
72	*Ceriops tagal*	
73	*Kandelia candel*	
74	*Rhizophora apiculata*	
75	*Rhizophora harrisonii*	

(continued)

Table 1.12 (continued)

Sl. No	Species	
76	*Rhizophora mangle*	
77	*Rhizophora mucronata*	
78	*Rhizophora racemosa*	
79	*Rhizophora samoensis*	
80	*Rhizophora stylosa*	

(continued)

Table 1.12 (continued)

Sl. No	Species	
81	*Scyphiphora hydrophyllacea*	
82	*Merope angulata*	
83	*Sonneratia alba*	
84	*Sonneratia apetala*	
85	*Sonneratia caseolaris*	

(continued)

Table 1.12 (continued)

Sl. No	Species	
86	*Sonneratia griffithii*	
87	*Sonneratia lanceolata*	
88	*Sonneratia ovate*	
89	*Sonneratia gulngai*	
90	*Sonneratia urama*	

(continued)

Table 1.12 (continued)

Sl. No	Species	
91	*Heritiera fomes*	
92	*Heritiera globosa*	
93	*Heritiera kanikensis*	
94	*Heritiera littoralis*	
95	*Brownlowia tersa*	

(continued)

Table 1.12 (continued)

Sl. No	Species	
96	*Clerodendrum inerme*	

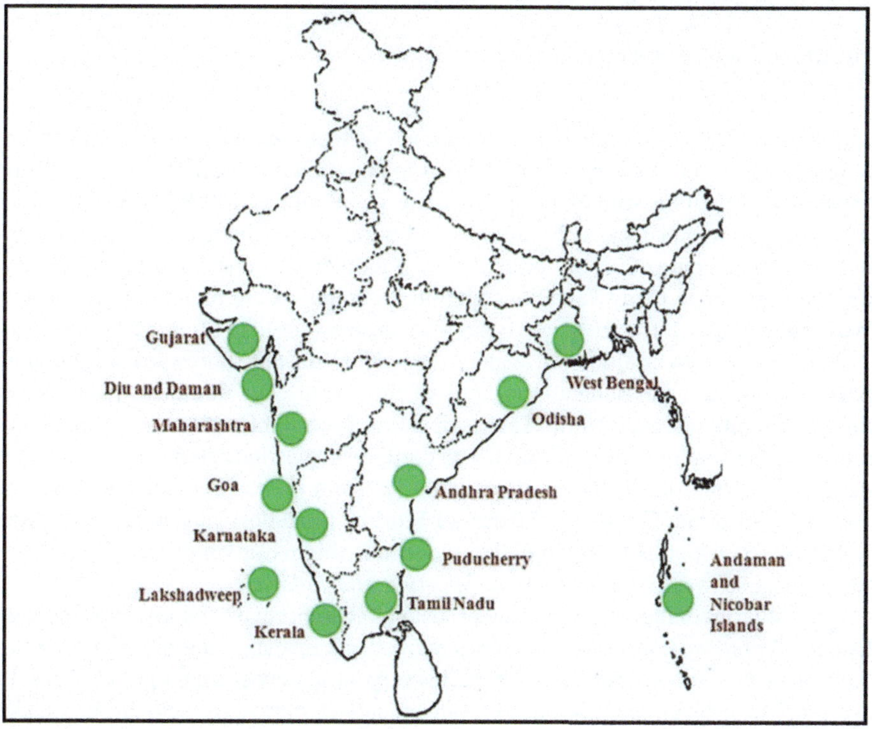

Fig. 1.15 Map showing mangrove habitats in India

Seagrasses and Saltmarsh Grasses

Seagrasses are considered as blue carbon like mangroves, which are usually restricted in shallow saline/brackishwater zone of the oceans, seas and estuaries. They have distinct differentiation of roots, stems and leaves and are characterized with the presence of flowers and seeds. The ecosystem service valuation models typically address only partial functions of seagrass ecosystems (Unsworth and Cullen-Unsworth 2014)

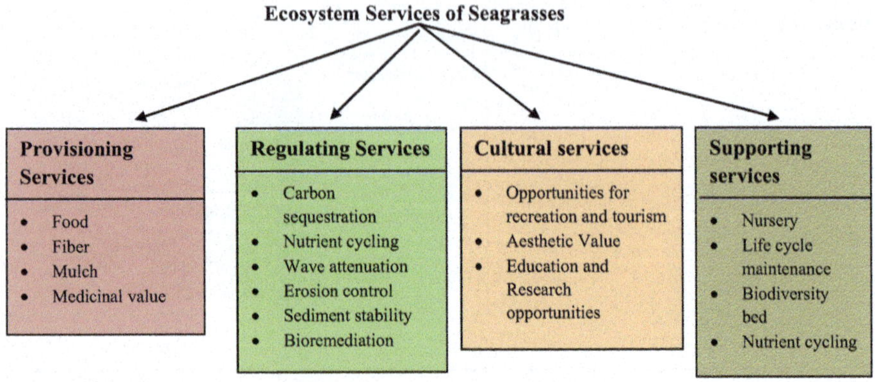

Fig. 1.16 Classification of ecosystem services of seagrasses

and mostly they are related to commercial fish production. Few literatures on seagrasses also focus on their bioremediation potential due to their nutrient cycling capability. Valuation studies often calculate economic value (Scientists HaSEEa 2005) based on total ecosystem capacity and often fail to isolate the seagrass contribution to this figure. Carbon sequestration (another non-consumptive value) also does not appear in many valuations although seagrasses are major sink of carbon. We present in Fig. 1.16 the various types of ecosystem services offered by seagrasses.

Salt marsh communities are widely distributed on the shoreward side of mud flats in temperate and subarctic regions of the world. The dominant plant life in this community on many North American shores is cordgrass (*Spartina* sp.) that has moved to the shallow intertidal area. The cordgrass *Spartina alterniflora* dominates the marshlands of the Atlantic coast, while *Spartina foliosa* is dominant along the coast of California. On the Gulf Coast of Florida, the tallest and most marsh grass is *Juncus* sp., which is found in higher elevations (higher up on shore). The species found most seaward is tall cordgrass, *Spartina alterniflora*.

The estuaries of Indian sub-continent sustains an important saltmarsh grass species namely *Porteresia coarctata*, which has a great role in stabilizing the islands. Their long network of roots binds the soil particles very intricately and retards the process of erosion. The growth, biomass and subsequent carbon storage capacity of the species is controlled by the ambient aquatic salinity. The species exhibits a luxuriant growth in relatively low saline regions (salinity range within 5–20 psu) with subsequently higher carbon content, but at salinity around 30 psu, the growth is stunted.

Marine biotechnology is a very prospective sector of blue economy that spins around the living resources of the marine and estuarine ecosystems. Several drugs are already in the market (Table 1.13).

Many drugs derived from marine flora and fauna are still under clinical trials. The use of mangroves to cure several types of diseases has also opened up a new chapter in the vertical of blue pharmacy although the supply chain management in this sector is still in an embryonic stage.

Table 1.13 Products developed from marine organisms and their uses

Species	Type of organisms	Product	Use
Cephalosporium sp.	Fungi	Cephalosporin	Antibiotic
Squalus acanthus	Spiny dogfish	Squalamine	Antibiotic
Tethya crypta	Caribbean sponge	Vidrabine/Ara-A	Antiviral
Tethya crypta	Caribbean sponge	Cytarabine/Ara-C	Anticancer
Conus magus	Cone snail	Ziconotide	Analgesic
Ecteinascidia turbinata	Caribbean tunicate	Trabectedin	Anticancer
Halichondria okadai	Marine sponge	Eribulin mesylate	Anticancer

A checklist of the true mangrove species under different families along with their respective curative actions for different diseases is presented here:

1. Under Acanthaceae family, two important species of mangrove namely, *Acanthus ilicifolius* and *Acanthus volubilis* are distributed abundantly in the intertidal zone and more preferably in the supra-littoral zone where there is minimum availability of tidal water. These species are mostly xerophytic in nature. *Acanthus* species is mostly used to cure the fracture of bones.
2. Under the family Myrsinaceae, a very common mangrove plant *Aegiceros corniculatum* is widely available in the intertidal zone of Sundarban. This species is used to cure inflammation, arthritis and asthma.
3. Under the family Avicenniaceae, three species namely *Avicennia alba, Avicennia marina* and *Avicennia officinalis* are commonly available in the intertidal zone. These species can thrive from moderate to high salinity. The species are known to contain several bioactive substances that can combat cancer, hepatitis, inflammation, HIV. etc.
4. Under the family Rhizophoraceae, four species of mangrove namely *Bruguiera cylindrical, Bruguiera gymnorrhiza, Bruguiera parviflora, Bruguiera sexangula* commonly grow in the intertidal zone. The bark of *Bruguiera* spp. is used to treat malaria, and the fruits have anti-viral properties.
5. Under the family Rhizophoraceae, two species of mangroves *Ceriops decandra, Ceriops tagal* are commonly available in the intertidal zone. *Ceriops decandra* is known for the treatment of gastrointestinal disorder, snake bites, inflammation and cancer. The phyto constituents of *Ceriops* species like ceriopsin A-G, lupeol, α amyrin, oleanolic acid, ursolic acid, catechin, procyanidins have potential to develop new drugs against various human diseases.
6. Under the family Rhizophoraceae, two species of mangroves namely *Rhizophora apiculata, Rhizophora mucronata* are commonly available in the intertidal zone of Indian Sundarbans. These species are known to possess antibacterial and antiviral properties.

7. Under the family Rhizophoraceae, another mangrove species *Kandelia candel* is also available in the intertidal zone. The species is known for its antibacterial activity.
8. Under the family Plumbaginaceae, a mangrove species *Aegialitis rotundifolia* is commonly found in the intertidal zone. The leaves of this halophyte are commonly used to get rid from insect bites and pains.
9. Under the family Tiliaceae, a mangrove species *Brownlowia tersa* is commonly found in the intertidal zone. This species is used to treat diahorrea, dysentery, wounds and boils.
10. Under the family Meliaceae, a mangrove species namely *Amoora cucullata* is found in the intertidal zone. The species is used to inhibit the spreading of enteropathogenic bacteria.
11. Under the family Combretaceae, a mangrove species namely *Lumnitzera racemosa* is found in the intertidal zone. This species is known for its anti-inflammatory properties.
12. Under the family Sterculiaceae, an important mangrove species namely *Heritiera fomes* is found in the intertidal zone. The species is gradually getting extinct from the Indian Sundarban part due to rise of aquatic salinity. This species contains several bioactive substances, which have antimicrobial and anticancer properties. Recently, this species has also been claimed to combat diabetes.
13. Under the family Arecaceae (Palmae) two mangrove species namely *Nypa fruticans* and *Phoneix paludosa* are found in the intertidal zone. *Nypa fruticans* is used to treat headache, toothache, ulcers and insect bites.
14. Under the family Meliaceae two mangrove species namely *Xylocarpus granatum* and *Xylocarpus mekongensis* grow in the intertidal zone. *Xylocarpus* spp. are used for curing diahorrea, cholera, fever and inflammation.
15. Under the family Leguminosae, three mangrove species namely *Cynometra ramiflora, Derris trifoliate* and *Derris umbrellatum* are distributed in the intertidal zone of Indian Sundarbans, which have unique anti-microbial properties.
16. Under the family Malvaceae, a mangrove species namely *Hibiscus tortuosus* is found in the intertidal zone of Indian Sundarbans. The species has antimicrobial properties and is often used to treat cough, tuberculosis, sore throat etc.
17. Under the family Tamaricaceae, two mangrove species of mangrove namely *Tamarix dioica* and *Tamarix gallica* grow in the intertidal zone. The root extracts of these species are known to fight against liver problems. They contain antioxidants (flavonoids) that prevent liver cell damage.
18. Under the family Euphorbiaceae, two important species of mangrove namely, *Excoecaria agallocha* and *Excoecaria bicolour* are found in the intertidal zone. *Excoecaria agallocha* has several pharmacological benefits and is used to treat ulcers, leprosy, epilepsy, and rheumatism.
19. Under the family Sonneratiaceae, two important species of mangrove that are found in Sundarbans are *Sonneratia apetala* and *Sonneratia caseolaris*. These two species are abundantly distributed in the intertidal zone inundated with

low saline water. The fruits of *Sonneratia* species are rich in vitamin C and are regularly consumed by the Sundarban island dwellers, which prevent them from cough and cold.

We isolated few phytochemical compounds from the common mangrove floral species collected from Indian Sundarbans during March, 2022 (Annexure 1A), which play crucial roles to cure several human diseases. Soxhlet extraction method using organic solvents (methanol, ethanol, chloroform, n-hexane, ethyl acetate etc.) were used followed by separation and purification through chromatography was employed to evaluate the compounds isolated from some true mangrove species that we collected during our field survey. Finally we used spectroscopic techniques to document their structures. These phytochemical compounds have great role in boosting up the marine/based pharmaceutical vertical in the domain of blue economy.

1.2.2 Marine Non-living Resources in Context to Blue Economy

The crucial role of non-living resources of the marine and estuarine ecosystems in upgrading the global economy is still under appreciated, although as per the estimates of the Economic Cooperation and Development (OECD) in 2010, the contribution in the financial domain from the ocean amounted to US$1.5 trillion, which is about 2.5% of the world's total financial wheel. About 30 million jobs sourced from the non-living resources of the marine and estuarine ecosystems are related directly to the blue economy and it is expected that this value of employment may touch 40 million in 2030. It has also been predicted that during 2030 the value of ocean economy contribution may reach US$3 trillion. This figure may deviate significantly due to COVID-19 pandemic, which has not been considered in the projection report.

One of the major missions in blue economy is to minimize the carbon footprint preferably from GHG emission and because of this vital component, thrust has been given on the energy derived from offshore wind, waves and tides. According to Ocean Energy System (OES), an intergovernmental organization, the zero carbon energy sources include waves and swells, tides and ocean currents. Energy can also be captured from the thermal sources and salinity gradients. OES estimates that between the time phase of 2030 to 2050, wave and tidal energy growth could match that of offshore wind energy sector.

Wave and tidal energy sectors are related to several spin-off units, as equipments used to capture these zero-carbon energies involves extensive supply chains stretching into a number of countries. The OECD reports that supply chains for ocean energy projects in Europe, for example, encompass "tidal turbines, hydro-turbines, steel spare parts" (Austria); "wave power plants and generators" (Germany); "wave power attenuators and overtopping devices" (Denmark).

Emission standards for ocean going vessels will become more stringent in future, and shipping companies would have to reduce their emissions of sulphur and nitrous oxides. Many may choose to meet these standards by employing cleaning systems known as "scrubbers," which reduce sulphur from exhaust gases. A number of providers offer this technology. Information and communications technology in shipping sector combined with "big data" analytics, can bring a revolution in the shipping industry. Satellite applications, mobile communications, remote sensing, and radar technologies could result in greater automation and data collection, leading to improved maritime management and decision-making.

For convenience of the readers, we have categorized the various prospective sectors of blue economy related to non-living resources in to six basic types, namely (i) Marine Transportation, (ii) Ship Breaking and Ship Waste Management, (iii) Marine and Coastal Constructions, (iv) Marine Based Energy, (v) Oil and Natural Gas Extraction and (vi) Deep Sea Minerals.

Marine Transportation

Marine transportation is primarily used for passenger transportation, military and defense, fishing and marine tourism. According to the Ministry of Shipping, around 95% of India's trading by volume and 70% by value is done through maritime transport. In November 2020, the Prime Minister of India, Mr. Narendra Modi re-coined the name of Ministry of Shipping as the Ministry of Ports, Shipping and Waterways.

Presently India has 13 major and 205 notified minor and intermediate ports. Under the National Perspective Plan for Sagarmala, six new mega ports will be developed in the country. The important ports in India are listed in Table 1.14.

Table 1.14 Important ports in India along with their state-wise location

Sl. No	Port	State
1	Chennai	Tamil Nadu
2	Kochi	Kerala
3	Ennore	Tamil Nadu
4	Kolkata	West Bengal
5	Kandla	Gujarat
6	Mangalore	Karnataka
7	Mormugao	Goa Maharashtra
8	Mumbai Port Trust	Goa Maharashtra
9	Jawaharlal Nehru Port Trust (JNPT) also known as Nhava Sheva, Navi Mumbai	Maharashtra
10	Paradip	Odisha
11	Tuticorin	Tamil Nadu
12	Visakhapatnam	Andhra Pradesh
13	Krishnapatnam	Andhra Pradesh

The Indian ports and shipping industry play a pivotal role in strengthening the growth in the country's trade and commerce. India is the sixteenth-largest maritime country in the world with a coastline of about 7,517 kms. The Government of India has taken important steps in upgrading the ports sector. It has allowed Foreign Direct Investment (FDI) of up to 100% under the automatic route for port and harbour construction and maintenance projects. It has also facilitated a 10-year tax holiday to enterprises that develop, maintain and operate ports, inland waterways and inland ports.

The role of Indian ports in hiking the GDP of the country is appreciable. The key ports had a capacity of 1534.91 million tonnes per annum (MTPA) in FY20. In FY21, all key ports in India handled 672.60 million tonnes (MT) of cargo traffic. Merchandise exports reached US$290.63 billion in FY21. In October 2021, India's merchandise exports grew 43.05% Year-over-Year to reach US$33.65 billion.

The Indian Government has taken several measures to improve operational efficiency through mechanization, deepening the draft and speedy evacuations. Between April 2021 and August 2021, cargo handled by India's 12 state-owned major ports recorded an increase of 19.54%, from 245.289 MT to 293.226 MT. In FY22 (until October 2021), cargo traffic handled by India's major ports reached 406.98 MT, a 14.59% increase over the same period last year.

Indian ports received cumulative FDI inflow worth US$1.63 billion between April 2000 and June 2021. In October 2021, the Shyama Prasad Mookerjee Port, Kolkata, gave importers the opportunity to bring in vessels at the deep drafted anchorages located at Sagar Island and Sandheads region located at the confluence of the River Hooghly and Bay of Bengal. In October 2021, Adani Group announced that it wants to make Adani Port a net-zero carbon emitter by 2025 and power all its data centres with renewable energy by 2030. Jawaharlal Nehru Port Trust (JNPT) Special Economic Zone (SEZ) became the first of its kind operational port-based multi-product SEZ in India. The Competition Commission of India (CCI) approved Adani Ports and Special Economic Zone's proposed acquisition of 10.40% equity investment in Gangavaram Port in September 2021. The 10.4% equity shareholding will be bought from the government of Andhra Pradesh. APSEZ (Adani Ports and Special Economic Zone) plans to become the world's largest private port company by 2030 and carbon neutral by 2025.

This sector, however also faces several adverse impacts. Maritime transport infrastructure and services came under severe stress as a result of the pandemic and the closure of the Suez Canal. The adverse impacts of climate change that causes extreme weather events, including floods, hurricanes and cyclones, have been causing frequent and intense disruptions for both coastal infrastructure and hinterland connections. It is therefore the need of the hour to focus on climate-resilient maritime transport infrastructure and services, as well as provide momentum to the development of related legal, policy and technical measures, and capacity-building.

Ship Breaking and Ship Waste Management

Ship breaking activity dates back to ship building activities. In this operation obsolete vessels are fully dismantled after anchoring down on shore to recover its useful components which are about 99% of its light displacement tonnage (LDT). Ship breaking was a major onshore business in USA and some European countries before 1960's. However, these activities have now been shifted to Asian countries and Turkey owing to possibility of more profitable business.

In India, full-fledged ship breaking units were initiated around early eighties in Mumbai and Kolkata. Indian Ship breaking sites are distributed all along the coastal states of West Bengal, Andhra Pradesh, Tamil Nadu, Karnataka, Kerala, Maharashtra and Gujarat (Table 1.15). The various types of ships broken in India are listed in Table 1.16.

In 1979, Government of India recognized ship breaking as manufacturing industry. This industry provides 10–15% of the demand for steel in India. Currently, Alang, Gujarat is the only site with considerable activities of ship breaking.

During the peak period from 1994–2004, this industry has provided employment for about 40,000 workers and dismantled approximately broke about 350 ships per

Table 1.15 Ship breaking locations in India (CPCB 2001)

State	Ship breaking location
Andhra Pradesh	Vishakhapatnam
Gujarat	Alang
	Sachna
Karnataka	Tadri
	Mangalore
	Malpe
Kerala	Bypore
	Cochin
	Azhical
Maharashtra	Mumbai
Tamil Nadu	Tuticorin
West Bengal	Kolkata

Table 1.16 Types of ships broken in India (HPC 2001)

Type of Ship	Percentage (approx.)
General Cargo and Bulk carrier	60–65
Refrigerator Cargo vessel	15–20
Oil tanker	10
Passenger ship and War-ship	5
Cruiser and Drill-ship	5

year on an average of almost one ship each day. Easy beaching of vessels, presence of considerable tidal amplitude, higher scrap value, and cheap workforce has made India a preferable country for ship breaking. However, there has been high competition from Bangladesh and Pakistan with higher demand for scrap and lower taxes. Due several environmental issues connected with ship breaking, this unit has been severely criticized by the environmentalists and hence there is a great demand for eco-friendly ship breaking facilities. In this scenario, an economically feasible environment management system is the need of the hour to make the Indian ship breaking industry environmentally sustainable.

Marine and Coastal Constructions

Blue economy in Indian sub-continent means an enormous scope of economic opportunities that are intricately related to sustaining livelihoods. With a coastline over 7,500 km spread across nine maritime states, four union territories (UTs) and two island UTs, 12 major, and 200 minor ports, India's blue economy supports 95% of the country's business through transportation and contributes an estimated 4% to its Gross Domestic Product (GDP). India is also the third largest fish producing and second largest aquaculture fish producing country in the world (NFDB 2020). Therefore, all the sectors across blue economy have the potential to engage a large workforce and have been doing so from the past many decades at least in sectors such as fishing, aquaculture, fish processing, marine tourism, shipping and port activities. In recent time new sectors such as offshore wind, marine biology, biotechnology, and other activities like ship building and ship breaking is also emerging extensively. On this background, there is great scope for construction of roadways, landing stations, ports and harbors. Even in the marine and coastal tourism sectors, corrosion is a major issue due to high salinity and hence corrosion resistant buildings, embankments and aquaculture need to be constructed.

Marine Based Energy

Increasing power demands and rapid deterioration of the air quality index across majority of the urban localities of India has compelled researchers to find sustainable and harmless alternative sources of energy. Oceans have been an untapped source of energy for India since a long time. Preliminary investigation and identification of potential sites for harvesting ocean energy (OE) began in 1988 and first pilot scale wave energy plant came up in 1990. Vast research has been conducted by many researchers in the field of OE till date but still the contribution of OE is negligible towards meeting the nation's needs. OE can contribute hugely to transform the coastal areas and islands into self reliant ecosystems in terms of energy demands. This unit focuses on reassessing the OE potential of India, the available technologies for harnessing OE and stimulates research on the topics that have not been in focus so far.

India has a coastline of 7516.6 km (Sanil Kumar et al. 2006). Widespread presence of such marine conditions is a great advantage for India to harvest marine energy. OE harvesting technologies in India can be broadly classified under the following 3 heads.

(1) Ocean Wave Energy (OWE)
(2) Tidal Energy (TE)
(3) Ocean Thermal Energy (OTE)

(1) **Ocean Wave Energy (OWE)**

Ocean waves originate as result of energy transfer from the sun to wind and subsequently to water. These waves have the capability to travel thousands of miles with little energy loss.

Measurement of the forces acting upon the wave energy convertor (WEC) is essential for the design process. There are mainly four categories of instruments for measuring wave data. They are wave staff, sub-surface sensor, buoys and ship borne systems.

The power (P) in a wave can be expressed by the formula (Raju and Ravindran 1997),

$$P = \frac{\rho g^2}{64\pi} H_s^2 T \tag{1.1}$$

where P is in kW/m crest length, ρ is the mass density of sea water, g is the acceleration due to gravity, H_s is the significant wave height in meter and T is the wave period in seconds for actual sea state.

For precise design of underwater WEC, determination of wave power at operational depth E(d) is essential. E(d) is calculated by (1.2) (Caroll et al. 2004), where d is depth below sea water level (SWL) and λ is wavelength.

$$E(d) = E(SWL) \cdot e^{\frac{2\pi d}{\lambda}} \tag{1.2}$$

Above equation stands valid only for d greater than (λ)/2.

(2) **Tidal Energy (TE)**

Tidal Energy harvesting under most conditions is not continuous but the timing of tides can be calculate very precisely which makes tidal energy a reliable source of energy.

Tidal energy refers to harnessing energy from fluctuation of sea surface under the influences of tides or from tidally driven currents.

Tidal motion is the result of gravitational pull of the sun and moon. It is the marine manifestation of energy fluxes existent in the earth, moon and sun system (Bryden 2013). Presence of land masses complicates these energy fluxes and result in local enhancement where conditions are befitting.

An initial high expense of setting up a tidal barrage for TE harvesting is one of the primary reasons for limited presence of TE plants in the nation. But local enhancement of tidal fluxes at certain locations across the nation makes TE an economical alternative source of energy. Researchers have been working in the domain of TE harvesting because of its long life and reliability. Because of high expenses, extensive theoretical estimation and hydrodynamic modeling is essential for any detailed discussion on TE.

Tides contain both potential as well as kinetic energy.

$$E_p = 0.5\rho g A_b \Delta h_b^2 \tag{1.3}$$

Equation 1.3 (Murali and Sundar 2017) gives potential energy contained in a water volume impounded in a basin, where E_p is the potential energy over a tide cycle, ρ is the density of sea water, g is the acceleration due to earth's gravity, A_b is the horizontal area of enclosed basin Δh_b is the mean tidal range in the basin.

$$E_k = 0.5 C_p g A_s V^3 \tag{1.4}$$

Equation 1.4 (Murali and Sundar 2017) gives the power generated by a tidal turbine from kinetic energy of the tides, where E_k the kinetic energy of the tidal stream is, C_p is the turbine power coefficient, g is acceleration due to earth's gravity, A_s is the swept area by turbine blades, V is the tidal current.

A study conducted to scan the tidal range, tidal currents along with the available potential and kinetic energies is highlighted in Table 1.17.

(3) **Ocean Thermal Energy (OTE)**

Table 1.17 Tidal potential across Indian coastline

Coastal region	Tidal range (m)	Typical tidal currents (m/s)	Mean potential energy available per square kilometer (MW)	Mean kinetic energy available per square meter (W)
Khambhat	5–11	2.5	10.9	2604.3
Kutch	4–9	3.0	7.2	4500.2
South Gujarat	2–4	2.0	1.5	1333.4
Maharashtra	2–4	1.5	1.5	562.5
Karnataka	1–1.5	1.5	0.2	562.5
Kerala	1–1.5	1.5	0.2	562.5
Tamil Nadu coast	1	0.8	0.1	85.3
Andhra coast	1–2	1	0.2	166.7
Orissa coast	2–4	1.5	1.5	562.5
Hooghly estuary (West Bengal)	4–7	2–3	7.2	2604.3

In a tropical country like India, the temperature difference between free surface of the ocean and depth of around 900 m is above 20 °C (Jayashankar et al. 1998). The fluctuation in this value is only 1 °C round the year. Thus OTEC can be considered as a source of alternative energy.

OTE conversion (OTEC) exploits the temperature difference between sea surface and deep sea or ocean in pursuit to drive a thermodynamic heat engine that can generate electricity.

Hotspots for OTE are the tropical regions which lie between 20° of the equator. In these regions the surface temperature can rise up to 33 °C. At a depth of about 1 km temperature 5 °C. India lies between 8°4′ north to 37°6′ north latitude. It is clearly visible that India is a hotspot for OTE. US National Renewable Energy Laboratory estimated that 60 million km^2 of tropical ocean absorbs an energy equivalent to 250 billion barrels of oil/day (Paul 2019).

Thermal efficiency of a heat engine can computed using (1.5), where W is work, T is surface water temperature, T° is deep water temperature and Q is the thermal value.

$$W = \frac{T - T_{\circ}}{T * Q} \tag{1.5}$$

India got its first breakthrough in OE sector when the OWC/WEC started operation in 1990. As the Table 1.18 summarizes vast developments had taken place in the OE sector in last 3 decades. In 2015, successful completion of trials of Backward Bent Duct Buoy (BBDB) empowered India for generation of offshore power as well. In this decade intense research and developments are under progress pertaining to optimization of hydrokinetic turbines and OWC turbines.

Oil and Natural Gas Extraction

Petroleum beds are common in the Indian Ocean Region (IOR). When viewed in an aggregated sense, the total oil and gas reserves held by IOR states as a percentage of the entire world's proven reserves are impressive (Table 1.19).

India is 85% dependent on imports to meet its oil needs, and a way to cut the high import bill is to increase domestic production.

Mumbai High, which was discovered in 1974, and Bassein and Satellites (B&S) that was put into production in 1988 are Oil and Natural Gas Corporation's (ONGC) mainstay assets, contributing two-thirds of its current oil and gas production. Without these assets, the company will be left with only smaller fields.

ONGC's consolidated crude oil output has fallen from 25.5 million metric tonnes (MMT) in FY17 to 22.5 MMT in FY21, while its natural gas production has fallen to 22.8 billion cubic meters (BCM) from a peak of 25.8 BCM in FY19 (https://indianexpress.com/article/business/economy/seek-foreign-partners-to-raise-output-from-mumbai-high-bs-fields-govt-to-ongc-7603318/).

Table 1.18 Major developments in OE sector in India

Decade	India's milestones in OE sector
1980–1990	• CEA had identified tidal potential in the Gulf of Kutch in Gujarat. The investigations were formally launched in 1982. The techno-economic feasibility study was completed in 1988
	• First wave power effort from IIT Madras Pilot plant by GoI
1990–2000	• First OWC type wave energy pilot plant was installed at Vizhinjam
	• Generation and grid connectivity demonstrated
	• Hybrid BW was proposed by DOD at Thangassery and Mus point
2000–2010	• Horizontal axis turbine by NIOT
	• BBDB program started
	• 1 MW OTEC-Installation difficulties
2010–2020	• Kalpasar Tidal
	• Double chamber OWC at IIT Madras-optimization
	• BBDB Demo by NIOT
	• Atlantis-Kutch for 50–200 MW-Tidal Stream
	• Establishment of 1 lakh litres per day LTTD plant powered by OTEC at Kavaratti Island
	• Establishment of waste heat recovery plant at Tuticorin Thermal Power Station
	• Development of turbines for harnessing energy from OTEC/Wave energy/Ocean currents which could be stand-alone units or used to power LTTD plants or small loads

Considering this downfall, the petroleum Ministry, Government of India has asked state-owned Oil and Natural Gas Corporation (ONGC) to consider offering up to a 60% stake and operating rights over parts of key Mumbai High and B&S oil and gas producing fields to foreign companies to boost production. The proposal aims to boost crude oil and natural gas recovery from the maturing fields.

Deep Sea Minerals

The physical, chemical and biological characteristics in the deep sea environment are unique and regulate the formation and distribution of minerals on the sea floor. The minerals extracted from the sea bed exhibit the presence of sulphur, cobalt, nickel, iron etc. and can be categorized into (i) Sea-floor Massive sulphides, (ii) Cobalt-rich Ferromanganese Crusts and (iii) Polymetallic (manganese) nodules. Manganese nodules have great potential in strengthening the economic backbone of the nation as they are composed of a wide range of elements (Table 1.20).

Data on the availability of the deep sea minerals in the South Asian region is very scanty. However India has gained a momentum in this sector and is in the way of building a processing plant for rare earth minerals by investing US$135 million on

Table 1.19 Proven oil and natural gas reserves in the Indian Ocean region

Country	Oil (BBL)	Percent of global total	Gas (TCM)	Percent of global total
Saudi Arabia	688.9	19.8	7.46	3.92
Iran	137.6	10.3	29.6	15.57
Iraq	115	8.6	3.17	1.67
Kuwait	104	7.6	1.79	0.95
UAE	97.8	7.3	6.07	3.19
Qatar	25.4	2	25.47	13.39
Sudan	6.8	0.5	0.85	0.04
India	5.8	0.5	1.07	0.57
Oman	5.5	0.4	0.85	0.45
Malaysia	5.5	0.4	2.35	1.24
Egypt	4.3	0.3	1.66	0.87
Australia	4.2	0.3	3.12	1.64
Indonesia	4.05	0.3	3	1.58
Yemen	3	0.2	0.48	0.25
Timor-Leste	0.55	–	0.2	0.11
Pakistan	0.44	–	0.84	0.44
Thailand	0.43	–	0.342	0.18
Bahrain	0.12	–	0.09	0.05
Myanmar	0.05	–	0.28	0.15
Bangladesh	0.02	–	0.12	0.1
South Africa	0.015	–	0.002	0
Israel	0.001	–	0.03	0.02
Jordan	0.001	–	0.006	0
Tanzania	0	–	0.0065	0
Somalia	0	–	0.0056	0

Table 1.20 Average elemental concentrations for Manganese Nodules from different seafloor provinces

Average elemental content (dry wt. %)	Atlantic Ocean	Pacific Ocean	Indian Ocean
Manganese	15.46	19.27	15.25
Iron	23.01	11.79	13.35
Nickel	0.308	0.846	0.534
Copper	0.141	0.706	0.295
Cobalt	0.234	0.290	0.247
Manganese/Iron	0.67	1.6	1.19

Source Cronan (1977, 1980, 2000)

a new exploration ship (http://www.deepseaminingoutofourdepth.org/india-backs-exploration-of-rare-earths-in-deep-sea/). The recent push for deep-sea mining exploration reflects India's concerns that China's deep-sea excavations will further increase China's dominance over rare-earth elements, which are used in aviation and defense manufacturing, as well as electronics. China Ocean Mineral Resources Research and Development Association (COMRA) obtained a license to mine polymetallic sulphides in the Southwest Indian Ridge (SWIR). The SWIR is a divergent tectonic plate boundary between the African and Antarctic plates, running from the South Atlantic to a junction—the place where two or more plates meet—in the Indian Ocean just south of Madagascar.

The International Seabed Authority (ISA), an autonomous international organisation established under the 1982 United Nations Convention on the Law of the Sea, allots the 'area' for deep-sea mining. India was the first country to receive the status of a 'Pioneer Investor' in 1987 and was given an area of about 1.5 lakh km^2 in the Central Indian Ocean Basin (CIOB) for nodule exploration. In 2002, India signed a contract with the ISA and after complete resource analysis of the seabed 50% was surrendered and the country retained an area of 75,000 km^2.

According to a release from the Ministry of Earth Sciences, the estimated polymetallic nodule resource potential in this area is 380 million tonnes (MT), containing 4.7 MT of nickel, 4.29 MT of copper, 0.55 MT of cobalt and 92.59 MT of manganese. Further studies have helped narrow the mining area to 18,000 km^2 which will be the 'First Generation Mine-site'.

India is the second most populated country and the seventh-largest country by land area (Schwartzberg et al. 2021).

Known for being the largest Democratic country globally and the highest number of official languages (Schwartzberg et al. 2021), India is also known for being the pioneer of its exploration of the deep sea (Aggarwal 2021).

India being the first to recover a polymetallic nodule in 1981 from the center of the Indian Ocean (Aggarwal 2021), began its exploration of the unknown.

The Indian government has been recently cleared to create a system in which they will mine polymetallic nodules, a mineral found in the Indian Ocean from about 6 km deep in the center of the ocean (Aggarwal 2021).

With the decrease of land-based minerals such as copper, nickel, aluminum, manganese, zinc, lithium, and cobalt to manufacture high-tech gadgets such as our smartphones (Aggarwal 2021), the need for deep-sea mining has increased. In this context, extraction of polymetallic nodule, also known as manganese nodule, is a great step, which is formed of concentrated minerals through concentric layers of iron and manganese hydroxide around the core (Hein et al. 2020).

The uses of different metals present in the polymetallic nodule is presented in Table 1.21

Deep sea mining operations have high probability to pose an adverse impact on the deep sea habitats preferably on the benthic organisms. We therefore suggest few precautionary approaches in connection to deep sea mineral extraction. These approaches are highlighted in Table 1.22.

Table 1.21 Uses of metals in polymetallic nodules

Sl. No	Metal/Alloy	Uses
1	Copper	1. Resistant to corrosion caused by salinity 2. Good conductor of electronic gadgets 3. Used in telecommunication and electronic gadgets
2	Titanium	1. Used in the aerospace industry 2. Used in the medical field for bone plates 3. Used in pacemakers
3	Nickel alloy	1. used in aerospace industry 2. Used in defense sector 3. Used for making electronic gadgets 4. Used in the power, oil and natural gas exploitation/extraction appliances
4	Manganese	1. Used in the steel industry 2. Used in making Mn batteries, Al alloys and micronutrients
5	Cobalt	1. Used greatly in the pharmaceutical industries 2. Used as a catalyst for chemical and petroleum industries 3. Used in making magnets

1.3 Take Home Messages

(A) The over exploitation of forest/land resources have pushed the human civilization to touch the boundary wall of the resource base of the planet Earth. Hence it is perhaps the need of the hour to think of the blue economy, which is exclusively marine based economy encompassing the living and non-living resources of the oceans, seas, bays and estuaries. Covering 71% of the Earth's surface, the saline water of the marine and estuarine ecosystems embedded in different compartments exhibits substantial area and volume with rich reservoir of living and non-living resources. The living resources provide opportunities for industries and livelihoods in the sector of fisheries, aquaculture, marine pharmaceutical products and chemicals, tourism and recreation, carbon sequestration and eco-restoration. The non-living resources of the marine and estuarine ecosystems act as the foundation stones of few important industries like salts and marine chemicals, oil and natural gas, desalinization for freshwater generation, shipping, ship building, ship wrecking, ship waste recycling, marine transport, power plants based on solar, wind, tidal energy and Ocean Thermal Energy Conversion.

(B) India has a coastline of 8,129 km encompassing 3,600 fishing villages. The EEZ covers an area of 2.03 million km^2. The marine and estuarine ecosystems of the country support rich taxonomic diversity of the flora and fauna along with several categories of non-living resources. The most prospective non-living resources related to blue economy of the country are (i) Marine Transportation, (ii) Ship Breaking and Ship Waste Management, (iii) Marine and Coastal Constructions, (iv) Marine Based Energy, (v) Oil and Natural Gas Extraction and (vi) Deep Sea Minerals. According to a note released by the Ministry

Table 1.22 Precautionary approaches for deep sea mining operations

Sl. No	Precautionary approach	Description
1	Gap area filling	1. On the basis of the baseline data, the gap areas should be identified and as per the need of the companies related to blue economy, the data need to be provided backed up with proper environmental impact management 2. Need for quantification of resource based coupled with improved understanding of the extraction process (CAPEX and OPEX in a DSM mine model per deposit type)
2	Long-term planning	1. Capacity building for efficient and effective exploration and extraction 2. Development of a financial model for resource distribution uncertainties and good revenue administration 3. Risk analysis 4. Linking the resource base with National economy
3	Integrated marine management	1. Environmental Impact Assessment 2. Development of ecofriendly Environment Management Plan (EMP)
4	Manpower	1. Trained manpower needs to be deployed to understand the dynamics of the ocean for exploration and extraction of deep sea minerals 2. Trained environmentalists need to be developed to ensure protection to deep sea biodiversity during and after the extraction process
5	Policy	1. Implement policy legislative, and institutional reforms on a country-by-country basis 2. Ensure that the prescribed regime is consistent with the International Law and the World Bank Group Safeguards and Performance Standards

of Earth Sciences, Government of India, the estimated polymetallic nodule resource potential in within Indian jurisdiction is some 380 million tonnes (MT), that contain 4.7 MT of nickel, 4.29 MT of copper, 0.55 MT of cobalt and 92.59 MT of manganese. The demands for these metals are gradually rising as they are widely used in aerospace industry, steel industry, different categories of pharmaceutical industries, medical equipment manufacturing units, electrical and electronic goods/gadgets manufacturing units and petroleum industries.

(C) Deep sea mining operations have high probability to pose an adverse impact on the deep sea habitats preferably on the benthic organisms. Hence eco-friendly technologies need to be developed along with capacity building. Trained manpower needs to be deployed to understand the dynamics of ocean for exploration and extraction of deep sea minerals along with environmentalists/marine biodiversity experts to ensure protection to deep sea biodiversity during and after the extraction process. It is also important to ensure that the mining operations are consistent with the International Law and the World Bank Group Safeguards and Performance Standards.

Annexure A Phytochemical Compounds Isolated from Mangrove Species

Species	Compound isolated	
Acanthus ilicifolius	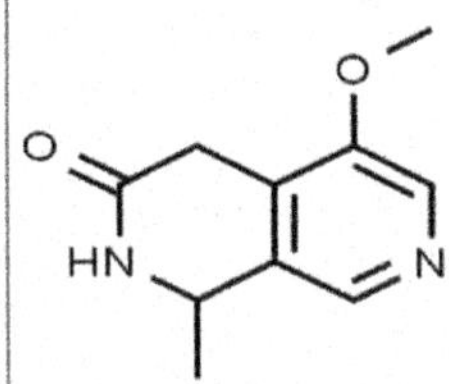 Acanthicifoline	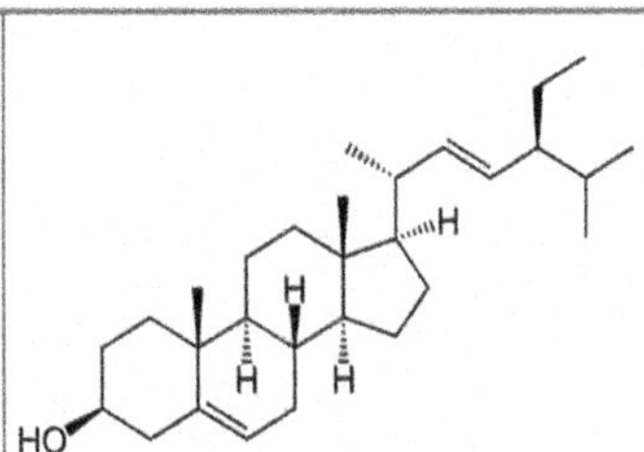 Stigmasterol
Aegiceras corniculatum	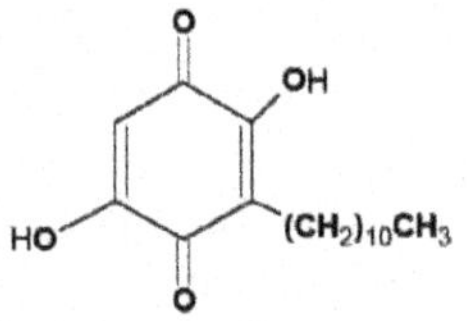 2,5-dihydroxy-3-undecyl-1,4 benzoquinone	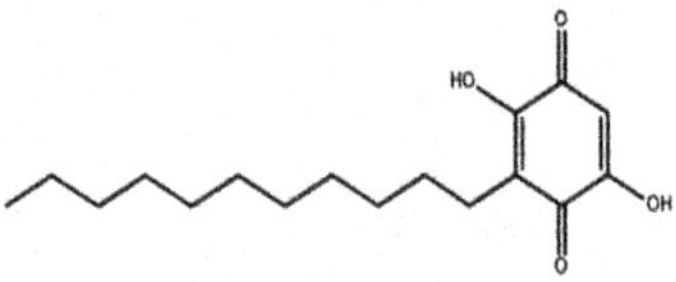 Methyl Embelin
Avicennia alba	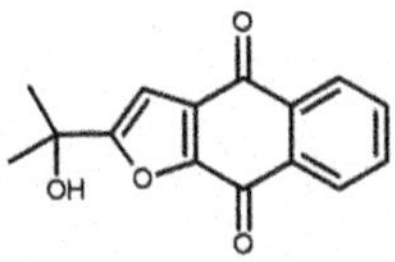 Avicequinone C	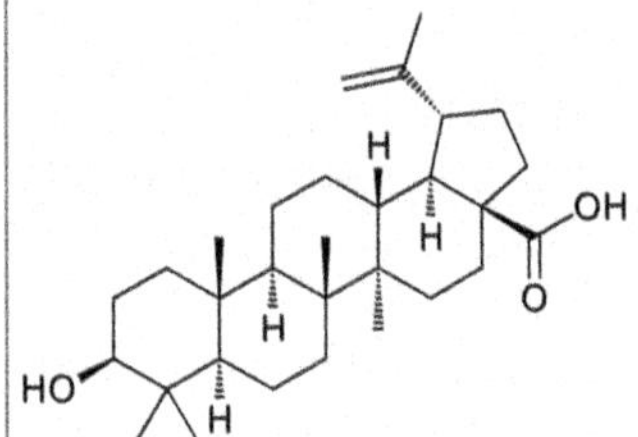 Betulinic acid

(continued)

(continued)

Species	Compound isolated	
Avicennia marina	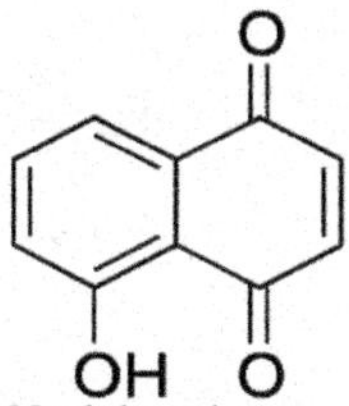 Naphthoquinone	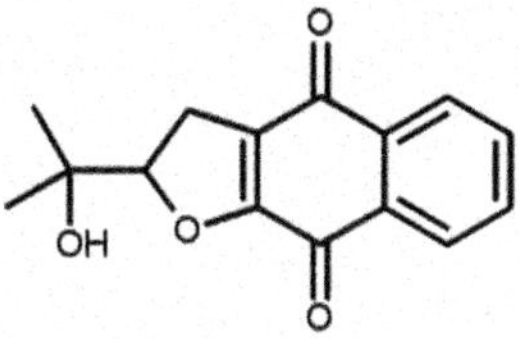 Stenocarpoquinone B
Avicennia officinalis	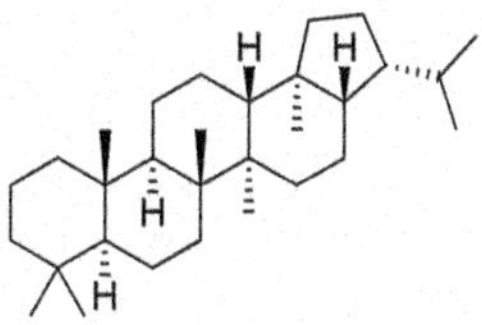 Triterpene	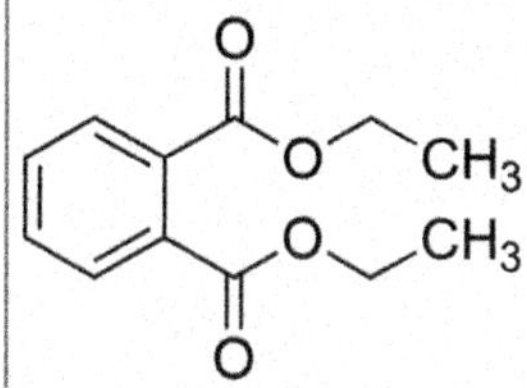 Diethyl phthalate
Bruguiera gymnorrhiza	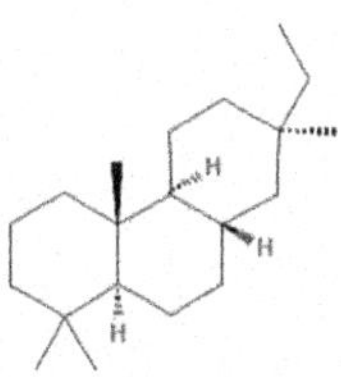 Pimaranes	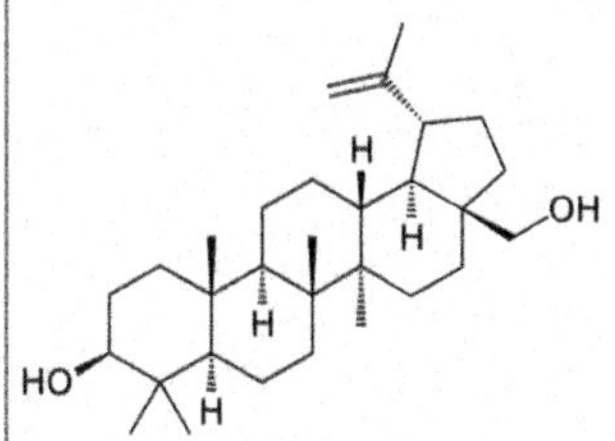 Betulin
Ceriops decandra	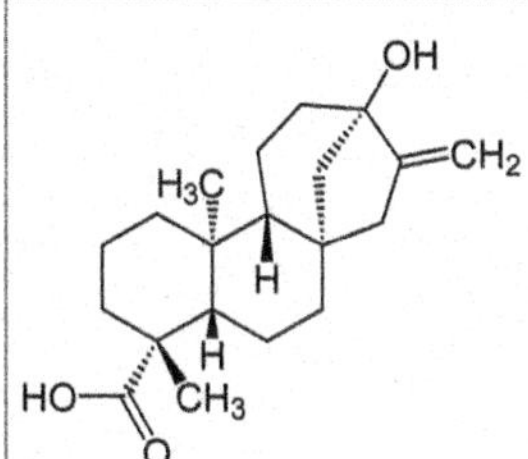 Steviol	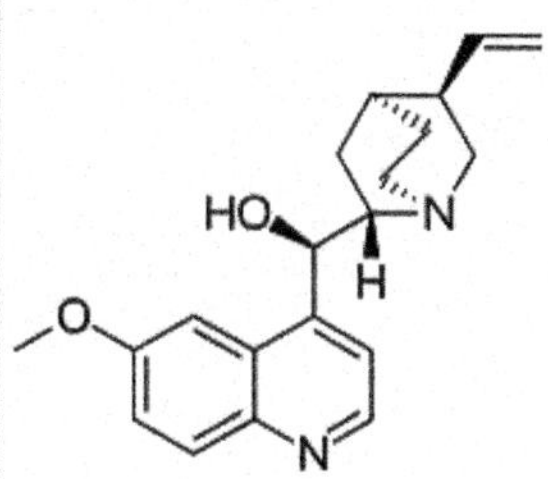 Quinine

(continued)

(continued)

Species	Compound isolated	
Excoecaria agallocha	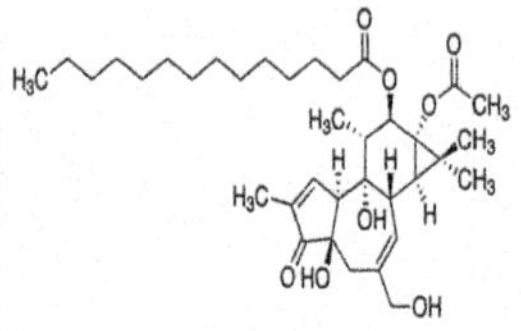 Phorbol ester	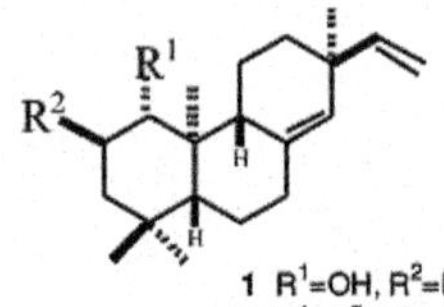 1 R^1=OH, R^2=H 2 R^1=R^2=OH 3 R^1=H, R^2=OH Ent-isopimarane diterpenoids
Heritiera fomes	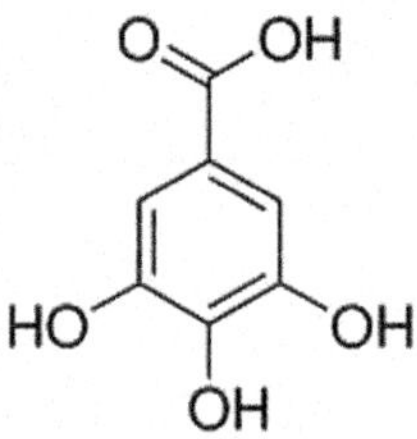 Tannin	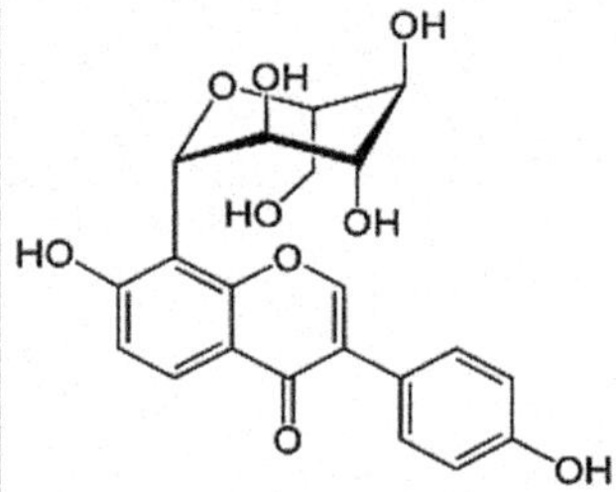 Polyphenol
Lumnitzera racemosa	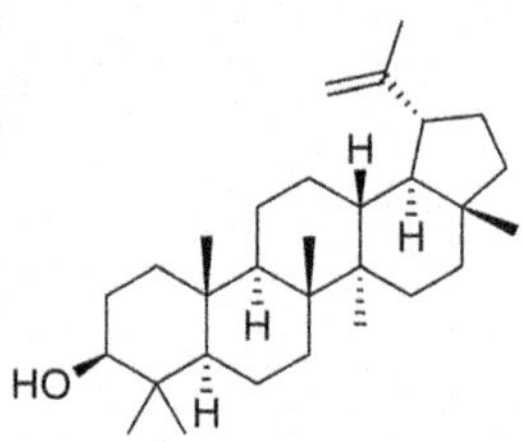 Lupeol	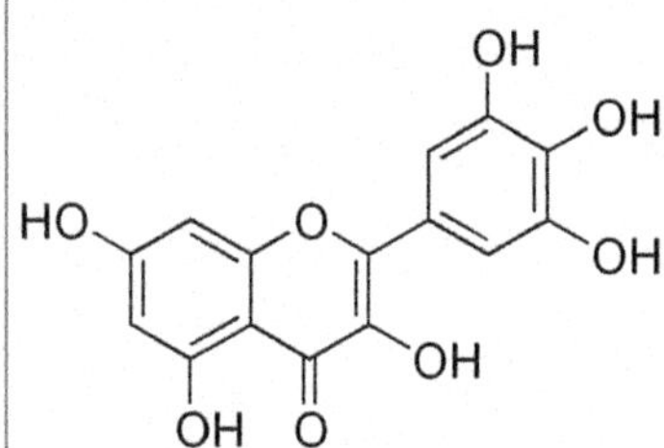 Myricetin
Nypa fruticans	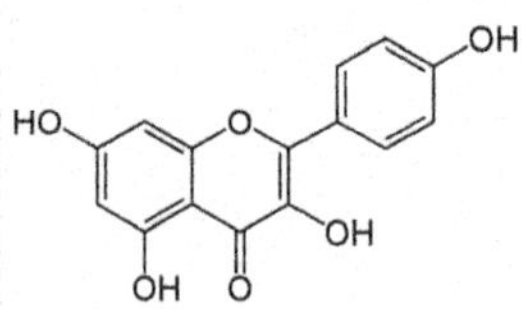 Kaempferol	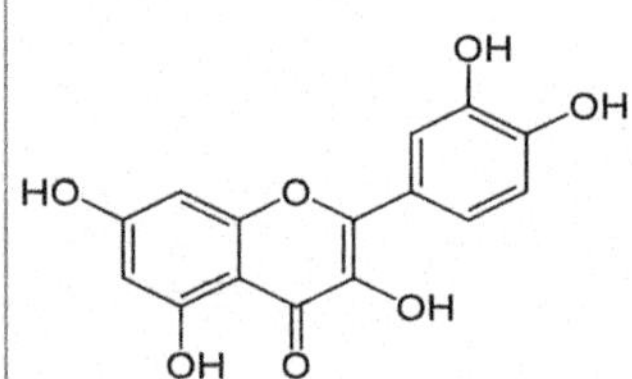 Quercetin
Phoenix paludosa	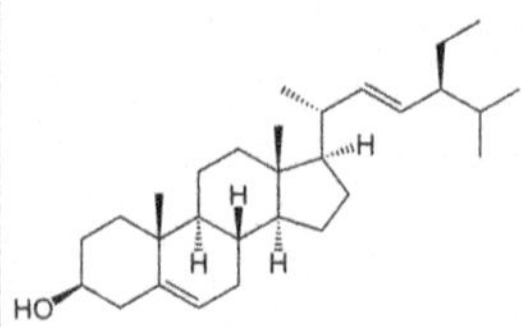 Stigmasterol	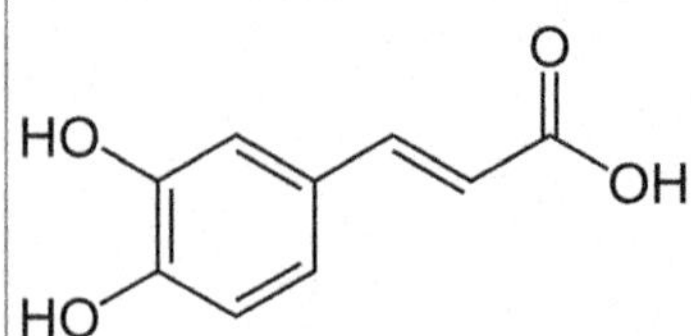 Caffeic acid

(continued)

(continued)

Species	Compound isolated	
Rhizophora apiculata	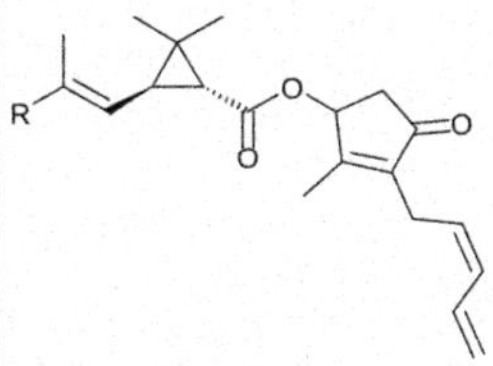 Pyrethrin	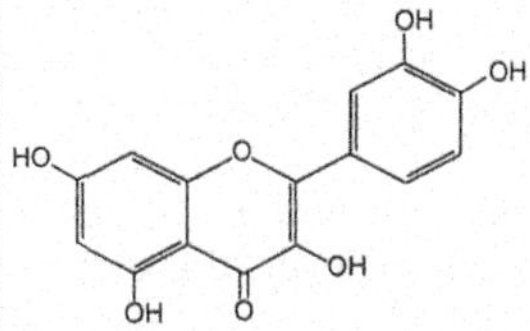 Quercetin
Sonneratia apetala	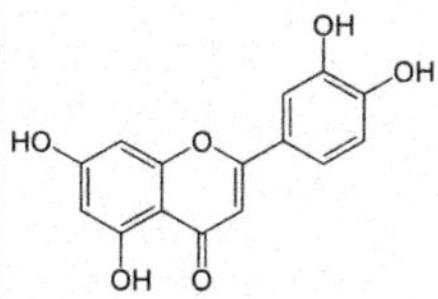 Luteolin	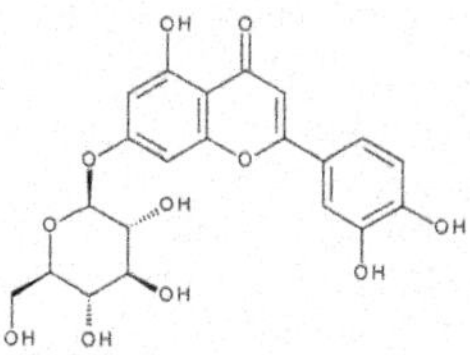 Luteolin 7-O-β-glucoside
Xylocarpus granatum	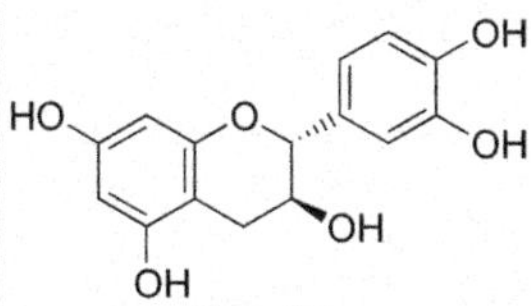 Catechin	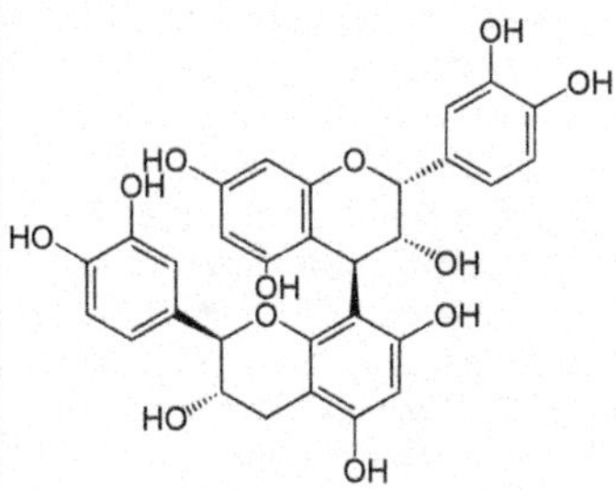 Procyanadin B1

References

Aggarwal M (2021) India's deep seabed mining plans gear up for a dive. Mongabay. https://india.mongabay.com/2021/07/indias-deep-seabed-mining-plans-gear-up-for-a-dive/.

Bryden IG (2013) Tidal energy. In: Kaltschmitt M, Themelis NJ, Bronicki LY, Söder L, Vega LA (eds) Renewable energy systems. Springer, New York, NY

Caroll C, Bell M (2004) Wave energy convertors utilizing pressure differences. US patent 6.933.6233. To Ocean Power Technologies, Inc., Patent and Trademark Office, Washington D.C.

CPCB Center Home Page of Central Pollution Control Board (2001) http://www.cpcb.nic.in/News%20Letters/Archives/Polychlorinated%20Biphenyls%20. (PCBs)/Ship%20Breaking%20Activities%20-%20Source_of_toxic.html

Cronan DS (1977) Deep-sea nodules: distribution and geochemistry. In: Glasby GP (ed) Marine manganese deposits. Elsevier oceanographic series 15. Elsevier (pubs.) Amsterdam, pp 11–44

Cronan DS (1980) Underwater minerals. Academic Press (pubs.), London, pp 121–195

Cronan DS (ED) (2000) Handbook of marine mineral deposits. CRC Press (pub.), pp 347–368
Dagar JC, Mongia AD, Bandhyopadhyay AK (1991) Mangroves of Andaman and Nicobar Islands. Oxford and IBH, New Delhi, p 166
Duke NC (1992) Mangrove floristics and biogeography. In: Robertson AI, Alongi DM (eds) Tropical mangrove ecosystems. American Geophysical Union, Washington DC, USA, pp 63–100
Duke NC, Ball MC, Ellison JC (1998) Factors influencing biodiversity and distributional gradients in mangroves. Global Ecol Biogeogr 7(1):27–47
Ellison AM, Farnsworth EJ, Merkt RE (1999) Origins of mangrove ecosystems and the mangrove biodiversity anomaly. Global Ecol Biogeogr 8(2):95–115
Field CD (1995) Impacts of expected climate change on mangroves. Hydrobiologia 295(1–3):75–81
FSI (2015) India State of forest report. Forest Survey of India (FSI). Dehra Dun, pp 62–67
Hamilton RS, Snedaker SC (1984) Handbook for mangrove area management. Gland, Switzerland. Commission on Ecology, IUCN
Hamilton SE, Casey D (2016) Creation of a high spatio-temporal resolution global database of continuous mangrove forest cover for the 21st century (CGMFC-21). Global Ecol Biogeogr 25(6):729–738
Hein JR, Koschinsky A, Kuhn T (2020) Deep-ocean polymetallic nodules as a resource for critical materials. Nature News. https://www.nature.com/articles/s43017-020-0027-0
HPC (2001) A brief note on CPCB's guidelines and present status of implementation at Alang ship breaking yard. Ministry of Environment and Forests, India. http://www.envfor.nic.in/cpcb/hpc report/vol3pdf/vol-III%20part%20I.pdf. Accessed Jan 2006
http://www.bsienvis.nic.in/Database/IndianMangroves_3941.aspx#WORLD DISTRIBUTION OF MANGROVES
http://www.deepseaminingoutofourdepth.org/india-backs-exploration-of-rare-earths-in-deep-sea/
https://indianexpress.com/article/business/economy/seek-foreign-partners-to-raise-output-from-mumbai-high-bs-fields-govt-to-ongc-7603318/
https://worldpopulationreview.com/countries
https://www.worldometers.info/world-population/world-population-countries.php
Jayashankar V, Purnima J, Ravindran M, Mitsumort M, lkegami Y, Uehara H (1998) The Indian OTEC Program, Ocean Technology Conference Texas.
Kathiresan K (2010) Importance of Mangrove forest of India. J Coast Environ 1:11–26
Mandal RN, Naskar KR (2008) Diversity and classification of Indian Mangroves: a review. Trop Ecol 49:131–146
Mitsch WJ, Gosselink JG (2007) Wetlands, 4th edn. Wiley, New York, USA, p 582
Murali K, Sundar V (2017) Reassessment of tidal energy potential in India and a decision-making tool for tidal energy technology selection. Int J Ocean Clim Syst 8(2):85–97
NFDB (National Fisheries Development Board) (2020) Introduction to fish and fisheries. NFDB: National Fisheries Development Board. http://nfdb.gov.in/Fish-and-Fisheries-of-India
Paul B (2019) Chapter 9—Tidal barrage power plants. Power generation technologies, 3rd edn. Newnes, pp 203–217. ISBN 9780081026311
Polidoro BA, Carpenter KE, Collins L, Duke NC, Ellison AM, Ellison JC, Farnsworth EJ, Fernando ES, Kathiresan K, Koedam NE, Livingstone SR, Miyagi T, Moore GE, Nam VN, Ong JE, Primavera JH, Salmo SG III, Sanciangco JC, Sukardjo S, Wang Y, Yong JWH (2010) The loss of species: Mangrove extinction risk and geographic areas of global concern. PLoS ONE 5(4):1–10
Raju VS, Ravindran M (1997) Wave energy: potential and programme in India. Renew Energy 10(2/3):339–345
Richards JF (1986) World environmental history and economic development in sustainable development of the biosphere. In: Clark WC, Munn RE (eds) International Institute for Applied Systems Analysis, Laxenburg, Austria
Ricklefs RE, Latham RE (1993) Global patterns of diversity in Mangrove Floras. In: Ricklefs RE, Schluter D (eds) Species diversity in ecological communities, historical and geographical perspectives. University of Chicago Press, pp 215–229

Saenger P (2002) Mangrove ecology, silviculture and conservation. Kluwer Academic Publishers, The Netherlands, p 360

Sanil Kumar V, Pathak KC, Pednekar P, Raju NSN, Gowthaman R (2006) Coastal processes along the Indian coastline. Curr Sci 91(4)

Schwartzberg JE, Alam M, Calkins PB, Spear TGP, Dikshit KR, Allchin FR, Subrahmanyam S, Wolpert SA, Srivastava AL, Thapar R, Champakalakshmi R (2021) India. Encyclopedia Britannica. https://www.britannica.com/place/India

Scientists HaSEEa (2005) Biscayne Bay economic study-task 3 report final Biscayne economic baseline and trends report (South Florida Water Management Disctrict)

Spalding M, Kainuma M, Collins L (2010) World Atlas of mangroves. ISME Publication, p 320

Tomlinson PB (1986) The botany of mangroves. Cambridge (Cambridge University Press), p 419

Unsworth RK, Cullen-Unsworth LC (2014) Biodiversity, ecosystem services, and the conservation of seagrass meadows. Coast Conserv 19:95

Wang BS, Liang SC, Zhang WY, Zan QJ (2003) Mangrove flora of the world. Acta Botanica Sinica 45(3):644–653

Chapter 2
Blue Economy in Context to the Mangrove Dominated Indian Sundarbans

Abstract Indian Sundarbans in the lower Gangetic delta with the crowns of World Heritage Site and Biosphere Reserve is blessed with unique genetic diversity. The region supports some 34 species of true mangroves and a wide spectrum of floral and faunal diversity that render several ecosystem services. It is the hotspot of biodiversity and serves as the home of a large number of endangered and vulnerable species. The present chapter highlights the ecosystem services of the natural resources of Sundarban mangrove ecosystem that can offer a wide range of jobs and livelihoods related to blue economy in this region. The most relevant sectors of blue economy in the mangrove dominated Indian Sundarbans are (i) fisheries and aquaculture, (ii) non-living resources like oil and natural gas, (iii) shipping and transportation of goods, (iv) non-conventional energy, (v) marine biodiversity and conservation and (vi) estuary/sea-based tourism. The present chapter highlights the magnitude of opportunities of each of these sectors that serve as the spokes of blue economy.

2.1 Biotic Resources of Indian Sundarbans

The estuaries of Indian Sundarbans along with the living and non-living resources are the basic sources of livelihoods in and around this World Heritage Site. Mangroves, mangrove associate species, seaweeds, fishes of various categories, molluscs, arthropods and even the big mammals like Royal Bengal Tiger (*Panthera tigris tigris*), deer, monkey, wild boar and birds of various types, reptiles and innumerable flora and fauna have potential to sustain the blue economy of the region. A sustainable blue economy encompasses activities in which ecological risks are mitigated and minimized. This is primarily achieved through application of 'circular economy' in which several eco-friendly and innovative technologies are involved. There are three major points to achieve sustainable blue economy from the living and non-living resources embedded in the dynamic martrix of mangrove dominated Indian Sundarbans and these are growth, risk management (through eco-friendly carbon neutral approaches) and Return On Investment (ROI).

The foundation of blue economy in the mangrove dominated Indian Sundarbans stands on the pillars of floral and faunal diversities, which are unique in this region.

A. Mitra et al., *Blue Economy in Indian Sundarbans*,
https://doi.org/10.1007/978-3-031-07908-5_2

2.1.1 Overview of Floral Biodiversity

Indian Sundarbans having the crowns of World Heritage Site and Biosphere Reserve supports a wide spectrum of flora that encompasses true mangroves, mangrove associate species, saltmarsh grass, seaweeds and phytoplankton. In this unit, apart from providing an overview of floral spectrum in the region, we have highlighted few floral species that have the possibility of strengthening and sustaining blue economy in the region.

True Mangroves

True mangroves are special types of halophytes having features like pneumatophores, prop roots, stilt roots, viviparous germination etc. that are abundantly available along the estuaries of the deltaic complex in Indian Sundarbans. 34 varieties of true mangrove species have been documented from this region, which exhibit significant spatial variation in response to salinity that vary widely between the western, central and eastern sectors of Indian Sundarbans (Mitra and Zaman 2015, 2021; Mitra 2020). These true mangroves (Table 2.1) have several economic importances like production of timber, fuel wood, thatching materials, honey, wax etc. Many mangrove species also have medicinal properties and are widely used by the local people to cure various diseases. All these products support the mangrove based alternative livelihood either directly or indirectly and can be linked with the wings of blue economy in the region (Tables 2.2 and 2.3).

It is to be noted that all the 34 varities of true mangrove flora do not flourish in the same site/region; rather their growth and survival are direct functions of salinity. Table 2.4 reflects the salinity preference of few dominant mangrove species in this deltaic complex along with their possible linkage with regional blue economy.

A study done at an interval of ten years in 24 stations in Indian Sundarbans showed relatively more above ground biomass (expressed in tonnes/ha) of *Sonneratia apetala* (a fresh water loving mangrove species) in the low saline zones like Lothian Island, Sagar Island, Muriganga etc. compared to high saline zone like Thakuran, Herobhanga, Ajmalmari etc. in the central Indian Sundarbans (Fig. 2.1), which confirms the regulatory role of salinity on the growth of mangrove flora in the coastal and estuarine regions of the maritime state of West Bengal, India (Mitra et al. 2011; Mitra 2013; Mitra and Zaman 2014, 2016, 2021; Mitra 2020).

Mangrove Associate Species

Mangrove associate floral species constitute an integral part of mangrove biodiversity in Indian Sundarbans. In Indian Sundarbans, the salt marsh grass *Porteresia coarctata* is a very common mangrove associate species, which can tolerate a wide range of salinity (Fig. 2.2).

Table 2.1 True mangroves in the Indian Sundarbans Biosphere Reserve along with their abundance (documented during our qualitative field survey from January, 2021—March, 2021)

Sl. No.	True mangrove flora	Abundance		
		W	C	E
1	*Acanthus ilicifolius*	+++	+++	+++
2	*Acanthus volubilis*	+++	+++	++
3	*Aegiceros corniculatum*	+++	+++	+++
4	*Aegialitis rotundifolia*	+++	+	+++
5	*Amoora cucullata*	+	+	++
6	*Avicennia alba*	++	+++	++
7	*Avicennia marina*	++	+++	+++
8	*Avicennia officinalis*	++	+++	+++
9	*Brownlowia tersa*	++	+	+++
10	*Bruguiera cylindrica*	++	+	++
11	*Bruguiera gymnorrhiza*	+++	++	+++
12	*Bruguiera parviflora*	+++	+	+++
13	*Bruguiera sexangula*	+++	+	+++
14	*Ceriops decandra*	+	+++	+++
15	*Ceriops tagal*	+	++	+++
16	*Cynometra ramiflora*	NS	NS	NS
17	*Derris trifoliate*	+++	+	+++
18	*Derris umbrellatum*	NS	NS	NS
19	*Excoecaria agallocha*	+++	+++	+++
20	*Excoecaria bicolor*	+++	+++	+++
21	*Heritiera fomes*	+	#	++
22	*Hibiscus tortuosus*	NS	NS	NS
23	*Kandelia candel*	NS	NS	NS
24	*Lumnitzera racemosa*	NS	NS	NS
25	*Nypa fruticans*	++	+	+++
26	*Phoneix paludosa*	–	+++	+++
27	*Rhizophora apiculata*	+	++	++
28	*Rhizophora mucronata*	+	++	++
29	*Sonneratia apetala*	+++	+	+++
30	*Sonneratia caseolaris*	+	+	+++
31	*Tamarix dioica*	NS	NS	NS
32	*Tamarix gallica*	NS	NS	NS
33	*Xylocarpus granatum*	++	++	++
34	*Xylocarpus mekongensis*	NS	NS	NS

'–' means absent, '+' means present in small number, '++' means present in moderate level and '+++' means high abundance, 'NS' means not seen at the sites during our survey period; W, C and E represent the western, central and eastern regions of Indian Sundarbans respectively in the study area

Table 2.2 Direct benefits from mangroves

Sl. No.	Item	Benefits
1	Fuel	Mangrove trees are the sources of firewood for cooking, heating and burning bricks; charcoal and alcohol are also obtained from mangrove vegetation
2	Construction	Mangroves are the sources of timber for boat building, dock pilings, fence, posts, chipboards and flooring and paneling, beams and poles for building; The leaves of *Nypa* are used for thatching the houses
3	Fishing	Mangrove trees can be used to make poles for fish traps, fishing floats, etc.; fishing boats are also made from mangrove wood
4	Textiles	Mangroves are the sources of dye for cloth, and tannin for leather preservation
5	Food, drugs and beverages	Mangroves can be used for extracting sugar, alcohol, cooking oil, vinegar, medicines from bark, leaves and fruits, tea substitute, fermented drinks, etc.
6	Household items	Mangrove wood can be used for making furniture, glue, hair dressing oil, tool handlers, rice mortars, toys, matchsticks and incense
7	Agriculture	Mangrove leaves are used as fodder and green manure. The cyanobacterial strains present on the forest floor/intertidal mudflat of mangrove ecosystem with high nitrogent content are important source of biofertilizer
8	Paper products	Mangrove wood are used for making paper and packing boxes

The species initiate the process of ecological succession and imparts stability to the intertidal mudflats of deltaic Sundarbans, which otherwise could face severe erosion. We extensively visited different islands of Indian Sundarbans and observed that embankments/intertidal mudflats without this vegetation cover face erosion of very high degree (Fig. 2.3).

Many of these mangrove associate species provide several types of ecosystem services that are intricately related to the regional blue economy. A list of mangrove associate species commonly available in Indian Sundarbans is highlighted in Table 2.5.

Seaweeds

Seaweeds are thallophytes with no differentiation of roots, stems and leaves and thrive luxuriantly in the marine and estuarine ecosystems.

A checklist of seaweeds commonly available in Indian Sundarbans is highlighted in Table 2.6.

Table 2.3 Indirect benefits from mangroves

Sl. No.	Item	Benefits
1	Food	Mangrove forests serve as the breeding and nursery grounds for fishes, prawns, shrimps, crabs, oysters, mussels and cockles which are the important food items for the local population. Many of these finfishes and shell fishes have high export values. In addition to these few mangroves are also used for making snacks, jellies, jams etc. (Pramanick et al. 2014; Biswas et al. 2019). The recent use of vitamin C-rich fruits of *Sonneratia apetala* for making sweet snacks locally called 'narkel naru' has the possibility to open up a mangrove based cottage industry in the region (*Vide* Annexure 3.2 in Chap. 3 for details). This sweet food item has high demand in the maritime state of West Bengal preferably during the festivals
2	Extracts and other products	Mangrove forests are the housing complexes for insects, birds, mammals, reptiles and several faunal species. Honey, wax, food, etc. are obtained from mangrove ecosystem. The molluscan species in the mangrove ecosystem (preferably the oysters) are the sources of lime
3	Recreational activities	The mangrove ecosystem can be used for several recreational activities such as power boating, canoeing, fishing, collecting molluscs and crustaceans, picnic spots, swimming, bird watching, wildlife observation, photography, nature education and tourism
4	Regulating services	Bioremediation, nutrient cycling, carbon sequestration, oxygen generation, wave attenuation etc. are the contributions of mangrove floral species that have not been valued with accuracy in the present geographical locale

Phytoplankton

Marine and estuarine phytoplankton are tiny floral entities distributed in the ocean, seas, bays and estuaries, which cannot move against the current of the water bodies. They are free floating in nature and have multiple uses that can be linked with blue economy as listed here.

1. Phytoplankton are the primary producers lying at the base of the food chain/web and support the entire animal kingdom in the marine and estuarine ecosystems. They provide nutrition and energy across the food pyramid.
2. Phytoplankton are the sources of fossil fuels. It has been reported that some species of phytoplankton store extra food as oil rather than starch to get an advantage in terms of buoyancy. When oil-storing forms of phytoplankton die and descend to the bottom of the sea and undergo microbial degradation, they are often covered by huge sediment load and subjected to enormous pressure. Such conditions are supposed to change the deposited phytoplankton into fossil fuel after hundreds and millions of years. Thus human world is benefitted from these tiny organisms even after their death.
3. Diatoms constitute an important group in phytoplankton community and are commercially important as 'diatomaceous earth' (diatomite or kieselghur), which

Table 2.4 Salinity preferences of most common mangrove species in Indian Sundarbans along with their possible linkages to regional blue economy

S. No.	Species	Salinity preference for optimum growth (psu)	Application in the blue economy vertical
1	 *Sonneratia apetala*	5–12	• Disaster management (protection against cyclone) • Land stabilization • Food for deer population in the mangrove forest • Medicinal values of fruits, seeds and barks of the species • Rich in vitamin C and has been used for making jellies (Pramanick et al. 2014) • Flowers are good sources of honey
2	 *Sonneratia caseolaris*	1–8	• Disaster management (protection against cyclone) • Land stabilization • Pneumatophores are used for making corks, fishing floats, shoe heels etc • Have been used in making jelly from the fruits rich in vitamin C (explained in details in Chap. 3 *Vide* Annexure 3.2)
3	 *Hertiera fomes*	2–7	• Timber production that are used in bridge building, house construction and boat and trawler manufacturing • Making hard board • Fuel wood • Medicinal properties *e.g.* the ethanol extract of bark has rich antioxidant value and exhibits antimicrobial properties

(continued)

Table 2.4 (continued)

S. No.	Species	Salinity preference for optimum growth (psu)	Application in the blue economy vertical
4	 *Nypa fruticans*	3–15	• Protection against cyclones • Leaves are used for thatching roofs • Plant parts are used for making umbrellas, raincoats, hats, mats, brooms, baskets, ropes etc • Sugary sap of the inflorescence is used as a source of molasses, amorphous sugar, vinegar and alcohol • Medicinal properties *e.g.*, ash of the stem is used to prevent toothache and headache
5	 *Avicennia marina*	12–30	• Prevents coastal erosion and helps in wave attenuation • Bioremediation as the species can absorb heavy metals from the ambient media • Bark is rich in tannin and is a source of reddish and brown dyes • Wood burn smoke is used as mosquito repellent • Timber, firewood and charcoal production • Medicinal properties *e.g.*, the wood ash is used to cure dermatological problems
6	 *Avicennia officinalis*	10–30	• Protection against cyclones and embankment erosion; wave attenuation • Wood is used to make boats, houses and furniture • Bark and roots are u sed for tanning • Medicinal property *e.g.*, the resinous substance exuded from the bark acts as contraceptive

(continued)

Table 2.4 (continued)

S. No.	Species	Salinity preference for optimum growth (psu)	Application in the blue economy vertical
7	 *Avicennia alba*	10–28	• Disaster management (against cyclonic depression and wave actions) • Wood of the species is used for smoking • Heart wood extract issed as tonic • Resin is used for birth control • Seeds of the species are often consumed as curry
8	 *Phoenix paludosa*	16–32	• The fruits are consumed in curries • The fruits can also be used as substitute of coffee • The leaves are used to make ropes, mats, umbrellas and also for thatching houses • The stems are used for making posts

(continued)

Table 2.4 (continued)

S. No.	Species	Salinity preference for optimum growth (psu)	Application in the blue economy vertical
9	 *Excoecaria agallocha*	5–30	• The roots of the species are used as floats • The species has bioremediation potential as the vegetative parts can absorb Zn, Cu and Pb from the ambient aquatic phase (Mitra and Zaman 2020) • Medicinal properties *e.g.*, the resin is used as purgative, and the oil extracted through wood distillation is used to cure dermatological problems
10	 *Bruguiera gymnorrhiza*	6–12	• Stabilization of the intertidal mudflats • Used for construction purpose, boat making, posts etc. • Used for making furniture • Used as fuel wood • Wood is used for making tool handles, paddles, digging sticks etc. and the chips are used for pulp production • Medicinal properties *e.g.*, the bark is often used to cure diarrhoea and fever

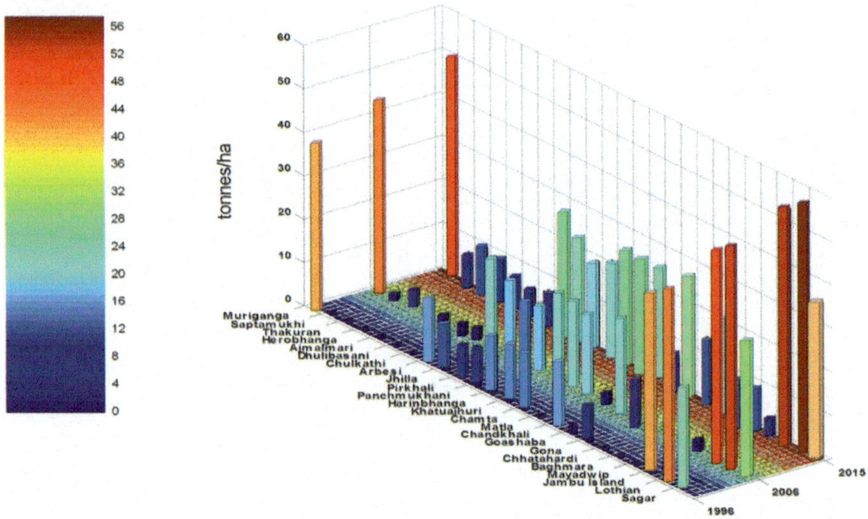

Fig. 2.1 Spatio-temporal variation of AGB in *Sonneratia apetala*

Fig. 2.2 *Porteresia coarctata* in the intertidal mudflats of Sundarbans

are congregation of dead silicon-rich diatom frustules in the sea beds like protozoan oozes. This material is employed in the filtration of fruit juice, syrup and varnish; in the boilers, electric ovens and refrigerators as insulators; as polishing agent for precious metals; in the separation of paraffin wax from petroleum and in the manufacture of concrete.

Fig. 2.3 Erosion of banks and intertidal mudflats in Sundarbans

4. The distribution and abundance of the commercially important fish and shellfish and their larvae are dependent on some species of the phytoplankton besides serving as their main food source. Among the species of diatoms, *Fragilaria oceanica* and *Hemidiscus hardmannianus* have been observed to indicate the abundance of clupeid fish, Hilsa in the Hooghly estuary and oil sardine, *Sardinella longiceps* in the west coast of India. The 'white water phenomenon' due to abundance of the coccolithophores has been a good omen for the herring fisheries in the British waters. Similarly, the colonial diatom, *Fragilaria antartica* is known to indicate the abundance of the Antarctic krill, *Euphausia superba.*

   The dinoflagellates also at times play an important role in the fishery sector as by virtue of their power of luminescence they can emit 'cold living light' which though lasts for about one-tenth of a second, helps in identifying certain fish shoals during night.

5. Phytoplankton also help to mitigate the rise of carbon dioxide in the atmosphere as they serve as potential sink of carbon. We made a survey in the Hooghly estuarine water adjacent to Sagar Island (21° 37′ 49.90″ N; 88° 04′ 00.51″ E, Fig. 2.4) during April, 2019 (before COVID-19 phase) and April, 2020 (COVID-19 lockdown phase) and came out with the interesting finding that the cell carbon content in the phytoplankton species were more during April, 2019 compared to April, 2020 (Table 2.7). This may be due to high level of carbon dioxide in the atmosphere during pre-COVID-19 phase compared to the COVID-19 lockdown phase. The table confirms that phytoplankton of the marine and estuarine ecosystems are unique sink of carbon and may be cultured in brackish water pond or aquaculture farms to serve dual benefits (i) food for cultured fishes and (ii) lowering of atmospheric carbon dioxide level at local scale.

Table 2.5 Commonly available mangrove associate species in Indian Sundarbans

Sl. No.	Species	Family	Application in the blue economy vertical
1	*Aglaia cucullata*	Meliaceae	• Fuelwood • Posts for making houses and building boats • Medicinal properties *e.g.*, leaves and fruits of the species are used to treat diarrhoea, dysentery and dermatological problems • Carbon sequestration
2	*Scyphiphora hydrophyllacea*	Rubiaceae	• Wood is dark brown and is used to make craft and small toys • Flowers of the species are used as cleansing agent or whitening agent used in laundry • Medicinal properties *e.g.*, leave extract is commonly used to treat stomach ache • Carbon sequestration
3	*Derris trifoliata*	Fabaceae	• Stem fibres are used for making ropes and fish nets • Leave extract is used as insecticide as it is rich in rotenone • Medicinal properties *e.g.*, the root is used as stimulant and against chronic paralysis; the root decoction is also used to prevent high fever • Carbon sequestration
4	*Sarcolobus globosus*	Apocynaceae	• It is a poisonous plant and the seeds are known to kill cat, dogs and wild animals • Medicinal properties *e.g.*, the plant extract is used to treat rheumatism and fever • Rich in antioxidant due to presence of barbigerone • Carbon sequestration
5	*Clerodendrum inerme*	Verbenaceae	• The species plays an important role in health care sector as it has anti-inflammatory activity, anti-maliarial activity and antimicrobial activity • The species is rich in antioxidant • Leaf extract of the species has insecticidal property and is used as mosquito repellent • Carbon sequestration

(continued)

Table 2.5 (continued)

Sl. No.	Species	Family	Application in the blue economy vertical
6	*Suaeda nudiflora*	Chenopodiaceae	• Tender shoots are used as vegetables • Leaf extract has several medicinal uses like prevention against weakness, dehydration etc • Leaves are used to make snacks (like samosa and kachuri and has wide scope for initiating mangrove associate based cottage industry in coastal areas, a source of alternative livelihood) (Mitra and Biswas 2021) • Carbon sequestration
7	*Suaeda maritima*	Chenopodiaceae	• Tender leaves are used for preparing curries • Ashes of the plants are used as the source of sodium carbonate used for glass making • Leaves are used to make snacks (like samosa and kachuri) and has wide scope for initiating mangrove associate based cottage industry in coastal areas (Mitra and Biswas 2021) • Carbon sequestration
8	*Salicornia brachiata*	Chenopodiaceae	• The plant is used as green salad; only the green tender parts are recommended for edibility, the reddish being too high in salinity and silica • The shoots are processed into beverages like nuruk (a type of fermentation starter), makgeolli (a Korean rice wine), or vinegar (Song et al. 2013; Kim et al. 2013) • Leaves are used to make snacks (like samosa and kachuri) and has wide scope for initiating mangrove associate based cottage industry in coastal areas (Mitra and Biswas 2021) • Carbon sequestration
9	*Sesuvium portulacastrum*	Aizoaceae	• Beach stabilization • Sand dune fixation • Phytoremediation as the species is known to accumulate heavy metals from the ambient water and soil • Carbon sequestration
10	*Trianthema portulacastrum*	Aizoaceae	• Raw leaves of the species with salty taste are used in curry preparation and also in salads • Used as biofertilizer as it contains considerable amount of N, P and K • Medicinal properties as the leaf extract is used to cure jaundice and oedema • Carbon sequestration

(continued)

Table 2.5 (continued)

Sl. No.	Species	Family	Application in the blue economy vertical
11	*Ruppia maritima*	Ruppiaceae	• Imparts stability to the intertidal mudflats/beaches • Bioremediation, since the species has been found to absorb nutrients (preferably nitrate and phosphate) from the ambient water • Serves as an important food for migratory birds and wintering water fowls • Reduces the pace of coastal acidification
12	*Ipomoea pes-caprae*	Convolvulaceae	• Sand binders and imparts stability to the sandy beach • Bioremediation, as it is known to absorb Zn, Cu, Pb, Hg, Co, Ni etc. from the ambient water and sediment • Leaf extract acts as mosquito repellent • Carbon sequestration
13	*Porteresia coarctata*	Poaceae	• Pioneer species in the process of ecological succession and imparts stability to the intertidal mudflats • Serves as fodder for cattle • Wave attenuation • Carbon sequestration
14	*Saccharum spontaneum*	Poaceae	• Root decoction of the species mixed with sugar syrup is used to treat burning parts of the body and excessive thirst • Used for thatching, houses and making carpets, mats and other decorative items • The aroma acts as insect repellent • Carbon sequestration
15	*Acrostichum aureum*	Pteridanceae	• The dried ferns are used for thatching roof of the house • Acts as agent of bioremediation as it is known to accumulate heavy metals from the ambient media • The rhizomes and leaves of the species are used to cure ulcers, wounds and worm infection • Carbon sequestration

Table 2.6 Commonly available seaweed species in Indian Sundarbans

Sl. No.	Species	Family	Application in the blue economy vertical
1	*Enteromorpha intestinalis*	Ulvaceae	• Used as food, fodder etc. • Can be used as source material for biofertilizer • Bioremediation, as the species is known to absorb nutrients and heavy metals from the ambient water • Carbon sequestration • Can be used as source material for preparing bread, cookies, ice cream (Pramanick et al. 2016), which can be an alternative livelihood for the people of Indian Sundarbans
2	*Enteromorpha compressa*	Ulvaceae	• Source of essential oil and antioxidant • Often used in beauty parlour for skin treatment • Used for making biofertilizer due to presence of high nutrient level in the thallus body • Used for making Pakoda and ingredients of Gol gappa (commonly called phuchka in West Bengal) • Bioremediation, as the species is known to absorb nutrients and heavy metals from the ambient water • Carbon sequestration
3	*Ulva lactuca*	Ulvaceae	• Used as salads and also served with soup, curry and fishes • Rich in antioxidant • Bioremediation as the species is known to absorb nutrients and heavy metals from the ambient water • Carbon sequestration • Used for making biofertilizer due to presence of high nutrient level in the thallus body • Used as ingredients of organic fish feed preferably for shrimps and prawns

(continued)

Table 2.6 (continued)

Sl. No.	Species	Family	Application in the blue economy vertical
4	*Catenella repens*	Rhabdoniaceae	• Rich in antioxidant • Rich in protein and hence used in organic fish feed preparation • Bioremediation as the species is known to absorb nutrients and heavy metals from the ambient water • Carbon sequestration • Used for making biofertilizer due to presence of high nutrient level in the thallus body
5	*Catenella nipae*	Rhabdoniaceae	• Rich in protein (13.50–17.65%) and hence used in organic fish feed preparation • Bioremediation as the species is known to absorb nutrients and heavy metals from the ambient water • Carbon sequestration • Used for making biofertilizer due to presence of high nutrient level in the thallus body • Used for making bread, cookies etc. and hence can be an alternative livelihood programme in Indian Sundarbans
6	*Gelidium pusillum*	Gelidiaceae	• Rich in antioxidant • Exhibits antimicrobial activities and hence has wide scope in aquaculture industry • Potential agent of bioremediation as the species is known to absorb nutrients and heavy metals from the ambient water • Used for making biofertilizer due to presence of high nutrient level in the thallus body • Carbon sequestration
7	*Ceramium manorensis*	Corallinaceae	• Can be a good source of biodiesel due to high lipid level • Can be a good source of fish feed • Carbon sequestration
8	*Bostrychia simpliciuscula*	Rhodomelaceae	• Deposits of manganese were documented on the surface • Can be a good source of fish feed and biofertilizer as the species is a rich source of phosphorus and potassium • Carbon sequestration

(continued)

Table 2.6 (continued)

Sl. No.	Species	Family	Application in the blue economy vertical
9	*Polysiphonia mollis*	Rhodomelaceae	• Usually consumed in raw or dried form • Used in salad, soups and sushi • Can be a good source of fish feed and biofertilizer • Carbon sequestration
10	*Compsopogon coeruleus*	Erythrotrichiaceae	• Used in salad, soups and sushi • Boiled and mixed with coconut milk and tinned fish to produce a pudding that sets and can be sliced up as a delicacy food • Can be a good source of fish feed and biofertilizer • Carbon sequestration

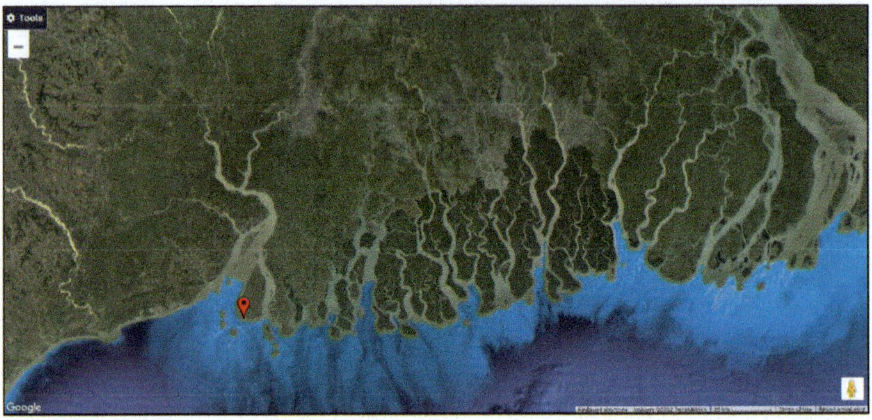

Fig. 2.4 Map of Sundarbans showing the location of Sagar Island along the Hooghly estuary. *Source* Google map using 'map.co' software

The main sources of carbon dioxide in and around the present study sites are (i) massive deforestation of mangroves for constructing aquaculture farms (Fig. 2.5), (ii) presence of large number of brick kilns that emit considerable amount of carbon dioxide (Fig. 2.6), (iii) mushrooming of tourism units at the cost of mangroves (Fig. 2.7), (iv) construction of fish landing stations at the cost mangroves (Fig. 2.8) and (v) emissions from fishing vessels and trawlers (Fig. 2.9).

6. Phytoplankton can also be used to reflect the water quality in terms of certain pollutants other than sewage (that contribute considerable amount of nitrate and phosphate in the ambient water). It has been reported that an increase in cell volume occurs in certain species (like *Dunaliella tertiolecta* and *Phaeodactylum tricornutum*) due to copper pollution. We made a detailed survey in the surrounding waters of Bali Island (22° 04′ 35.17″ N; 88° 44′ 55.70″ E), Sagar Island (21° 37′ 49.90″ N; 88° 04′ 00.51″ E) and Haldia port-cum-industrial complex (22° 02′ 30.48″ N; 88° 05′ 26.52″ E). All these sites are located in and around the World Heritage Site of Indian Sundarbans. We studied the dissolved copper level in the aquatic phase around these selected stations during December, 2021 and documented that Haldia port-cum-industrial complex had the highest level of dissolved copper 336.85 ppb compared to Sagar Island 223.69 ppb and Bali Island 164.55 ppb. We also studied simultaneously the cell volume of the sampled phytoplankton species and observed that in all the 47 species cell volume was highest in Haldia followed by Sagar Island and Bali Island (Table 2.8).

Table 2.7 Phytoplankton cell carbon content (in pg) before COVID-19 phase during April, 2019 and COVID-19 lockdown phase in April, 2020 in the Hoogly estuarine water adjacent to Sagar Island

Sl. No.	Species	April, 2019	April, 2020
1	*Coscinodiscus eccentricus*	1040.03	797.39
2	*Coscinodiscus jonesianus*	1099.79	760.25
3	*Coscinodiscus lineatus*	758.74	374.3
4	*Coscinodiscus radiates*	2202.61	1835.65
5	*Coscinodiscus gigas*	762.12	382
6	*Coscinodiscus oculusiridis*	741.34	363.57
7	*Planktoniella sol*	7714.02	7442.9
8	*Cyclotella striata*	1572.02	1226.67
9	*Thalassiosira subtilis*	572.14	195.02
10	*Ceratium tripos*	602.78	236.28
11	*Skeletonema costatum*	508.95	96.8
12	*Paralia sulcata*	617.59	219.02
13	*Rhizosolenia crassipina*	1389.93	1024.65
14	*Rhizosolenia setigeara*	737.3	371.87
15	*Rhizosolenia alata*	765.71	406.27
16	*Ceratium teres*	735.78	354.97
17	*Ceratium trichoceros*	759.37	434.9
18	*Bacteriastrum delicatulum*	508.95	96.8
19	*Bacteriastrum varians*	725.62	363.41
20	*Bacteriastrum comosum*	508.95	96.8
21	*Chaetoceros didymus*	2050.95	1711.66
22	*Chaetoceros peruvianus*	2001.27	1724.57
23	*Chaetoceros compressus*	1967.14	1673.21
24	*Ditylum sol*	763.09	417.23
25	*Triceratium favus*	741.6	371.94
26	*Triceratum reticulatum*	707.94	347.5
27	*Biddulphia sinensis*	627.76	236.44
28	*Biddulphia mobiliensis*	675.82	315.24
29	*Hemidiscus hardmanninus*	734.72	357.04
30	*Climacosphenia elongate*	1274.61	962.86
31	*Fragilaria oceanic*	1649.71	1321.13
32	*Rhaphoneis amphiceros*	519.71	124.38
33	*Thalassionema nitzschioides*	949.14	578.38
34	*Thalassiothrix longissima*	1862.11	1546.43
35	*Thalassiothrix fraunfeldii*	730.43	363
36	*Asterionella japonica*	542.83	148.29
37	*Ceratium extensum*	650.31	266.95
38	*Gyrosigma balticum*	542.79	142.8
39	*Pleurosigma normanii*	520.9	118.79
40	*Pleurosigma elongatum*	557.28	174.38
41	*Diploneis smithii*	508.95	96.8
42	*Cymbella marina*	691.71	298.85
43	*Nitzschia sigma*	611.3	219.72

(continued)

Table 2.7 (continued)

Sl. No.	Species	April, 2019	April, 2020
44	*Nitzschia closterium*	2346.45	2117.5
45	*Ceratium furca*	667.94	283.14
46	*Trichodesmium erythraea*	545.4	150.47
47	*Chlorella marina*	546.42	154.84

Fig. 2.5 Aquaculture farms constructed at the cost of mangroves in Indian Sundarbans, which otherwise could be a potential sink of carbon

2.1.2 Overview of Faunal Biodiversity

Sundarban ecosystem is the breeding, nursing and survival ground of a variety of fauna that represent almost all the major phyla of animal kingdom. Starting from microscopic protozoa to chordata all the phyla are represented by the faunal members of Indian Sundarbans (Table 2.9). However, as per the blue economy of the region is concerned, members of class pisces, reptilia, aves and mammalia are the main pillars for sustainable income in the region. The fishes caught from the estuaries of Indian Sundarbans are not only consumed in the local markets, but they are exported in different foreign countries and are the sources of foreign exchange. Shrimps, prawns and crabs are the major exportable species that fetch considerable foreign currencies for the country.

It is a fact that Sundarban ecosystem is still one of the most biologically productive and taxonomically diverse ecosystems of the Indian sub-continent, although about 6

Fig. 2.6 Brick kilns in Indian Sundarbans: A point source of carbon dioxide emission in Indian Sundarbans

vertebrates have disappeared from this ecosystem since the last 200 years and about 20 species are in the endangered species list (Table 2.10).

Many representatives of invertebrate phyla have great possibility to uplift the mangrove-centric economy of the region. The molluscan community in Indian Sundarbans can be a potential source of alternative livelihood for the local population not only for their edible value (still not known by the local population), but also as a source of lime. The edible oysters (*Saccostrea cucullata* and *Crassostrea madrasensis*), clams, mussels etc. can be brought under the umbrella of commercial culture to provide alternative source of income to the coastal population (Fig. 2.10). This approach can reduce the adverse pressure on mangrove forest and its natural resources to a great extent, which is a positive road map to preserve the biodiversity in this dynamic deltaic ecosystem.

We also want to mention in this context about two important arthropod species available on the mangrove swamps and sandy beaches of coastal West Bengal, which are commonly referred to as horseshoe crabs (Fig. 2.11). The two horseshoe crab species namely *Carcinoscorpius rotundicauda* and *Tachypleus gigas* are observed frequently during the premonsoon period of high salinity. These species have excellent biomedical values and have been established as potential source of bioactive substance, the Carcinoscorpius Amoebocyte Lysate (CAL) and Tachypleus Amoebocyte Lysate (TAL). The reagents are highly sensitive and useful for the rapid and

Fig. 2.7 Mushrooming of tourism units in Indian Sundarbans at the cost of mangroves destroyed the natural carbon reservoir

accurate assay of Gram negative bacteria even if they are present in a very minute quantity up to the level of 10^{-10} gm. Hence, these reagents have wide application in the medical sphere as they have immense role in detecting endotoxins in several pharmaceutical products and other life saving drugs like interferon, insulin etc. We presume that this detection method can also be used to assess the quality of COVID-19 vaccines.

The list of macro-invertebrate species found in this deltaic ecosystem exhibits unique species diversity although the magnitude of diversity (as reflected through Shannon Weiner Index or Index of Dominance) varies markedly with season (Mitra 2000). The invertebrate phyla act as magnificent "conveyer belts" in this ecosystem for the transferring energy to the higher vertebrates. The scat analysis of Indian Tiger (*Panthera tigris tigris*) revealed the presence of *Toxocara carti*, which proves the direct dependency of higher vertebrates on invertebrate species.

Apart from sustaining a wide spectrum of invertebrate species in the benthic substratum, this deltaic region also sustains niches for minor phyla. A relatively small phylum of marine worms, often overlooked by biologists, the Sipunculus, (number slightly more than 300 species in the globe with 16 recongnised genera) is also a representative of Indian Sundardans. The benthic substratum of deltaic Sundarbans supports one species of sipunculus namely *Phascolosoma arcuatum* (Gray), which

Fig. 2.8 Construction of fish landing stations in mangrove ecosystem reduced the sink of carbon in Sundarbans

Fig. 2.9 Carbon dioxide emission from trawlers is a common scene in Indian Sundarbans

Table 2.8 Spatial variation of cell volume (in μm^3) of phytoplankton at three selected stations with different dissolved copper level in and around Indian Sundarbans

Sl. No.	Species	Bali Island	Sagar Island	Haldia
1	*Coscinodiscus eccentricus*	10,539.29	14,957.48	16,722.7
2	*Coscinodiscus jonesianus*	12,034.12	13,984.9	15,750.12
3	*Coscinodiscus lineatus*	4098.6	4766.87	6532.09
4	*Coscinodiscus radiates*	44,353.09	45,906.96	47,672.18
5	*Coscinodiscus gigas*	4168.8	4930.84	6696.06
6	*Coscinodiscus oculusiridis*	3741.2	4540.07	6305.29
7	*Planktoniella sol*	264,896.66	271,394.4	273,159.62
8	*Cyclotella striata*	24,933.26	26,973.55	28,738.77
9	*Thalassiosira subtilis*	672.65	1316.32	3081.54
10	*Ceratium tripos*	259.8	569.52	2334.74
11	*Skeletonema costatum*	1405.33	1727.2	3492.42
12	*Paralia sulcata*	19,757.08	21,154.33	22,919.55
13	*Rhizosolenia crassipina*	3666.81	4715.34	6480.56
14	*Rhizosolenia setigeara*	4243.59	5454.46	7219.68
15	*Rhizosolenia alata*	3628.18	4359.75	6124.97
16	*Ceratium teres*	1088.5	1700.76	3465.98
17	*Ceratium trichoceros*	896.26	1269.75	3034.97
18	*Bacteriastrum delicatulum*	39,497.65	41,903.77	43,668.99
19	*Bacteriastrum varians*	37,930.42	42,317.29	44,082.51
20	*Bacteriastrum comosum*	36,861.23	40,676.86	42,442.08
21	*Chaetoceros didymus*	4188.9	5693.97	7459.19
22	*Chaetoceros peruvianus*	3746.38	4716.73	6481.95
23	*Chaetoceros compressus*	3072.61	4204.48	5969.7
24	*Ditylum sol*	1580.68	2037.55	3802.77
25	*Triceratium favus*	2453.96	3545.83	5311.05
26	*Triceratum reticulatum*	3606.69	4403.15	6168.37
27	*Biddulphia sinensis*	16,602.74	19,430.28	21,195.5
28	*Biddulphia mobiliensis*	27,207.75	29,781.62	31,546.84
29	*Hemidiscus hardmanninus*	111.21	253.66	2018.88
30	*Climacosphenia elongate*	8341.51	9417.38	11,182.6
31	*Fragillaria oceanic*	33,606.75	36,680.62	38,445.84
32	*Rhaphoneis amphiceros*	3520.17	4528.04	6293.26
33	*Thalassionema nitzschioides*	259.26	587.47	2352.69
34	*Thalassiothrix longissima*	1981.87	2602.74	4367.96
35	*Thalassiothrix fraunfeldii*	258.77	509.87	2275.09

(continued)

Table 2.8 (continued)

Sl. No.	Species	Bali Island	Sagar Island	Haldia
36	*Asterionella japonica*	0.79	198.47	1963.69
37	*Ceratium extensum*	61.1	272.49	2037.71
38	*Gyrosigma balticum*	2756.85	3219.72	4984.94
39	*Pleurosigma normanii*	1298.61	1739.48	3504.7
40	*Pleurosigma elongatum*	49,053.25	55,251.12	57,016.34
41	*Diploneis smithii*	2306.2	2912.83	4678.05
42	*Cymbella marina*	292.95	618.86	2384.08
43	*Nitzschia sigma*	306.52	682.82	2448.04
44	*Nitzschia closterium*	327.4	707.68	2472.9
45	*Ceratium furca*	344.64	712.99	2478.21
46	*Trichodesmium erythraea*	3569.86	4221.54	5986.76
47	*Chlorella marina*	470.44	731.31	2496.53

is characterized by the presence of a spacious, unsegmented coelom, filled with fluid continuing free haemocytes.

Zooplankton commonly found in the pelagic system comprises a heterogeneous assemblage of animals covering many taxonomic groups which include copepod, mysids, lucifer, gammarid amphipods, cladocera, ostracod, hydromedusae, ctenophore and chaetognath among haloplankters and larval stage of polychaete, molluscs, echinoderm, crustacean and fin fish among meroplankters (Fig. 2.12).

The estuarine tourism sector of the mangrove dominated deltaic complex attracts tourists from all over the world to witness the wilderness and wild beauties of crocodiles basking on the sunny intertidal mudflats; birds chirping and roosting within the mangrove forests; monkeys, deer, wild boars and tigers passing through the dense halophytic forests along the estuaries. The beauty of mangrove swamps with migratory birds, the monkeys on the intertidal mudflats feeding their young ones or the tigers swimming along the estuaries are special bonus for the tourists visiting the region (Figs. 2.13, 2.14 and 2.15). Mangrove based tourism in this region can not only fetch revenues, but also serve the base of edu-tourism and scientific researches preferably for the researchers of life science, oceanography, marine biology, climate science, geography, fishery science, environmental science and many other related disciplines.

The watch towers in the midst of the mangrove habitat have been constructed to observe the behaviour of wild animals in the mangrove forest (Fig. 2.16).

The estuarine tourism is thus one of the most promising sectors in the domain of regional blue economy, which has large prospect to expand and generate revenues if organized with proper management action plan and institutional support.

The updated checklists of the members under class pisces, amphibia, reptilia, aves and mammalia available in the mangrove creeks, estuaries and bays in and around

Table 2.9 Faunal diversity of Sundarban mangrove ecosystem

Sl. No.	Chaudhuri and Chaudhuri (1994) Genera/species	Mandal and Nandi (1989) Genera/species
I. Invertebrates		
1. Protista	13/21	1/1
2. Porifera	1/1	7/7
3.Cnidaria	16/17 (9)	1/1
4. Ctenophore	2/2	1/1
5. Platyhelminthes	8/8 (6)	1/1
6. Rotifera	–	4/5
7.Nematoda	50/55 (45)	2/2
8. Sipuncula	–	4/4
9. Mollusca	45/57 (5)	57/94
10. Echiura	1/3	2/2
11. Annelida	39/49 (2)	37/18
12.Crustacea	37/69 (10)	92/169
13. Insecta	59/98 (29)	126/187
14. Arachunida	8/8	28/35
15. Xiphosura	2/2	2/2
16. Bryozoa	1/1	3/3
17. Entopoda	1/1	-
18. Chaetognatha	–	1/2
19. Echinodermata	4/6 (5)	12/14
20. Hemichordata	1/1	1/1
II. Vertebrates Chordates		
1. Pisces	96/250	100/141
2. Amphibia	4/7	4/8
3. Reptilia	49/57	40/57
4. Aves	120/300	122/161
5. Mammalia	33/40	32/40

Note Figures within parenthesis indicate unidentified species or spp. groups

the Sundarban mangrove forest are highlighted as Annexures 2A, 2B, 2C, 2D and 2E respectively.

Table 2.10 Endangered species of Sundarban mangrove ecosystem

Scientific name of the endangered species	Common name
A. Mammals	
Felis viverrina	Fishing Cat
Panthera tigris tigris	Indian Tiger
Platanista gangatica	Gangetic dolphin
Orcaella brevirostris	Snub-nosed dolphin
Neophocaena phocaenoides	Little porpoise
B. Birds	
Ardea goliath	Giant heron
Leptoptilos javanicus	Lesser adjutant
Haliacetus leucogastar	White bellied sea eagle
Pandion haliaetus	Osprey
C. Reptiles	
Crocodylus porosus	Saltwater or Estuarine crocodile
Lepidochelys olivacea	Olive Ridley turtle
Batagur baska	Batagur terrapin
Kachugatacta tecta	Indian tent turtle
Lissemys punctate	Indian flap shelledturtle
Aspideretus gangeticus	Indian soft shelled turtle
Aspideretus burun	Peacock marked soft shelled turtle
Varanus bengalensis	Common Indian monitor
Varanus flavescens	Yellow monitor
Varanus salvator	Water monitor
Oythonmolurus bivittatus	Indian monitor
Ophiophagus hannah	King cobra

Source Chaudhuri and Chaudhuri (1994)

2.2 Abiotic Resources of Indian Sundarbans

The non-living resources of deltaic Sundarbans do not include rich mineral based resources like phosphorite deposits, polymetallic sulphides, polymetallic nodules and gas hydrates or other major categories of commercially viable non-living components that are extracted from the marine environment (Fig. 2.17).

Evaporite deposits like salts were usually procured in the Digha-Mondarbani region in coastal West Bengal in early eighties (Fig. 2.18), but now the practice exists no more and instead hotels and shrimp culture farms have flourished in the region.

The blue economy in the lower Gangetic delta region is getting momentum through construction of a deep sea port off Sundarban mangrove forest near Sandhead Island.

Fig. 2.10 Oyster colony on hard substratum

Fig. 2.11 Horseshoe crab on the Sundarban mudflat

The primary aim of this establishment is to offer cargo handling and trans-shipment facility. Handing and trans-shipment of hazardous cargo is the priority of the project. However, several environmental issues have cropped up in connection to this project like adverse impact on the fish breeding and nursery ground in and around the mangroves and deterioration of estuarine water quality. A survey conducted on the awareness of the local stakeholders on the deep sea port project reveal that 76% of the fisher folks, 89% of the agriculturists and 65% of the common people are not aware of the project (Figs. 2.19, 2.20 and 2.21).

Fig. 2.12 Zooplankton community in Sundarban water

Fig. 2.13 Monkey (*Macaca mulatta*)—a prey base of Sundarban tiger (*Source* Mitra 2020)

We also carried out a stakeholder's analysis to assess the pulse of fisher folks, agriculturists and common people on the deep sea port project and its possible impacts.

73% of the fisher folks opined that the project, if installed, will greatly hamper their fish catch, while 20% stated that the project will moderately hamper their fish catch and 7% did not respond at all (Fig. 2.22).

Fig. 2.14 Deer (*Axis axis*)—a lucrative prey of Sundarban tiger (*Source* Mitra 2020)

Fig. 2.15 Tiger in the midst of *Phoneix paludosa* (local name of the tree is Henthal) (*Source* Mitra 2020)

Fig. 2.16 Watch Tower in the heart of mangroves (*Source* Mitra 2020)

The agriculturist of the deltaic complex mostly did not respond to the impact of the project (70%), while 25% stated that the fish resources will be moderately affected and 5% stated that the fish resource of the region will be greatly affected (Fig. 2.23).

The common people of the region who are mostly shrimp culturists, school teachers, businessman and daily labours were also interrogated. 57% of them expressed their views towards adverse impact of high degree on fish catch if the project is installed, while 33% opined that there will be a moderate impact and 10% did not respond at all (Fig. 2.24).

Oil and Natural Gas Corporation Limited (ONGC) has explored a gas field in the inshore region of Bay of Bengal, 55 km off Indian Sundarbans and estimated a reserve around 21 trillion cubic feet. The exploration of oil has several environmental issues like increase of SPM, lowering of DO level, increase of heavy metal level, oil and grease level etc., which have direct impact on the overall biodiversity of the geographical locale that includes plankton, nekton and benthos of the region. On this background, we carried out a respondent analysis considering the stakeholders like policymakers, fishermen, agriculturists, researchers and local inhabitants through the products of Threat Ranking (TR) and % of votes for every possible threat as per the standard practice (Mitra 2020; Mitra and Zaman 2021).

Table 2.11 clearly reflects that all the stakeholders exhibited perfect synchronization in their opinion on the stress due to oil and natural gas exploration, which is as per the order suspended particulate matter > oil and grease level in the aquatic phase

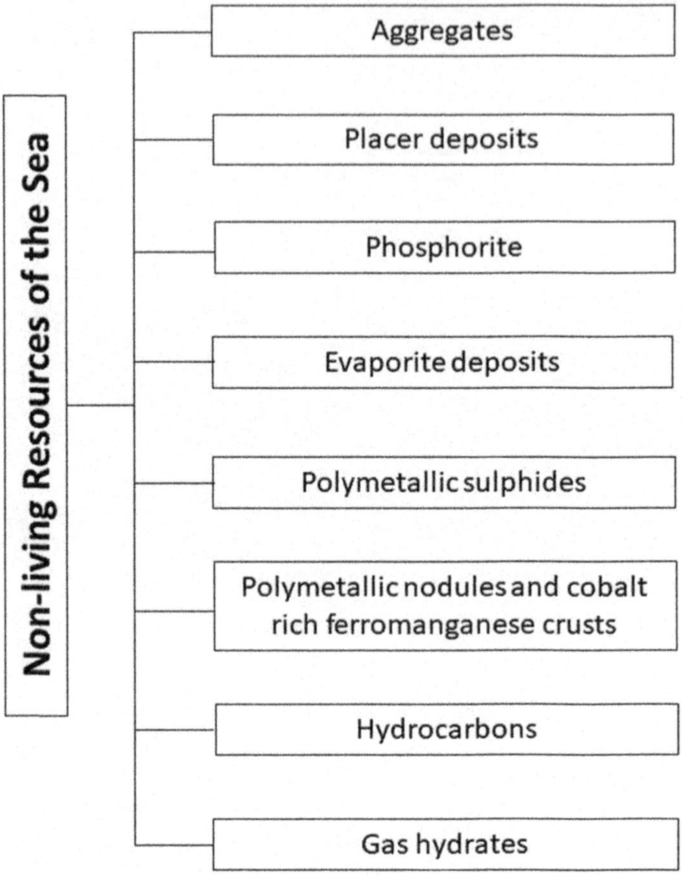

Fig. 2.17 Mineral resources of the marine environment

> impact on nekton > impact on phytoplankton > impact on benthos > dissolved heavy metals > dissolved oxygen > salinity.

Ship breaking is an important sector in the domain of blue economy and generates employment at several tiers. Most of the ships posses a shelf life of few decades and after the stipulated period if it is repaired then the entire process loose commercial viability. Ship breaking allows several materials obtained after demolition of the ships to be reused. Steel is one of such materials. Cobalt cables and aluminum may also be recycled and put to various uses. However, the present mode of demolition and scrapping as carried out in countries like India, Bangladesh and Pakistan are detrimental to the ecology and environment of the region. In addition to deterioration of the ambient environment, the involvement of the poor labourers without any proper safety measures often pose and adverse impact on their health and hygiene. The fine metallic dust affects the respiratory system of the workers who are involved in breaking and scrapping the ships and vessels. The methods by which the antifouling

Fig. 2.18 Salt pan in Mondarmoni (in the Midnapore district of West Bengal, India) during early eighties

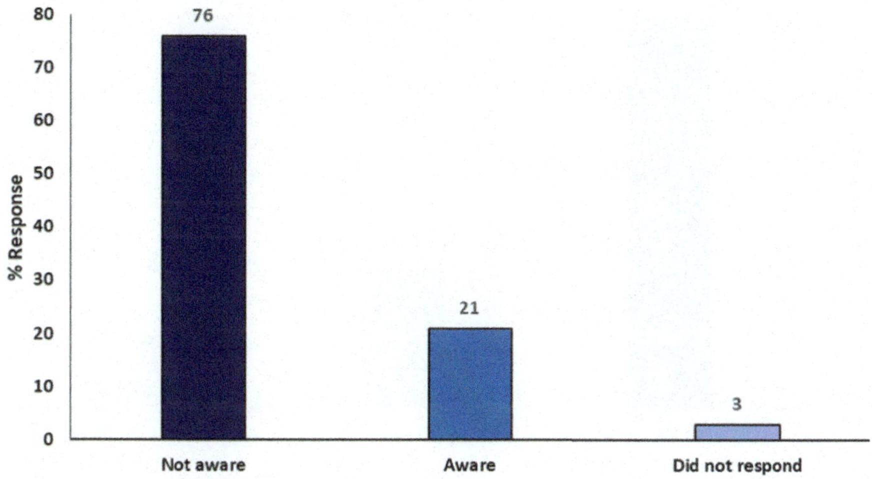

Fig. 2.19 Results of respondent analysis carried out with the inputs from 845 fisher folks on the deep sea port project off Indian Sundarbans

paints on the body of ship and vessels are burnt and removed often result in high emission of green house gases (Fig. 2.25).

In the maritime state of West Bengal, ship breaking is carried out at Netaji Subhas dock, Garden Reach at Kolkata. However, another ship breaking and ship recycling unit is in the pipeline. It is proposed to set up this unit at Tengra-Char village on the eastern bank of the Hooghly River estuary (the western-most distributary of the

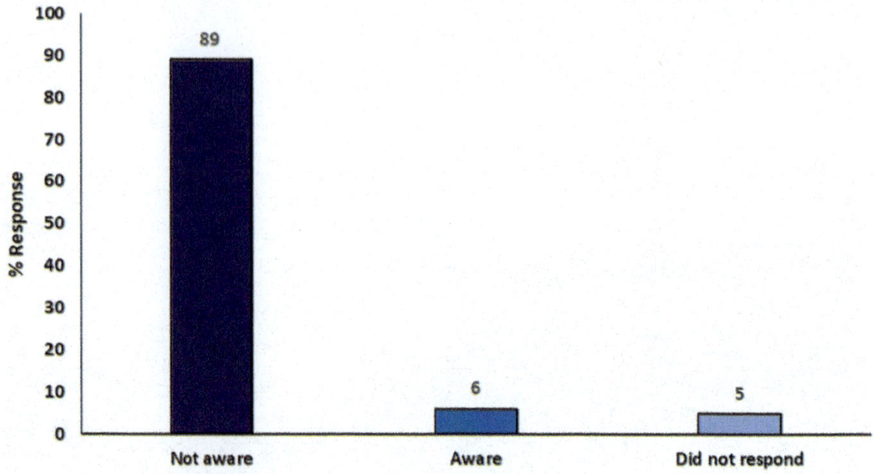

Fig. 2.20 Results of respondent analysis carried out with the inputs from 285 local agriculturists on the deep sea port project off Indian Sundarbans

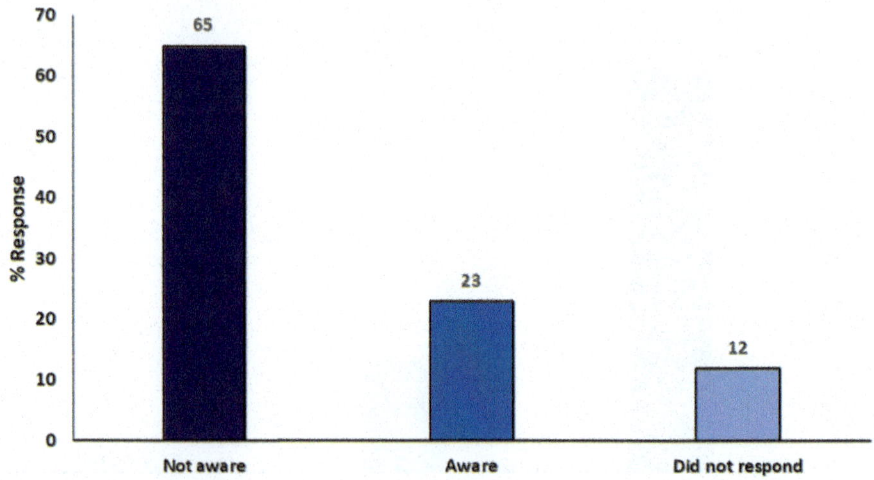

Fig. 2.21 Results of respondent analysis carried out with the inputs from 5230 common people on the deep sea port project off Indian Sundarbans

Ganges River) in Kulpi Block of South 24-Paragans District, West Bengal. The ship recycling yard shall be set up by Bengal Shipyard. It is expected that ships of total Light Displacement Tonnage (LDT) of about 60,000 tonnes will be recycled annually.

Non-conventional sources of energy are strong pillars of blue economy as it is not only a substitute of fast depleting coal reservoir, but also a zero-carbon emission energy source. India may have its first and the world's fifth tidal energy power plant at Durgaduari creek in Sundarbans. With the coastal regulatory zone clearance and

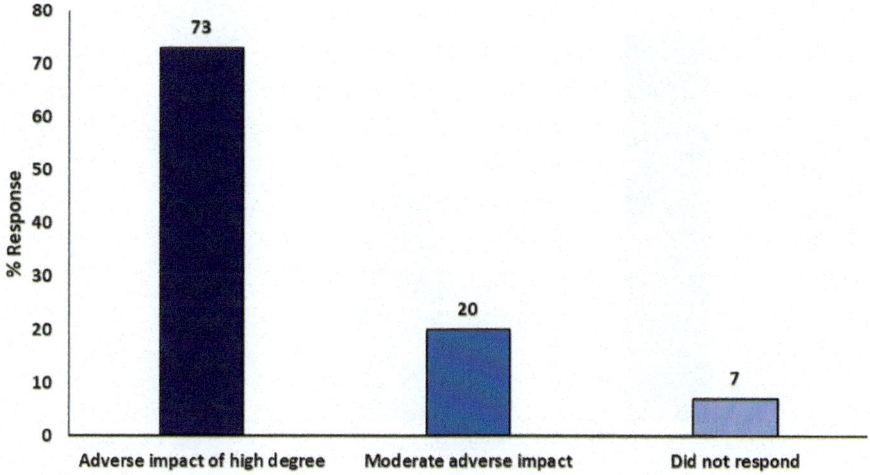

Fig. 2.22 Results of respondent analysis of fisherfolk on the impact of deep sea port project off Indian Sundarban on fish catch

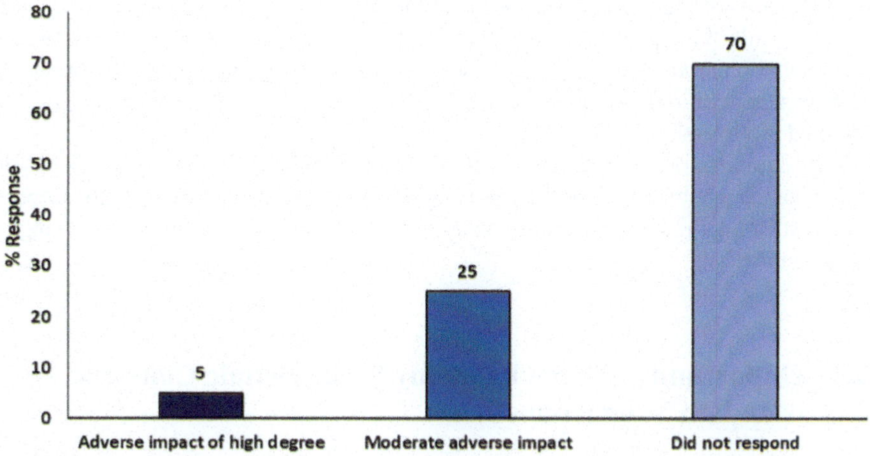

Fig. 2.23 Results of respondent analysis of agriculturists on the impact of deep sea port project off Indian Sundarbans on fish catch

environmental clearance in hand, the ministry of non-conventional energy sources (MNES) is all set to execute the tidal project.

According to MNES, India has a combined potential of 9,000 mw in Kutch, Sundarbans and Gulf of Cambay and if the project could see the face of success, India would expand its venture into exploiting tidal waves.

The estimated cost of the project is about Rs 40 crore. On the other hand, a hydro plant with a similar capacity would cost about Rs 20 crore. A tidal energy power plant, though more expensive to set up, would do away with controversies like over

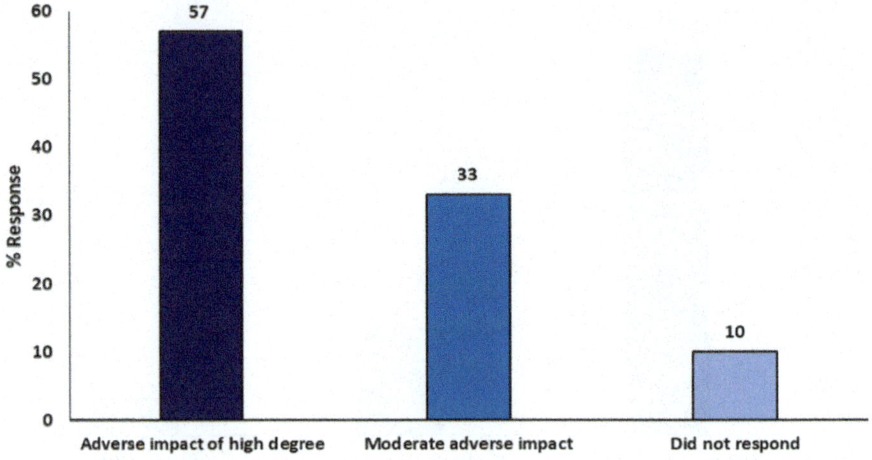

Fig. 2.24 Results of respondent analysis of common people on the impact of deep sea port project off Indian Sundarban on fish catch

flooding, adverse environmental problems and problems of rehabilitation. Also, tides being a result of the gravitational pull of the moon towards the earth make it a highly predictable phenomenon. This would help eliminate uncertainity factor in the sector of power generation. The project will also have minimum adverse impact on the regional biodiversity.

The saline atmosphere that would affect the equipment is something the operation maintenance would have to deal with. With 90% financial assistance from the Centre, the West Bengal renewable energy development agency is gradually progressing to execute the plan.

2.3 Opportunity of Blue Economy in the Deltaic Complex of Indian Sundarbans

The deltaic complex of Indian Sundarbans with rich mangrove biodiversity, fish resources, wild life, congenial aquatic environment and skilled human resources (specially in the field of aquaculture, deep sea fishing, mangrove oriented tourism, agriculture, vessel repairing and building etc.) has considerable scope to boost up the marine and estuarine based livelihoods and job opportunities. There is also great space to institutionalize the holy tourism in the region due to presence of Kapil Muni temple that attracts 10–15 lakhs people every year from different parts of the country to take the holy bath during Makar Shankranti (celebration in mid January). We have focused in this section only the most prospective sectors of blue economy considering the resource base of the region.

Table 2.11 Threats operating on Sundarban mangrove ecosystem in context to oil and natural gas exploration

Threat	Policymaker (Respondent Type 1)		
	TR	% of Vote	TS 1
Suspended particulate matter	8	16.1	128.8
Oil and grease level in the aquatic phase	7	13	91
Dissolved heavy metals	5	11.7	58.5
Dissolved Oxygen	4	8.8	35.2
Salinity	2	10.4	20.8
Impact on phytoplankton	6	13.4	80.4
Impact on nekton	6	14.1	84.6
Impact on benthos	6	12.5	75
Threat	**Fisherman (Respondent Type 2)**		
	TR	**% of Vote**	**TS 2**
Suspended particulate matter	8	16	128
Oil and grease level in the aquatic phase	8	14.3	114.4
Dissolved heavy metals	4	10.1	40.4
Dissolved Oxygen	4	7.4	29.6
Salinity	3	3.8	11.4
Impact on phytoplankton	6	16.1	96.6
Impact on nekton	6	16.7	100.2
Impact on benthos	6	15.6	93.6
Threat	**Agriculturist (Respondent Type 3)**		
	TR	**% of Vote**	**TS 3**
Suspended particulate matter	8	16.3	130.4
Oil and grease level in the aquatic phase	8	13.2	105.6
Dissolved heavy metals	5	9.8	49
Dissolved Oxygen	4	8.6	34.4
Salinity	4	6.6	26.4
Impact on phytoplankton	6	15.7	94.2
Impact on nekton	6	17.1	102.6
Impact on benthos	6	12.7	76.2
Threat	**Researcher (Respondent Type 4)**		
	TR	**% of Vote**	**TS 4**
Suspended particulate matter	7	16.9	118.3
Oil and grease level in the aquatic phase	8	13.4	107.2
Dissolved heavy metals	3	11.9	35.7
Dissolved Oxygen	4	7	28

(continued)

Table 2.11 (continued)

Threat	Policymaker (Respondent Type 1)		
	TR	% of Vote	TS 1
Salinity	3	7.9	23.7
Impact on phytoplankton	6	14.7	88.2
Impact on nekton	6	15.7	94.2
Impact on benthos	6	12.5	75
Threat	**Local inhabitant (Respondent Type 5)**		
	SR	% of Vote	**TS 5**
Suspended particulate matter	8	16.8	134.4
Oil and grease level in the aquatic phase	7	15.2	106.4
Dissolved heavy metals	4	11.6	46.4
Dissolved Oxygen	2	10.7	21.4
Salinity	2	3.7	7.4
Impact on phytoplankton	6	14.1	84.6
Impact on nekton	6	15	90
Impact on benthos	6	12.9	77.4

Fig. 2.25 Antifouling paints on the vessels are removed by burning that emit considerable GHGs

1. **Fisheries and Aquaculture**

Marine fish landings in India for the year 2019 were estimated as 3.58 million tonnes which was about 2.2% more than the estimate of 2018. Marine fish catches declined in Gujarat, Maharashtra, Goa, Kerala and Puducherry and increased in Daman and Diu, Karnataka, Tamil Nadu, Andhra Pradesh, Odisha and West Bengal.

In the economic sustainability and trade section, marine fish landings during 2019 were valued at the fish landing centre level and retail level. State-wise valuation across value chains was also performed for Andhra Pradesh, Goa, Gujarat, Karnataka, Kerala, Maharashtra, Odisha, Puducherry, Tamil Nadu, West Bengal and Daman and Diu. Species wise valuation was also done. Macro-economic indicators at the national level were, total operating cost INR 34,702.24 crores and the net operating income INR 261,78.89 crores. The average capital productivity worked out to 0.57 and the gross value added to the marine fisheries sector was estimated at 'INR 40,790.36 crores, which is 67 per cent of the gross revenue.

It is a fact that the fishery and aquaculture sector has strong contribution to blue economy in coastal West Begal, but the opportunity is often shadowed by several stressors. Primarily of two kinds, these stressors influence the aquatic ecosystem and its biotic community almost directly. The first kind is the unregulated and often illegal harvest of marine resources. For example, the sharks, skates, and rays in the Indian Sundarbans are overexploited for their body parts such as meat, liver, fins and skin. Similarly, noncompliance of rules with regard to the prohibition of small mesh size (<60 mm) results in the frequent capture of juvenile fish of commercially valuable species such as hilsa. Fishers also capture the female hilsa at the mouth of the estuary during upstream migration for breeding, adversely affecting population recruitment.

The second kind of in-situ stressor is the toxic spill of fly ash in the estuarine waters of the Sundarbans. Annually, India exports three million tonnes of fly ash to Bangladesh, where it is used in the cement industry. Fly ash is highly toxic and contains various lethal components such as lead, arsenic, mercury, cadmium and uranium. The environmental impact of these accidental spills is immense, yet they remain understudied from the perspecive of their impact on the fragile Sundarbans ecosystem. Natural disasters like cyclonic depressions that originate primarily from the Bay of Bengal region also pose severe adverse impact on the fishery and aquaculture sectors.

Considering the prospect/opportunity and intricate relationship of Sundarban fish resources with the regional blue economy a survey was conducted through an online invitation (via googlemeet) to feel the pulse of stakeholders (n = 980–1350; only the common people were considered who are mostly school teachers, graduate students, ice factory owners, agriculturists and members of local fisherman association) on the Fisheries and Aquaculture sector of blue economy (Annexure 2F).

The major findings that emerged from this questionnaire are presented in Figs. 2.26, 2.27, 2.28, 2.29 and 2.30.

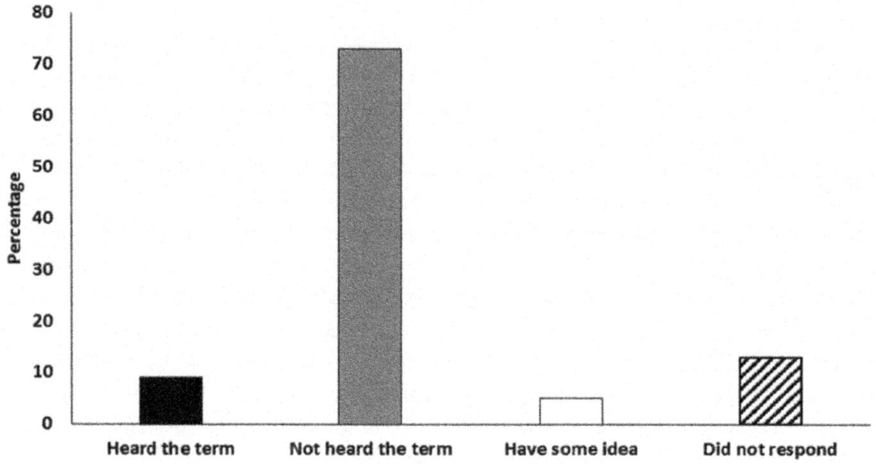

Fig. 2.26 Results of respondent analysis on the awareness of Blue Economy (n = 1067)

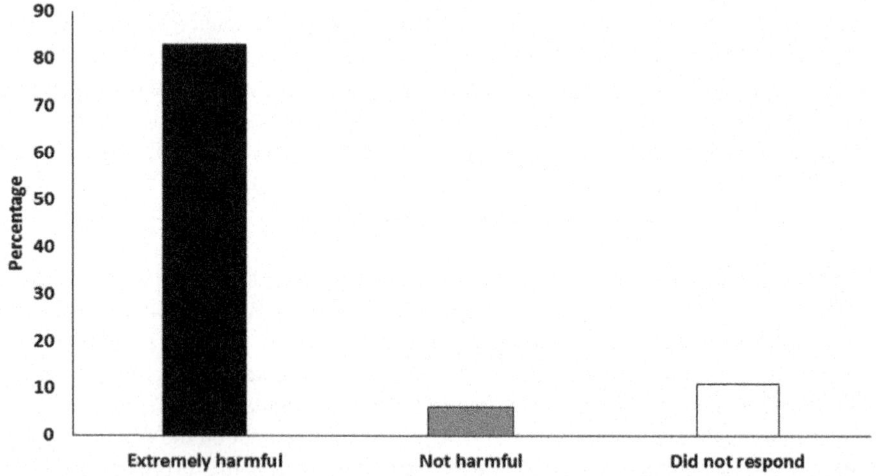

Fig. 2.27 Results of respondent analysis on the impact of aquaculture on mangrove ecosystem (n = 1350)

In addition to the above questionnaire, we also carried out a survey in the major fish landing stations of West Bengal to evaluate the temporal variation of catch volume of the major species in the lower Gangetic estuarine water (Table 2.12 and Fig. 2.31).

The figures in Table 2.12 points out an optimistic picture for all types of species except Indian shad, *Tenualosa ilisha*, which has a high export value and is noted for its delicacy. The primary reasons for such dip in the catch volume of the species in 2019 compared to 2010 (57.6%) may be attributed to three major reasons namely (i) alteration of salinity, (ii) pollution and (iii) illegal catching of the gravid females as well as small sized fishes (Mitra 2013).

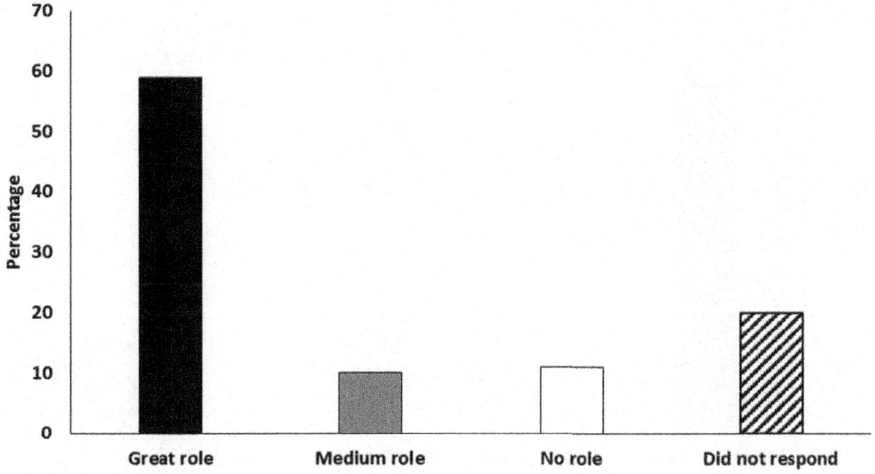

Fig. 2.28 Results of respondent analysis on the role of fishery in Indian Sundarban economy (n = 980)

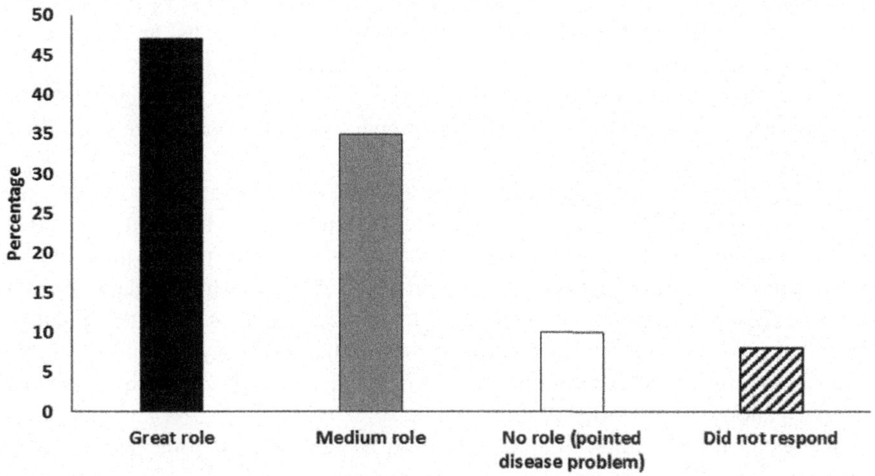

Fig. 2.29 Results of respondent analysis on the role of aquaculture in local economy (n = 1025)

2. **Shipping and Transportation of Goods**

In 2017 and early 2018, the global shipping industry saw a marked improvement of fundamentals in most market segments, with the exception of the tanker market. Key drivers were the combined strengthening in global demand, on the one hand, and the reduced fleet growth, on the other. Overall, freight rates improved across all markets in 2017, with the exception of tankers.

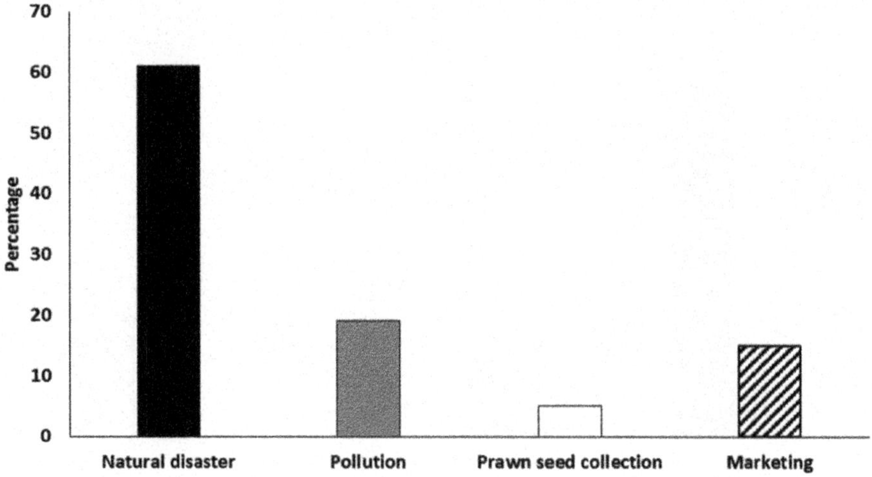

Fig. 2.30 Results of respondent analysis on the threats to fisheries and aquaculture (n = 1350)

Container freight rate levels increased, and averages surpassed performance in 2016. A better supply–demand balance in container ship markets, underpinned by stronger demand, was the main driver. The container shipping industry ended 2017 with a total profit of roughly $7 billion, driven mainly by a significant increase in transported volumes, freight rates and revenue, as well as proactive operational management discipline.

During the year, consolidation, whether in the form of alliances or mergers and acquisitions, persevered in the container industry in response to the negative environment of the industry a pessimistic scenario has been observed, which soraded high during the covid-19 lockdown phase of 2020. While outright negative impacts on trade and costs have not been reported, there exist concerns about the impact of growing market concentration on competition and the level playing field.

The shipping sector also exhibited a great dip in income due to closure of industrial activities and associated transport of materials.

The estuaries of Indian Sundarbans serve as navigational route for ships, oil tankers and vessels with their destinations at Kolkata or Haldia port. This sector, being a strong pillar of regional blue economy needs to be strengthened keeping the ambient environment ecologically sound.

In context to lower Gangetic delta, a questionarrie developed to feel the pulse of respondents (n = 1500; only the tanker owners, marine engineers/captains, drivers of passenger vessels and deep sea vessels were considered) to understand the viability of shipping and transportation in Sundarban estuaries are presented as Annexure 2G.

The major findings that emerged from the questionnaire are represened in Figs. 2.32, 2.33, 2.34, 2.35 and 2.36.

3. **Estuary/Sea-Based Recreation and Tourism**

Table 2.12 Marine fish landing (unit in tonnes) by species in West Bengal coast

Fish variety	2010	2011	2012	2013	2014	2015	2016	2017	2018	2019
Arius spp.	19,415	22,451	18,566	20,743	22,120	23,945	25,770	27,553	29,110	30,811
Sciaenidae (croakers)	19,732	29,000	19,845	21,933	23,010	24,875	25,841	27,641	29,257	31,025
Tenualosa ilisha	55,300	18,010	10,325	12,474	14,615	16,430	18,046	19,775	21,490	23,487
Harpodon nehereus	12,026	13,425	11,464	13,575	14,997	16,756	18,555	20,141	21,753	23,489
Other marine fish	13,990	15,702	12,535	14,321	15,423	17,321	19,025	20,856	22,614	24,410
Rastrelliger kanagurta	11,314	11,598	10,129	12,007	13,489	15,102	16,869	18,753	20,489	22,347
Penaeid Prawn	12,191	10,143	9260	11,345	12,753	14,563	16,325	18,141	19,913	21,698
Lates spp. and *Lutjanus* spp. (snapper and parch)	11,843	12,567	12,283	14,157	15,990	17,827	19,641	21,157	23,192	250.14
Stromateidae (butter fishes)	7999	8714	11,959	13,119	12,857	14,672	16,489	18,226	20,157	21,057
Other clupeoides	8012	7216	9724	10,955	12,931	14,509	16,432	18,055	19,873	21,478
Carangidae (bangada)	9150	10,057	8202	9913	11,819	13,473	15,124	16,983	18,696	20,453
Trichiuridae (hairtail ribbon fish)	6841	6605	3302	5148	7245	8843	10,775	12,690	14,650	15,741
Mugilidae (mullets)	6947	7581	11,470	12,875	15,119	16,789	1539	3371	5369	7259
Elasmobranchii (shark, rays, skates etc.)	9519	10,364	8759	10,523	12,358	14,009	15,836	17,559	19,478	21,317
Non-penaeid prawns	5610	5311	4597	6254	7965	9075	10,994	12,853	14,775	16,475
Indian oil sardine	3190	7956	5174	7001	8531	10,123	11,867	13,641	15,421	17,240
Engraulidae (anchovies)	7332	8490	10,574	11,963	13,869	15,489	17,429	19,103	20,069	21,459
Caranx spp. (Kala bangada)	4008	4261	3549	5189	6885	8753	10,488	12,278	14,108	15,745
Polynemidae (Indian salmon)	3483	3251	2356	4272	6247	7863	9687	11,451	13,247	14,921
Gerreidae (common silver belly)	2456	2508	2128	4005	6198	7894	9721	11,408	13,213	15,075

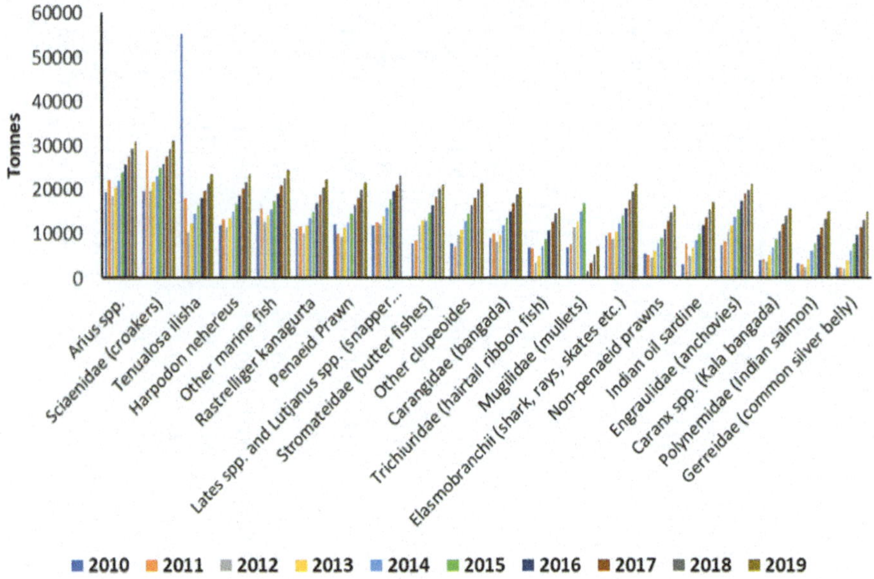

Fig. 2.31 Temporal variation of catch volume of the major species in the lower Gangetic delta region

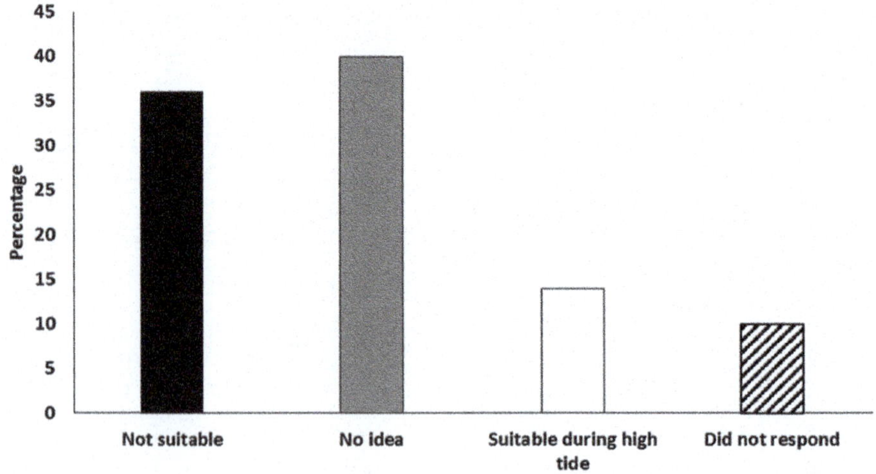

Fig. 2.32 Results of respondent analysis on the suitability of draft for navigation in Sundarban estuaries

Fig. 2.33 Results of respondent analysis on the purpose of using navigation channel in Sundarbans

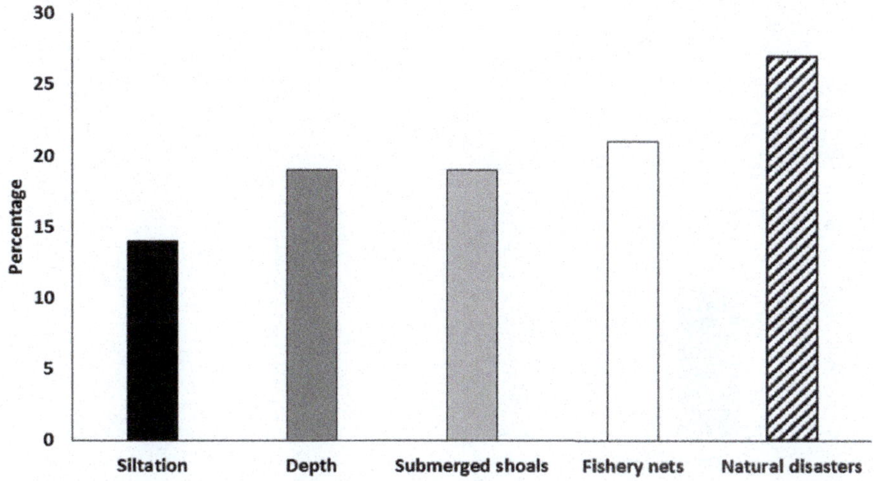

Fig. 2.34 Results of respondent analysis on the hindrances in navigation in Sundarban estuaries

Tourism sector in coastal West Bengal has the scope of earning huge profit. The major tourist spots are Digha (21° 37′ 00.12" N; 87° 30′ 10.54" E) in East Midnapore district, Sagar Island (21° 38′ 03.28" N; 88° 05′ 20.23" E) in south 24 Parganas district, Bakkhali (21° 33′ 15.98" N; 88° 16′ 00.45" E) in South 24 Parganas district and various areas within Sundarban mangrove forests.

It has been documented that Digha receives about 45% of the tourist flow in the maritime state of West Bengal. Lakhs of foreign tourists along with more than twenty five lakhs of Indian tourists visit this coastal zone. Mandarmoni (21° 39′ 37.48" N;

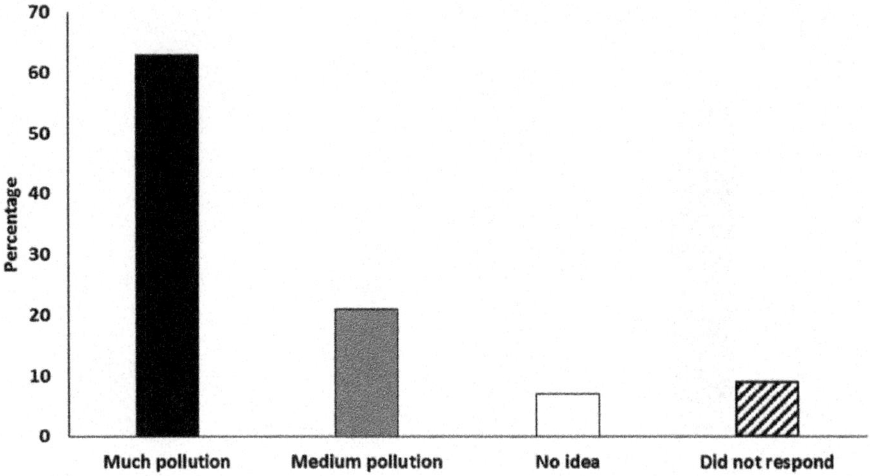

Fig. 2.35 Results of respondent analysis on pollution from ships, fishing vessels and trawlers

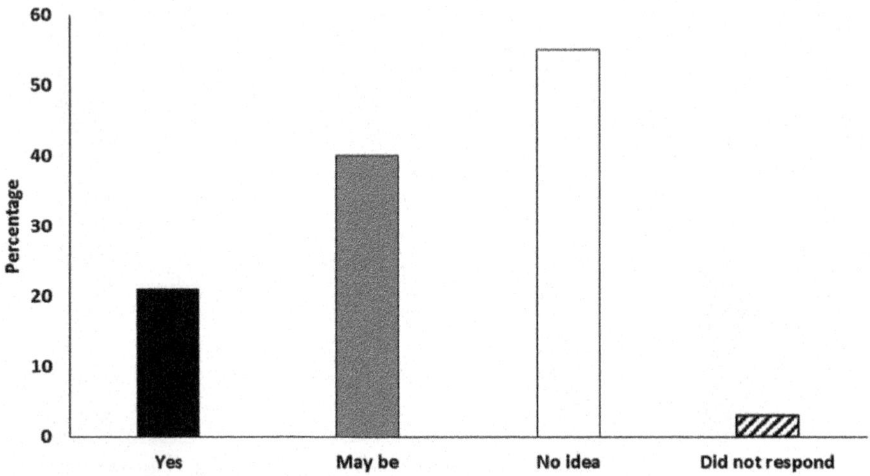

Fig. 2.36 Results of respondent analysis on pollution from ballast discharge

87° 42′ 43.14" E), Shankarpur (21° 37′ 49.49" N; 87° 34′ 33.70" E) and Talsari (21° 36′ 01.96" N; 87° 27′ 44.19" E) surrounding the Digha coastal regions are important places of interest that attract large number of tourists every year.

The Kapil Muni Ashram (21° 38′ 15.37" N; 88° 04′ 30.50" E) at Sagar Island attracts about ten to fifteen lakhs pilgrims during Makar shankranti in mid-January every year. The devotees take holy bath in the estuary adjacent to Kapil Muni Ashram and thus make the island a remarkable spot in the domain of pilgrimage. However, the main bottlenecks of problems for this pilgrimage are lack of frequent communication while crossing the Hooghly River by passenger vessels, absence of efficient grid electric power etc.

Bakkhali in the western Indian Sundarbans is noted for its unique beach; the shrimp farms in the Henry's island (21° 34′ 19.18" N; 88° 17′ 54.71" E) and fish landing stations at Frasargunge (21° 34′ 54.42" N; 88° 15′ 33.31" E) are the other attractions surrounding Bakkhali tourist spot.

The dense mangrove forests of the Indian Sundarbans are the primary attractions for wild life enthusiasts. It is the only spot in the World map where the Royal Bengal tiger (*Panthera tigris tigris*) stay symbiotically with mangrove floral community. Deer, crocodiles, wild boars, monkeys and birds of various species have made the ecosystem a unique hub for observing nature, education and research and learning the age-old tradition and folk tales of the region.

We made an extensive survey amongst the local inhabitants to assess the viability of estuarine tourism in the region during 2019. The reflections of respondent analysis ($n = 1285$), who include local people, researchers, NGOs, and policy makers are presented in Figs. 2.37, 2.38, 2.39, 2.40 and 2.41. It is to be noted in this context that about 74% of the total respondents were the local inhabitants. We preferred the dominancy of local population in our survey only to get the ground reality, as the other categories of respondents (like researchers, NGOs, and policy makers are mostly migratory in nature).

The respondent analysis reflects that most of the local people and island dwellers understands tourism as a source of revenue generation that come from forest visits/wild life citations and religious activities (preferably the holy bath in mid January in the Hooghly estuary) surrounding the Kapil Muni temple at Sagar Island (Fig. 2.42). However, the concept of eco-toirism is not very clear to majority of the population (*vide* Fig. 2.41) although they have deep sense of knowledge on the

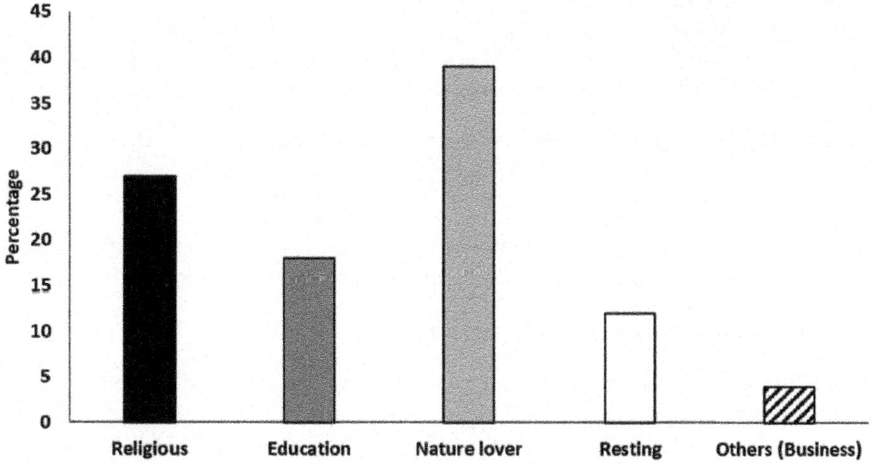

Fig. 2.37 Results of respondent analysis on the purpose of estuarine tourism

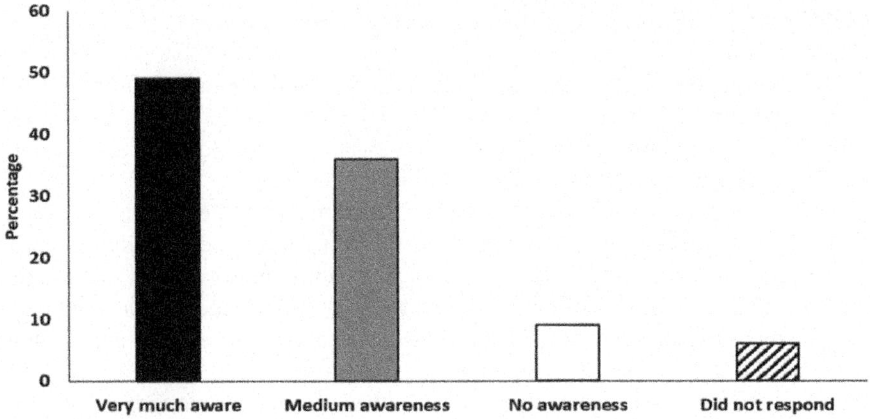

Fig. 2.38 Results of respondent analysis on the awareness of tourists on biotic resources on Indian Sundarbans

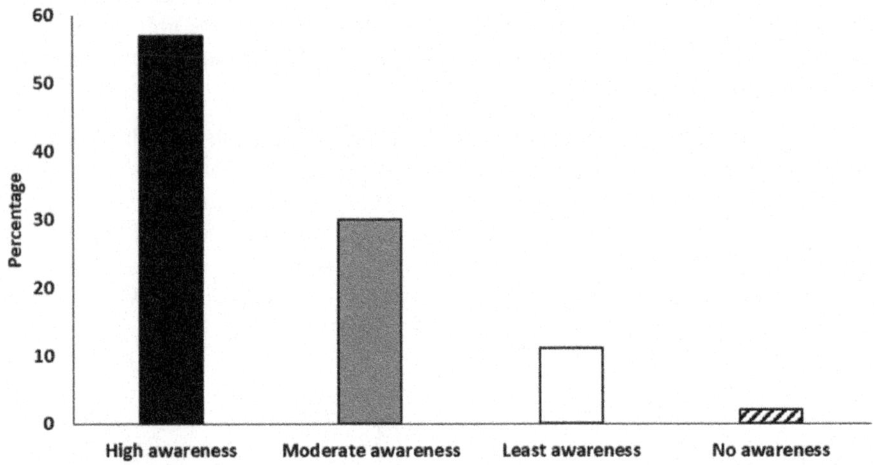

Fig. 2.39 Results of respondent analysis on the awareness of tourists towards the ecology of Indian Sundarbans

ecology of mangrove forests and their biotic resources. This may be possibly because of their dependency on the forest resources for their livelihoods and daily needs.

4. **Marine Biodiversity and Conservation**

Blue Economy will see a sustainable face if and only if the production sectors adopt a zero-carbon emission policy/technology. The producer community of the ecosystem acts as a sink of carbon and can serve to reduce the carbon dioxide level at local scale (Agarwal et al. 2021). In the Sundarban deltaic complex, mangroves are the most potential reservoir of carbon storage (Mitra et al. 2011; Banerjee et al. 2013;

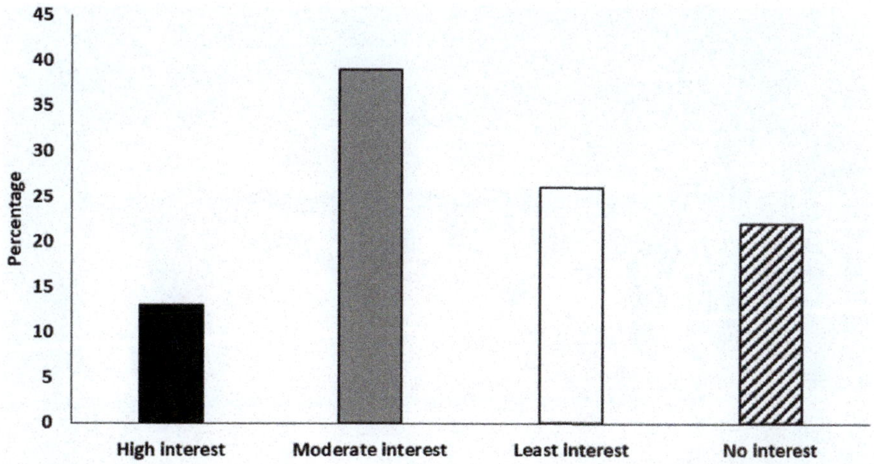

Fig. 2.40 Results of respondent analysis on the interest of tourists to know local culture and folk tales

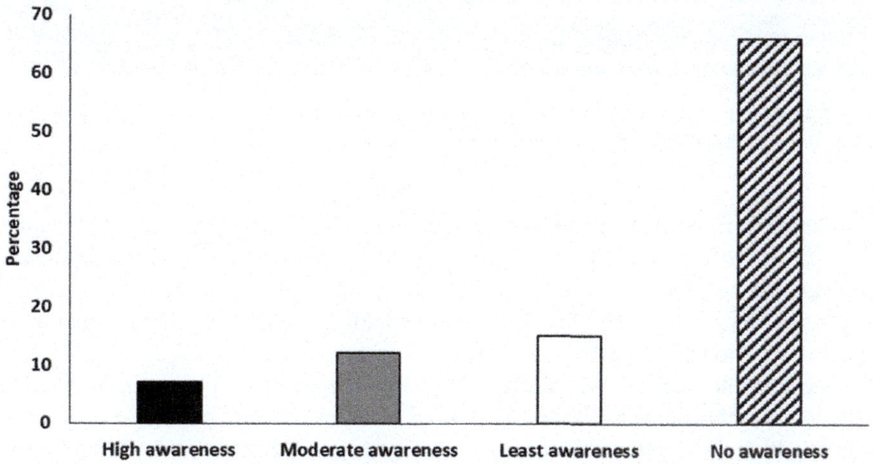

Fig. 2.41 Results of respondent analysis on the concept of eco-tourism by stakeholders

Raha et al. 2013; Sengupta et al. 2013; Mitra 2013, 2020; Mitra and Zaman 2014, 2015, 2016, 2020, 2021; Mitra and Gati 2015). Hence the survival, growth and diversity of this unique halophytic community need to be monitored with perfection so as to assess their carbon sequestration potential. On this background, we carried out a survey during 206, 2017 and 2018 to evaluate the health of commonly available mangrove species in Indian Sundarbans. A total of 25 mangrove species (*Acanthus ilicifolius, Acrostichum aureum, Aegiceras corniculatum, Aegialites rotundifolia, Avicennia alba, Avicennia marina, Avicennia officinalis, Bruguiera cylindrical, Bruguiera gymnorhiza, Bruguiera hexangular, Bruguiera parviflora, Ceriops*

Fig. 2.42 Kapil Muni temple at Sagar Island attracts about 10–15 lakhs people to take the holy bath during Makar Shankranti (in mid January every year)

decandra, Ceriops tagal, Exocoecaria agallocha, Heritiera fomes, Kandelia candel, Lumnitzera racemosa, Nypa fruticans, Phoenix paludosa, Rhizophora apiculata, Rhizophora mucronata, Sonneratia apetala, Sonneratia caseolaris, Xylocarpus granatum, Xylocarpus mekongensis) was recorded from the selected sites during our study period (2016, 2017 and 2018).

The total number of individuals of all the documented species per unit area, which represents the dense/sparse growth of the trees exhibits pronounced variation due to which the Shannon Weiner species diversity is affected. Highest numbers of individuals (N) are observed in the quadrate of Lothian island, a pristine forest patch almost with no human intervention, whereas lowest value is observed in Canning, the sampling station with maximum anthropogenic stress (Table 2.13 and Fig. 2.43).

The results of Shannon-Weiner Index are shown in Table 2.14 and Fig. 2.44. The spatial order of the index is Lothian island (2.6344, 2.6290 and 2.6086 in 2016, 2007 and 2018 respectively) > Prentice island (2.4509, 2.4739 and 2.4766 in 2016, 2007 and 2018 respectively) > Sajnekhali (2.3901, 2.3852 and 2.3852 in 2016, 2007 and 2018 respectively) > Pakhiralaya (2.3454, 2.3659 and 2.3068 in 2016, 2007 and 2018 respectively) > Harinbari (2.3328, 2.3313 and 2.3007 in 2016, 2007 and 2018 respectively) > Chemaguri (2.3017, 2.2587 and 2.2469 in 2016, 2007 and 2018 respectively) > Sagar South (2.2459, 2.2533 and 2.2409 in 2016, 2007 and 2018 respectively) >

Table 2.13 Spatio- temporal variation of N (No./100 m^2)

Sectors	Sl. No	Name	2016	2017	2018
Western	1	Harinbari (Stn. 1)	64.2	59.0	55.8
	2	Chemaguri (Stn. 2)	58.0	52.1	48.9
	3	Sagar South (Stn. 3)	58.5	55.1	52.0
	4	Lothian island (Stn. 4)	80.1	76.5	73.2
	5	Prentice island (Stn. 5)	73.6	68.7	65.4
Central	6	Canning (Stn. 6)	46.9	45.8	41.8
	7	Sajnekhali (Stn. 7)	51.4	47.6	43.2
	8	Chotomollakhali (Stn. 8)	53.2	48.5	44.3
	9	Satjelia (Stn. 9)	53.9	50.2	48.5
	10	Pakhiralaya (Stn. 10)	64.0	58.5	55.2

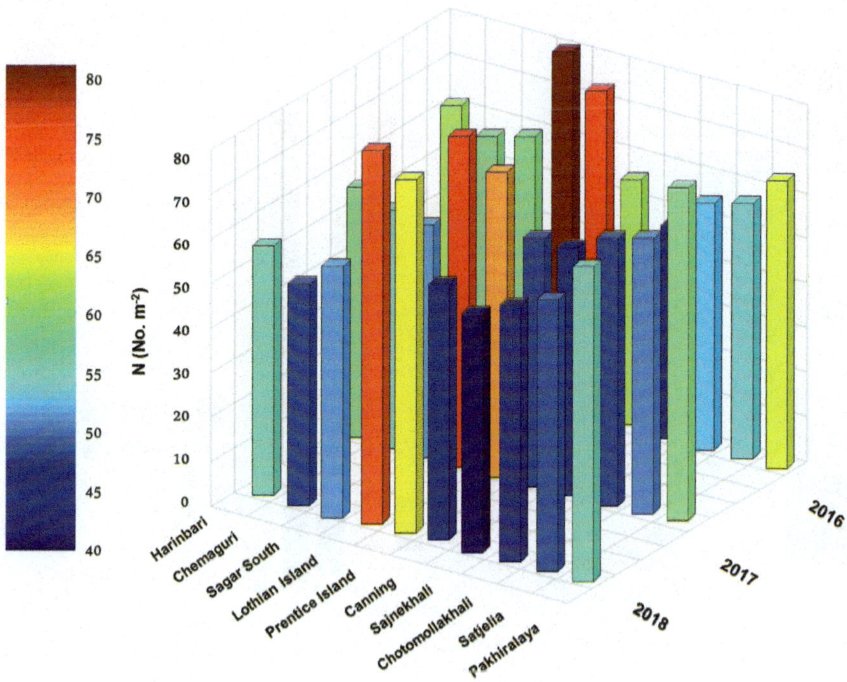

Fig. 2.43 Spatio-temporal variation of population density of mangroves (in No./100m^2) in the 10 selected stations during 2016, 2017 and 2018

Table 2.14 Spatio-temporal variation of $\overline{H}$

Sectors	Sl. No	Name	2016	2017	2018
Western	1	Harinbari (Stn. 1)	2.3328	2.3313	2.3007
	2	Chemaguri (Stn. 2)	2.3017	2.2587	2.2469
	3	Sagar South (Stn. 3)	2.2459	2.2533	2.2409
	4	Lothian island (Stn. 4)	2.6344	2.6290	2.6086
	5	Prentice island (Stn. 5)	2.4509	2.4739	2.4766
Central	6	Canning (Stn. 6)	2.0661	2.0661	2.0524
	7	Sajnekhali (Stn. 7)	2.3901	2.3852	2.3852
	8	Chotomollakhali (Stn. 8)	2.0967	2.0922	2.0662
	9	Satjelia (Stn. 9)	2.1284	2.1405	2.1307
	10	Pakhiralaya (Stn. 10)	2.3454	2.3659	2.3068

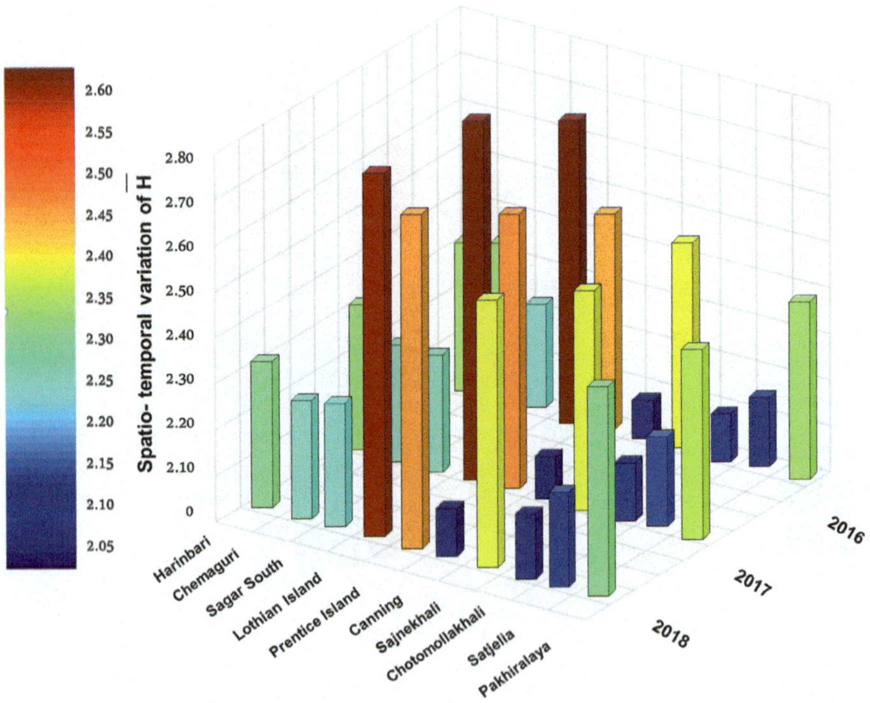

Fig. 2.44 Spatio-temporal variations of Shannon Weiner Species Diversity Index computed on the basis of relative abundance of true mangrove floral species

Satjelia (2.1284, 2.1405 and 2.1307 in 2016, 2007 and 2018 respectively) > Chotomollakali (2.0967, 2.0922 and 2.0662 in 2016, 2007 and 2018 respectively) > Canning (2.0661, 2.0661 and 2.0524 in 2016, 2007 and 2018 respectively).

Table 2.15 Spatio-temporal variation of Evenness Index

Sectors	Sl. No	Name	2016	2017	2018
Western	1	Harinbari (Stn. 1)	0.9488	0.9545	0.9607
	2	Chemaguri (Stn. 2)	0.9363	0.9411	0.9457
	3	Sagar South (Stn. 3)	0.9211	0.9273	0.9310
	4	Lothian island (Stn. 4)	0.9728	0.9794	0.9823
	5	Prentice island (Stn. 5)	0.9703	0.9755	0.9798
Central	6	Canning (Stn. 6)	0.8956	0.9004	0.9064
	7	Sajnekhali (Stn. 7)	0.9618	0.9677	0.9702
	8	Chotomollakhali (Stn. 8)	0.9002	0.9069	0.9119
	9	Satjelia (Stn. 9)	0.9031	0.9111	0.9165
	10	Pakhiralaya (Stn. 10)	0.9549	0.9603	0.9657

The variation in the value of the Shannon Weiner Species Diversity Index reflects (i) the degree of stress (both natural and anthropogenic) (ii) conditions of the ambient environment (in terms of hydrological parameters and soil quality). Greater value of the index represents a more congenial environment which usually occurs due to the survival of more number of species or even distribution of the number of individuals amongst different species in the quadrate.

Evenness index is a measure of favourable environmental conditions similar to Shannon Weiner Species Diversity Index.

The results of Evenness Index are shown in Table 2.15 and Fig. 2.45. The spatial order of the index is Lothian island (0.9728, 0.9794 and 0.9823 in 2016, 2007 and 2018 respectively) > Prentice island (0.9703, 0.9755 and 0.9798 in 2016, 2007 and 2018 respectively) > Sajnekhali (0.9618, 0.9677 and 0.9702 in 2016, 2007 and 2018 respectively) > Pakhiralaya (0.9549, 0.9603 and 0.9657 in 2016, 2007 and 2018 respectively) > Harinbari (0.9488, 0.9545 and 0.9607 in 2016, 2007 and 2018 respectively) > Chemaguri (0.9363, 0.9411 and 0.9457 in 2016, 2007 and 2018 respectively) > Sagar South (0.9211, 0.9273 and 0.9310 in 2016, 2007 and 2018 respectively) > Satjelia (0.9031, 0.9111 and 0.9165 in 2016, 2007 and 2018 respectively) > Chotomollakali (0.9002, 0.9069 and 0.9119 in 2016, 2007 and 2018 respectively) > Canning (0.8956, 0.9004 and 0.9064 in 2016, 2007 and 2018 respectively).

The results of Index of Dominance are shown in Table 2.16 and Fig. 2.46. The spatial order of the index is Canning (0.1407, 0.1412 and 0.1457 in 2016, 2007 and 2018 respectively) > Pakhiralaya (0.1352, 0.1365 and 0.1400 in 2016, 2007 and 2018 respectively) > Chotomollakhali (0.1299, 0.1311 and 0.1372 in 2016, 2007 and 2018 respectively) > Satjelia (0.1287, 0.1302 and 0.1336 in 2016, 2007 and 2018 respectively) > Chemaguri (0.1208, 0.1233 and 0.1301 in 2016, 2007 and 2018 respectively) > Harinbari (0.1154, 0.1167 and 0.1196 in 2016, 2007 and 2018 respectively) > Sagar South (0.1116, 0.1150 and 0.1169 in 2016, 2007 and 2018 respectively) > Prentice

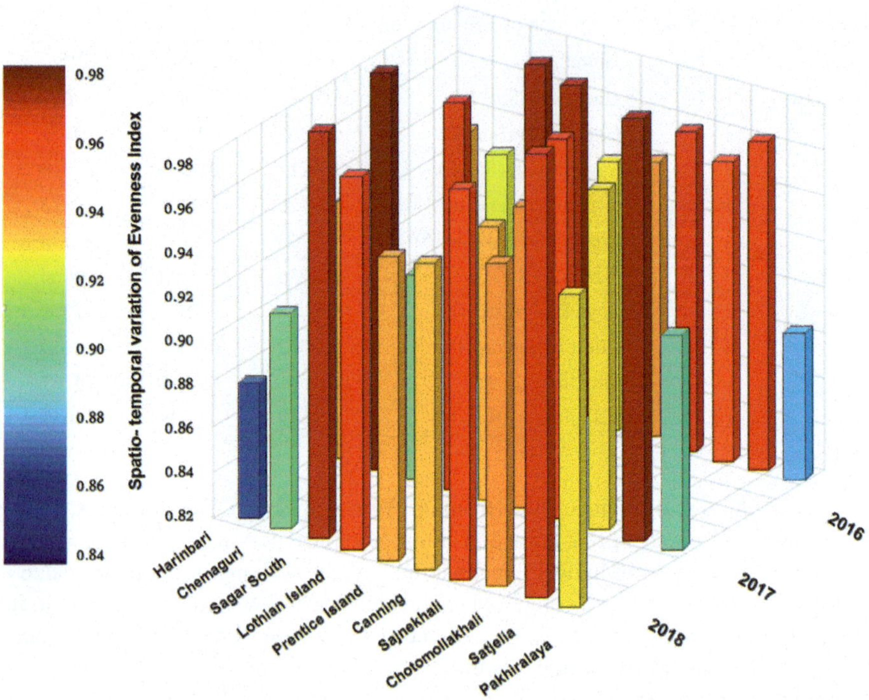

Fig. 2.45 Spatio-temporal variation of Evenness Index computed on the basis of relative abundance of true mangrove floral species

Table 2.16 Spatio-temporal variation of Index of Dominance

Sectors	Sl. No	Name	2016	2017	2018
Western	1	Harinbari (Stn. 1)	0.1154	0.1167	0.1196
	2	Chemaguri (Stn. 2)	0.1208	0.1233	0.1301
	3	Sagar South (Stn. 3)	0.1116	0.1150	0.1169
	4	Lothian island (Stn. 4)	0.0772	0.0779	0.0808
	5	Prentice island (Stn. 5)	0.1062	0.1073	0.1088
Central	6	Canning (Stn. 6)	0.1407	0.1412	0.1457
	7	Sajnekhali (Stn. 7)	0.0993	0.1015	0.1027
	8	Chotomollakhali (Stn. 8)	0.1299	0.1311	0.1372
	9	Satjelia (Stn. 9)	0.1287	0.1302	0.1336
	10	Pakhiralaya (Stn. 10)	0.1352	0.1365	0.1400

island (0.1062, 0.1073 and 0.1088 in 2016, 2007 and 2018 respectively) > Sajnekhali (0.0993, 0.1015 and 0.1027 in 2016, 2007 and 2018 respectively) > Lothian island (0.0772, 0.0779 and 0.0808 in 2016, 2007 and 2018 respectively).

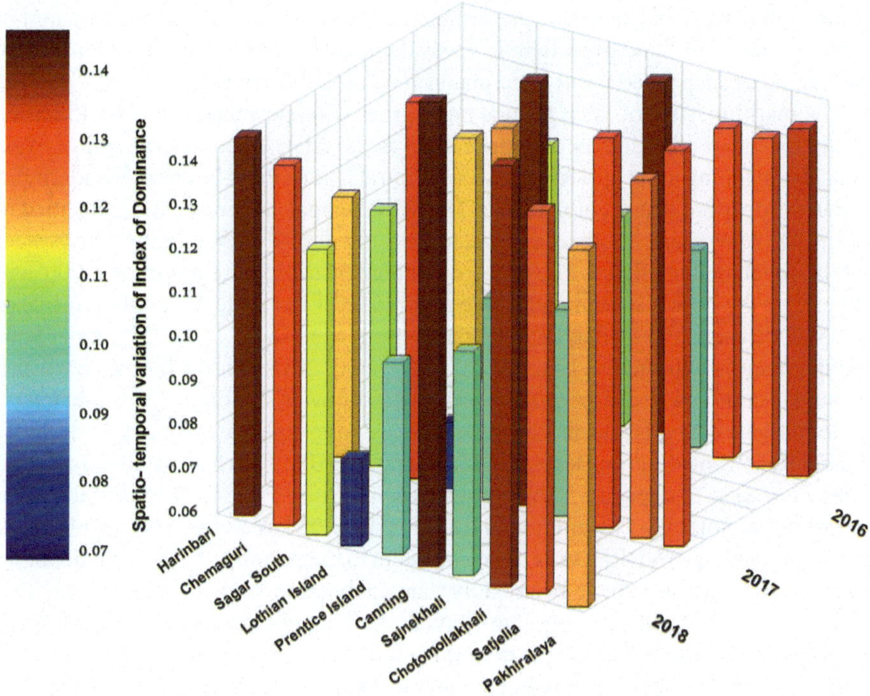

Fig. 2.46 Spatio-temporal variation of Index of Dominance computed on the basis of relative abundance of true mangrove floral species

Index of Dominance (ID) is a measure of environmental stress. The greater value of ID is an indicator of extreme environmental stress. Thus in this study is clear that the stress is minimum at Lothian island and the environment is congenial for the growth of the floral species.

5. **Non-conventional Renewable Energy Sector**

For islands and mangrove forest dominated ecosystems like Sundarbans, tidal energy is one of the best options for generation of electric current. The mechanism of working of the tide-based energy conversion is not very complicated. During ingression of the water, the tidal energy works via a turbine in which the blade rotates 12–20 times per minute. The rotation of the blade is, however, a function of the tidal strength. The turbine, which is connected to a gear box, turns the generator that finally produces electric current. In this context, we would like to state that the high upfront cost is one of the greatest hindrances to implement tidal-based electric generation. Because of the high density of the water, the turbine need to bear high strength compared to wind turbines. For Sundarban deltaic ecosystem, siltation is a major problem to run the turbine as suspended particulate matter may clog the system.

The maritime state of West Bengal has a gross wind power potential of 450 MW, but the installed capacity, as of March 2013 is 2.5 MW. The Government of West Bengal is planning to set up a wind power station at Dadan Patra Bar near Mandarmoni in East Midnapore district of the state. The West Bengal Government has decided to set up 100 MW project with an estimated investment of INR 500 crore. The project will be developed in phases. The state government has already earmarked 1,000 acre of land for the power station and in the first phase of the project a total of 700 acre land will be used. The Dadan Patra Bar area has a wind speed of 19–25 km per hour and the plant is expected to generate 50 MW power in its phase I (https://www.projectstoday.com/news/west-bengal-to-set-up-wind-power-station).

To mitigate the emission and subsequent rise of atmospheric carbon dioxide, West Bengal Renewable Energy Development Agency (WBREDA) has installed four aero generators each of capacity 5 kW at two existing Solar PV Power Plants at Gayen Bazar 25 kW Solar PV Power Plant and at Moushuni 55 kW Solar PV Power Plant of Indian Sundarbans. WBREDA is facing problem in maintaining the Aero Generators. Three Aero Generators have been badly damaged due to Aila, a super cyclone that hit the lower Gangetic delta on 25th May, 2009. Performances of Aero Generators are found well particularly for the period from March to September. It is to be noted in this context that Sundarban is a cyclone prone zone and this can pose considerable damage to installed structures related to non-conventional energy generation.

To conclude, it can be stated that reduction of CO_2 emission through implementation of non-conventional energy can see the light of success if few actions as highlighted here are adopted on fast track basis.

1. **Universal access to energy**

While most part of the Indian sub-continent is on track to reach near universal access to energy, there are several islands in Indian Sundarban region with extremely low access rates. Almost all the islands in the central and eastern sectors of Indian Sundarbans do not have the power supply. These islands require particular attention and institutional support from the side of their respective Governments, as well as from international development partners. Issues of quality and reliability of energy access, as well as availability of reliable data for monitoring of this indicator need to be controlled by the policy makers.

2. **Clean energy**

National targets for clean energy and technologies need to be established on priority basis and clean energy is required to be integrated into energy policy frameworks of the country.

3. **Increasing the share of renewable in the energy mix**

Considering the quasi stagnation of the share of renewable energy in the energy matrix of the region, efforts at promoting renewable sources of energy should be the need of the hour. This is primarily because of the rapid rate of industrialization and urbanization, which are presently fully fuel dependent in India. Moreover the effort can be a practical step to curb emission of GHGs.

4. **Energy efficiency**

Energy efficiency measures should be considered with high priority in the transport, industry, tourism, aquaculture and building sectors.

5. **Intergovernmental and Trans-boundary cooperation**

Member states with the support of Economic and Social Commission for Asia and the Pacific (ESCAP) and other relevant regional institutions need to develop an intergovernmental mechanism that would promote trans-boundary electricity power trade and connectivity as one of the building blocks to attain sustainable development. This is especially applicable for single large mangrove chunk like Sundarbans, in which 62% lies in Bangladesh and the remaining 38% in India.

6. **Removing policy barriers and promoting funds to innovative technology based research**

To facilitate renewable energy and energy efficiency expansion, member States need to address existing policy barriers such as restrictive permit procedures, monopolistic utility positions and restrictions in energy prices. After successful implementation of these measures, the introduction of financial incentives for clean energy projects will bring positive results. While a number of member States have made strong efforts to incentivize clean energy and have committed to phasing out fossil fuel subsidies, more improvements need to be made to reach the respective targets. Despite significant technological innovation in many industries, international energy cooperation efforts to deploy innovative technologies are limited due to lack of a coherent approach among the member states. Cutting edge innovations in the sphere of solar energy, wind energy, biomass energy, geo-thermal energy, tidal energy need to be encouraged through adequate funding after a comprehensive SWOT analysis.

The pulse of ramification of different sectors of blue economy in the deltaic complex of Indian Sundarbans has remained underexplored so far, although the region supports congenial environment, biotic and abiotic resources that have considerable potential to provide momentum to regional blue economy. The critical role of estuarine-cum-mangrove based resources in sustaining the regional blue economy is under appreciated no doubt, but there are immense opportunities in this sector which are highlighted in Table 2.17.

Table 2.17 Prospective sectors of blue economy in Indian Sundarbans along with their respective opportunity

Sector	Major actor	Opportunity	Remark
Fishery	• State Fishery Department • State Forest Department • Local NGOs • Local Fisherman Association	+++	Organized sectors, but during fish catch conservation related laws are hardly followed
Aquaculture	• State Fishery Department • State Forest Department • Local NGOs • CRZ Authority (State Level Committee) • State Environment Department • State Pollution Control Board	+++	Unorganized sector; institutional support required; treatment plant (preferably bio-treatment plant) needs to be installed
Fish drying	• State Fishery Department • State Forest Department • Local NGOs • Local Fisherman Association	++	Unorganized sector; institutional support required; strong supply chain needs to be established with poultry and fish feed manufacturing units
Mangrove based tourism	• State Tourism Department • State Fishery Department • State Forest Department • Local NGOs • Universities & Academic institutions • State Environment Department	+++	Partially organized sector; Awareness on eco-tourism needs to be generated; Strict laws and policies need to be implemented to prohibit tourism related pollutants like plastics, left over food items, drink cans etc. from throwing in the estuaries or mudflats
Shipping and Marine transport	• Kolkata Port Trust • Coast Guard • Ministry of Water ways • State Fishery Department	++	Draft; Ballast discharge problems
Ship breaking	• Kolkata Port Trust • Coast Guard • Ministry of Water ways • State Environment Department • State Pollution Control Board	+	Ecofriendly technology needs to be adopted

(continued)

Table 2.17 (continued)

Sector	Major actor	Opportunity	Remark
Biodiversity & Carbon sequestration	• State Tourism Department • State Fishery Department • State Forest Department • Local NGOs • Universities & Academic institutions • State Environment Department	+++	Awareness on every tiers of the society needs to be generated; carbon credit needs to implemented through coordination with international agencies keeping local people and island dwellers in the loop

'+++' means maximum opportunity, ++' means moderate opportunity and '+' means less opportunity to boost up blue economy in Indian Sundarbans

2.4 Take Home Messages

[A] The estuaries of Indian Sundarbans along with the living and non-living resources are the basic sources of livelihoods in and around this World Heritage Site. Mangroves, mangrove associate species, seaweeds, fishes of various categories, molluscs, arthropods and even the big mammals like Royal Bengal Tiger (*Panthera tigris tigris*), deer, monkey, wild boar and birds of various types, reptiles and innumerable flora and fauna have potential to sustain the blue economy of the region. The rich biotic resources of Indian Sundarbans can not only support the fishery and aquaculture sectors of the region, but they have great potential to flourish the eco- and edu-tourism in the deltaic complex. Blue Economy will see a sustainable face if and only if the production sectors adopt a zero-carbon emission policy/technology. The producer communities of the ecosystem (blue carbon) act as a major sink of carbon and serve to reduce the carbon dioxide level at local scale. In the Sundarban deltaic complex, mangroves are the most potential reservoir of carbon storage.

[B] The non-living resources of deltaic Sundarbans do not include rich mineral based resources like phosphorite deposits, polymetallic sulphides, polymetallic nodules and gas hydrates or other major categories of commercially viable non-living components that are extracted from the marine environment, but there is great scope of procuring evaporate deposits like salts from the estuaries, preferably in central Indian Sundarbans, which is a hypersaline sector compared to western and eastern sectors of Indian Sundarbans. Evaporite deposits like salts were usually procured in the Digha- Mondarbani region in coastal West Bengal in early eighties, but now the practice exists no more and instead hotels and shrimp culture farms have flourished in the region.

[C] The blue economy in the lower Gangetic delta region is getting momentum through construction of a deep sea port off Sundarban mangrove forest near Sandhead Island. The primary aim of this establishment is to offer cargo handling and trans-shipment facility. Handing and trans-shipment of hazardous cargo is the priority of the project. However, several environmental issues have cropped up in connection to this project like adverse impact on the fish breeding and nursery ground in and around the mangroves and deterioration of estuarine water quality. In the maritime state of West Bengal, ship breaking is carried out at Netaji Subhas dock, Garden Reach at Kolkata. However, another ship breaking and ship recycling unit is in the pipeline. It is proposed to set up this unit at Tengra-Char village on the eastern bank of the Hooghly River estuary (the western-most distributary of the Ganges River) in Kulpi Block of South 24-Paragans District, West Bengal. The ship recycling yard shall be set up by Bengal Shipyard. It is expected that ships of total Light Displacement Tonnage (LDT) of about 60,000 tonnes will be recycled annually. Non-conventional sources of energy are strong pillars of blue economy as it is not only a substitute of fast depleting coal reservoir, but also a zero-carbon emission energy source. India may have its first and the world's fifth tidal energy power plant at Durgaduari creek in Sundarbans. With the coastal regulatory zone clearance and environmental clearance in hand, the ministry of non-conventional energy sources (MNES) is all set to execute the tidal project.

[D] The deltaic complex of Indian Sundarbans with rich mangrove biodiversity, fish resources, wild life, congenial aquatic environment and skilled human resources (specially in the field of aquaculture, deep sea fishing, mangrove oriented tourism, agriculture, vessel repairing and building etc.) has considerable scope to boost up the marine and estuarine based livelihoods and job opportunities. There is also great space to institutionalize the holy tourism in the region due to presence of Kapil Muni temple that attracts 10–15 lakhs people every year from different parts of the country to take the holy bath during Makar Shankranti celebrated in mid January.

Annexure 2A: A Checklist of Members of Class Pisces (Both Chondrichthyes and Osteichthyes) in the Brackish Water and Fresh Water Ponds of Indian Sundarbans (*Source* Mitra 2020)

1. **Order: Carcharhiniformes (Ground Sharks)**

Family: *Carcharhinidae* (Requiem sharks)

White cheek shark *Carcharhinus dussumieri*

Bull shark *Carcharhinus leucas*

Black tip shark *Carcharhinus limbatus*

Black tip reef shark *Carcharhinus melanopterus*

Tiger shark *Galeocerdo cuvier*

Ganges shark *Glyphis gangeticus*

Milk shark *Rhizoprionodon acutus*

Indian dog shark or Spadenose shark *Scoliodon laticaudus* Whale shark *Rhincodon typus*

Grey bamboo shark *Chiloscyllium griseum*

Slender bambooshark *Chiloscyllium indicum*

Family: *Sphyrnidae* (Hammerhead, bonnethead or scoophead sharks)

Winghead shark, *Eusphyra blochii*

Scalloped hammerhead *Sphyrna lewini*

Great hammerhead *Sphyrna mokarran*

2. **Order: Pristiformes (Sawfishes)**

Family: *Pristidae*

Knifetooth sawfish *Anoxypristis cuspidata*

Largetooth sawfish *Pristis microdon*

Smalltooth sawfish *Pristis pectinata*

3. **Order: Rajiformes (Skates and Rays)**

Family: *Dasyatidae* (Stingrays)

Small-eye stingray *Dasyatis microps*

Pale-edged stingray *Dasyatis zugei*

Bleeker's whipray *Himantura bleekeri*

Ganges stingray *Himantura fluviatilis*

Scaly whipray *Himantura imbricata*

Pointed-nose stingray *Himantura jenkinsii*

Blackedge whipray *Himantura marginatus*

Honeycomb stingray *Himantura uarnak*

Family: *Myliobatidae* (Eagle and manta rays)

Spotted eagle ray *Aetobatus narinari*

Banded eagle ray *Aetomylaeus nichofii*

Family: *Rhinobatidae* (Guitarfishes)

Annandale's guitarfish *Rhinobatos annandalei*

Smoothback guitarfish *Rhinobatos lionotus*

Giant guitarfish *Rhynchobatus djiddensis*

4. **Order: Torpediniformes (Electric Rays)**

Family: *Narcinidae* (Numbfishes)

Brown numbfish *Narcine brunnea*

5. **Order: Osteoglossiformes (Bony Tongues)**

Family: *Notopteridae* (Featherbacks or knifefishes)

Clown knifefish *Chitala chitala*

Bronze featherback *Notopterus notopterus*

6. **Order: Elopiformes (Tarpons and Tenpounders)**

Family: *Elopidae* (Tenpounders and ladyfishes)

Tenpounder *Elops machnata*

Family: *Megalopidae* (Tarpons)

Indo-Pacific tarpon *Megalops cyprinoides*

7. **ORDER: ALBULIFORMES (BONE FISHES)**

Family: *Albulidae* (Bonefishes)

Roundjaw bonefish *Albula glossodonta*

8. **ORDER: ANGUILLIFORMES (EELS AND MORAYS)**

Family: *Anguillidae* (Freshwater eels)

Indian mottled eel *Anguilla bengalensis*

Indonesian shortfin eel *Anguilla bicolor*

Mottled eel *Anguilla nebulosa*

Family: *Moringuidae* (Worm and spaghetti eels)

Moringua macrocephalus

Purple spaghetti eel *Moringua raitaborua*

Family: *Muraenidae* (Moray eels)

Freshwater moray *Gymnothorax tile*

Family: *Ophichthidae* (Snake eels and worm eels)

Finny snake eel *Caecula pterygera*

Rice paddy eel *Pisodonophis boro*

Greenspot snake eel *Pisodonophis hijala*

Maimed snake eel *Muraenichthys schultzei*

Family: *Muraenesocidae* (Pike conger eels)

Yellow pike conger *Congresox talabon*

Indian pike conger *Congresox talabonoides*

Common pike conger *Muraenesox bagio*

Daggertooth pike conger *Muraenesox cinereus*

9. **Order: Clupeiformes (Herrings)**

Family: *Clupeidae* (Herrings, shads, sardines and menhadens)

Chacunda gizzard shad *Anodontostoma chacunda*

Thai gizzard shad *Anodontostoma thailandiae*

Ganges river sprat *Corica soborna*

White sardine *Escualosa thoracata*

Ganges river gizzard shad *Gonialosa manmina*

Indian river shad *Gudusia chapra*

Kelee shad *Hilsa kelee*

Bloch's gizzard shad *Nematalosa nasus*

Tardoore *Opisthopterus tardoore*

Raconda *Raconda russeliana*

Indian oil sardine *Sardinella longiceps*

Hilsa shad, River shad *Tenualosa ilisha*

Toli shad *Tenualosa toli*

Family: *Pristigasteridae*

Elongate ilisha *Ilisha elongata*

Coromandel ilisha *Ilisha filigera*

Kampen's ilisha *Ilisha kampeni*

Bigeye ilisha *Ilisha megaloptera*

Indian ilisha *Ilisha melastoma*

Indian pellona *Pellona ditchela*

Family: *Engraulidae* (Anchovies)

Goldspotted grenadier anchovy *Coilia dussumieri*

Ramcarat grenadier anchovy *Coilia ramcarati*

Reynald's grenadier anchovy *Coilia reynaldi*

Shorthead hairfin anchovy *Setipinna breviceps*

Dusky hairfin anchovy *Setipinna melanochir*

Gangetic hairfin anchovy *Setipinna phasa*

Scaly hairfin anchovy *Setipinna taty*

Indian anchovy *Stolephorus indicus*

Dussumier's thryssa *Thryssa dussumieri*

Hamilton's thryssa *Thryssa hamiltonii*

Malabar thryssa *Thryssa malabarica*

Moustached thryssa *Thryssa mystax*

Oblique jaw thryssa *Thryssa purava*
Bengal thryssa *Thryssa spinidens*

Slender thryssa *Thryssa stenosoma*

Orangemouth anchovy *Thryssa vitrirostris*

Family: *Chirocentridae* (Wolf herrings)

Dorab wolf herring *Chirocentrus dorab*

Whitefin wolf herring *Chirocentrus nudus*

10. **Order: Gonorhynchiformes (Milk Fish)**

Family: *Chanidae*

Milkfish White mullet *Chanos chanos*

11. **Order: Cypriniformes (Carps)**

Family: *Cyprinidae* (Minnows, carps, barbs)

Mola carplet *Amblypharyngodon mola*

Mrigal *Cirrhinus cirrhosus*

Sind danio *Devario devario*

Silver razorbelly minnow *Salmostoma acinaces (Chela argentea)*

Large razorbelly minnow *Salmostoma bacaila*

Finescale razorbelly minnow *Salmostoma phulo*

Bengala barb *Bengala elanga*

Zebra danio *Danio rerio*

Flying rasbora (barb) *Esomus danricus*

Slender rasbora *Rasbora daniconius*

Silver hatchet danio *Chela cachius*

Indian glass barb *Chela laubuca*

Catla *Catla catla*

Reba *Labeo ariza*

Bata *Labeo bata*

Kalbosu (Orange-fin labeo) *Labeo calbasu*

Rohu *Labeo rohita*

Swamp barb *Puntius chola*

Rosy barb *Puntius conchonius*

Golden barb *Puntius gelius*

Olive barb *Puntius sarana*

Pool barb *Puntius sophore*

Onespot barb *Puntius terio*

Ticto barb *Puntius ticto*

12. **Order: Siluriformes (Cat Fish)**

Family: *Bagridae* (Bagrid cat fishes)

Menoda cat fish *Hemibagrus menoda*

Day's mystus *Mystus bleekeri*

Gangetic mystus *Mystus cavasius*

Long whiskers cat fish *Mystus gulio*

Striped dwarf cat fish *Mystus vittatus*

Rita *Rita rita*

Long whiskered cat fish *Sperata aor*

Giant river cat fish *Sperata seenghala*

Family: *Siluridae* (Sheat fishes)

Butter cat fish *Ompok bimaculatus*

Pabdah cat fish *Ompok pabda*

Wallago *Wallago attu*

Family: *Schilbeidae* (Schilbeid cat fishes)

Gangetic ailia *Ailia coila*

Garua cat fish *Clupisoma garua*

Sharpnose cat fish *Eutropiichthys vacha*

Indian potasi *Pseudeutropius atherinoides*

Silond cat fish *Silonia silondia*

Family: *Pangasiidae* (Shark cat fishes)

Yellowtail cat fish, River pangus *Pangasius pangasius*

Family: *Sisoridae* (Sisorid cat fishes)

Dwarf goonch *Bagarius bagarius*

Goonch *Bagarius yarrelli*

Indian gagata *Gagata cenia*

Gagata gagata

Glyptothorax botius

Glyptothorax telchitta

Gogangra viridescens

Kosi nangra *Nangra nangra*

Sisor cat fish *Sisor rabdophorus*

Family: *Clariidae* (Airbreathing cat fishes)

Walking cat fish *Clarias batrachus*

Family: *Chacidae* (Squarehead cat fishes)

Squarehead cat fish *Chaca chaca*

Family: *Ariidae* (Sea cat fishes)

T hreadfin sea cat fish *Arius arius*

Engraved cat fish *Arius caelatus*

Blacktip sea cat fish *Arius dussumieri*

Gagora cat fish *Arius gagora*

Blackfin sea cat fish *Arius jella*

Spotted cat fish *Arius maculatus*

Smooth headed cat fish *Arius nenga*

Flat mouth cat fish *Arius platystomus*

Sagor cat fish *Arius sagor*

Sona sea cat fish *Arius sona*

Shovelnose sea cat fish *Arius subrostratus*

Beardless sea cat fish *Batrachocephalus mino*

River cat fish *Hemipimelodus jatius*

Soldier cat fish *Osteogeneiosus militaris*

Family: *Heteropneustidae* (Airsac cat fishes)

Stinging cat fish *Heteropneustes fossilis*

Family: *Plotosidae* (Eeltail cat fishes)

Gray eel cat fish *Plotosus canius*

Striped eel cat fish *Plotosus lineatus*

13. **Order: Aulopiformes (Grinners)**

Family: *Synodontidae* (Lizard fishes, Bombay duck)

Greater lizard fish *Saurida tumbil*

Brushtooth lizard fish *Saurida undosquamis*

Bombay duck *Harpadon nehereus*

14. **Order: Batrachoidiformes (Toad Fishes)**

Family: *Batrachoididae*

Grunting toad fish *Allenbatrachus grunniens*

15. **Order: Gadiformes (Cods)**

Family: *Bregmacerotidae* (Codlets)

Spotted codlet *Bregmaceros mcclellandi*

16. **Order: Cyprinodontiformes (Rivulines, Killi Fishes and Live Bearers)**

Family: *Aplocheilidae* (Killi fishes)

Blue panchax *Aplocheilus panchax*

17. **Order: Beloniformes (Needle Fishes)**

Family: *Belonidae.*

Banded needle fish *Strongylura leiura*

Spottail needle fish *Strongylura strongylura*

Freshwater gar fish *Xenentodon cancila*

Family: *Hemiramphidae*

Gangetic half beak *Dermogenys brachynotopterus*

Wrestling half beak *Dermogenys pusilla*

Jumping half beak *Hemiramphus archipelagicus*

Congaturi half beak *Hyporhamphus limbatus*

Long billed half beak *Rhynchorhamphus georgii*

Ectuntio half beak *Zenarchopterus ectuntio*

Family: *Adrianichthyidae* (Ricefishes)

Rice fish *Oryzias carnaticus*

18. **Order: Syngnathiformes (Pipefishes and Seahorses)**

Family: *Syngnathidae*

Freshwater pipefish *Ichthyocampus carce*

Crocodile tooth pipefish *Microphis cuncalus*

Deocata pipefish *Microphis deocata*

19. **Order: Synbranchiformes (Swamp and Spiny Eels)**

Family: *Synbranchidae* (Swamp eels)

Cuchia, Gangetic mud eel *Monopterus cuchia*

Bengal eel *Ophisternon bengalense*

Family: *Mastacembelidae* (Spiny eels)

Lesser spiny eel *Macrognathus aculeatus*

One-stripe spinyeel *Macrognathus aral*

Barred spiny eel *Macrognathus pancalus*

Zig-zag eel, Tire-track spiny eel *Mastacembelus armatus*

20. **Order: Scorpaeniformes (Scorpion Fishes & Flat Heads)**

Family: *Platycephalidae* (Flatheads)

Bartail flathead *Platycephalus indicus*

21. **Order: Perciformes (Perch-Like)**

Family: *Latidae* (Perches)

Barramundi, Giant seaperch *Lates calcarifer*

Waigieu seaperch *Psammoperca waigiensis*

Family: *Ambassidae* (Glass fishes)

Bald glassy Ambassis *gymnocephalus*

Elongate glass-perchlet *Chanda nama*

Highfin glassy perchlet *Parambassis lala*

Indian glassy fish *Parambassis ranga*

Family: *Serranidae* (Sea basses: groupers and fairy basslets)

Orange spotted grouper *Epinephelus coioides*

Giant grouper *Epinephelus lanceolatus*

Family: *Terapontidae* (Grunters and tigerperches)

Fourlined terapon *Pelates quadrilineatus*

Jarbua terapon *Terapon jarbua*

Small-scaled terapon *Terapon puta*

Largescaled therapon *Terapon theraps*

Family: *Sillaginidae* {Sillagos (Smelt-whitings)} Flathead sillago *Sillaginopsis panijus*

Clubfoot sillago *Sillago chondropus*

Silver sillago, *Sillago sihama*

Soringa sillago *Sillago soringa*

Family: *Carangidae* (Jacks and pompanos)

Indian threadfish *Alectis indicus*

Razorbelly scad *Alepes kleinii*

Longnose trevally *Carangoides chrysophrys*

Malabar trevally *Carangoides malabaricus*

Bigeye trevally *Caranx sexfasciatus*

Japanese scad *Decapterus maruadsi*

Golden trevally *Gnathanodon speciosus*

Torpedo scad *Megalaspis cordyla*

Black pomfret, Brown pomfret *Parastromateus niger*

Barred queen fish *Scomberoides tala*

Bigeye scad *Selar crumenophthalmus*

Yellowstripe scad *Selaroides leptolepis*

Family: *Menidae* (Moonfishes, bat fishes)

Moonfish *Mene maculata*

Family: *Leiognathidae* (Pony fishes)

Goldstripe pony fish *Leiognathus daura*

Common pony fish *Leiognathus equulus*

Striped pony fish *Leiognathus fasciatus*

Splendid pony fish *Leiognathus splendens*

Pugnose pony fish *Secutor insidiator*

Deep pugnose pony fish *Secutor ruconius*

Family: *Lutjanidae* (Snappers)

Mangrove red snapper *Lutjanus argentimaculatus* Humpback red snapper *Lutjanus gibbus*

John's snapper *Lutjanus johnii*

Malabar blood snapper *Lutjanus malabaricus*

Russell's snapper *Lutjanus russellii*

Family: *Datnioididae*

Fourstripe perch *Datnioides polota*

Family: *Lobotidae* (Tripletails)

Atlantic tripletail *Lobotes surinamensis*

Family: *Gerreidae* (Mojarras)

Whipfin silver-biddy *Gerres filamentosus*

Saddleback silver-biddy *Gerres limbatus*

Slender silver-biddy *Gerres oblongus*

Common silver-biddy *Gerres oyena*

Strong spine silver-biddy *Gerres phaiya*

Small Bengal silver-biddy *Gerres setifer*

Family: *Haemulidae* (Grunts)

Bluecheek silver grunt *Pomadasys argyreus*

Silver bream *Pomadasys hasta*

Family: *Sparidae* (Porgies and sea breams)

Yellowfin seabream *Acanthopagrus latus*

King soldierbream *Argyrops spinifer*

Goldlined seabream *Rhabdosargus sarba*

Family: *Nemipteridae* (Threadfin breams and spinycheeks)

Japanese threadfin bream *Nemipterus japonicus*

Family: *Sciaenidae* (Croakers and drums)

Chaptis bahaba *Bahaba chaptis*

Reeve's croaker *Chrysochir aureus*

Bengal corvina *Daysciaena albida*

Goatee croaker *Dendrophysa russelii*

Belanger's croaker *Johnius belangerii*

Karut croaker *Johnius carutta*

Coitor croaker *Johnius coitor*

Cuja croaker *Macrospinosa cuja*

Soldier croaker *Nibea soldado*

Bronze croaker *Otolithoides biauritus*

Pama croaker *Otolithoides pama*

Hooghly croaker *Panna heterolepis*

Pennahia ovata

Blackspotted croaker *Protonibea diacanthus* Blotched tiger-tooth croaker *Pterotolithus maculatus*

Family: *Polynemidae* (Threadfins)

Fourfinger threadfin *Eleutheronema tetradactylum*

Indian threadfin *Leptomelanosoma indicum*

Striped threadfin *Polydactylus plebeius*

Sixfinger threadfin *Polydactylus sexfilis*

Paradise threadfin *Polynemus paradiseus*

Family: *Mullidae* (Goat fishes)

Sulphur goatfish *Upeneus sulphureus*

Family: *Toxotidae* (Archerfishes)

Largescale archerfish *Toxotes chatareus*

Family: *Drepaneidae* (Sicklefishes)

Concertina fish *Drepane longimana*

Spotted sicklefish *Drepane punctata*

Family: *Monodactylidae* (Moonyfishes or fingerfishes)

Silver moony *Monodactylus argenteus*

Family: *Nandidae* (Asian leaf fishes)

Gangetic leaf fish *Nandus nandus*

Family: *Badidae*

Badis *Badis badis*

Family: *Kurtidae* (Nurseryfishes)

Indian hump head *Kurtus indicus*

Family: *Mugilidae* (Mullets)

Largescale mullet *Liza macrolepis*

Gold-spot mullet *Liza parsia*

Greenback mullet *Liza subviridis*

Tade mullet *Liza tade*

Flathead mullet *Mugil cephalus*

Corsula *Rhinomugil corsula*

Cascasia mullet, Yellowtail mullet *Sicamugil cascasia*

Bluetail mullet *Valamugil buchanani*

Bluespot mullet *Valamugil seheli*

Speigler's mullet *Valamugil speigleri*

Family: *Cichlidae* (Cichlids)

Green chromide *Etroplus suratensis*

Family: *Uranoscopidae* (Stargazers)

Uranoscopus guttatus

Family: *Callionymidae* (Dragonets)

River dragonet *Callionymus fluviatilis*

Arrow dragonet *Callionymus sagitta*

Family: *Eleotridae* (Sleepers)

Duckbill sleeper *Butis butis*

Gangetic sleeper *Odonteleotris macrodon*

Dusky sleeper *Eleotris fusca*

Lutea sleeper *Eleotris lutea*

Family: *Gobiidae* (Gobies)

Tropical sand goby *Acentrogobius caninus*

Acentrogobius cyanomos

Spotted green goby *Acentrogobius viridipunctatus*

Dragon goby *Apocryptes bato*

Scribbled goby *Awaous grammepomus*

Largesnout goby *Awaous melanocephalus*

Bathygobius ostreicola

Bumblebee goby *Brachygobius nunus*

Boddart's goggle-eyed goby *Boleophthalmus boddarti* Mudskipper *Boleophthalmus dussumieri*

Tank goby *Glossogobius giuris*

Glass goby *Gobiopterus chuno*

Rubicundus eelgoby *Odontamblyopus rubicundus*

Maned goby *Oxyurichthys microlepis*

Taileyed goby *Parachaeturichthys polynema*

Giant mudskipper *Periophthalmodon schlosseri*

Periophthalmodon septemradiatus

Atlantic mudskipper *Periophthalmus barbarus*

Pearse's mudskipper *Periophthalmus novemradiatus*

Elongate goby *Pseudapocryptes elongatus*

Many-finned eelgoby *Pseudotrypauchen multiradiatus*

Walking goby *Scartelaos histophorus*

Knight goby *Stigmatogobius sadanundio*

Eel worm goby *Taenioides anguillaris*

Burmese gobyeel *Taenioides buchanani*

Bearded worm goby *Taenioides cirratus*

Burrowing goby *Trypauchen vagina*

Family: *Scatophagidae* (Scats)

Spotted scat *Scatophagus argus*

Family: *Siganidae* (Rabbit fishes)

Streaked spinefoot *Siganus javus*

Family: *Sphyraenidae* (Barracudas)

Great barracuda *Sphyraena barracuda*

Family: *Trichiuridae* (Cutlass fishes and scabbard fishes)

Longtooth hairtail *Eupleurogrammus glossodon*

Smallhead hairtail *Eupleuroarammus muticus*

Coromandel hairtail *Lepturacanthus pantului*

Savalani hairtail *Lepturacanthus savala*

Gangetic hairtail *Trichiurus gangeticus*

Largehead hairtail *Trichiurus lepturus*

Family: *Scombridae* (Mackerels, tunas and bonitos)

Kawakawa *Euthynnus affinis*

Indian mackerel *Rastrelliger kanagurta*

Narrow-barred Spanish mackerel *Scomberomorus commerson* Indo-Pacific king mackerel *Scomberomorus guttatus*

Family: *Stromateidae* (Butterfishes)

Silver pomfret *Pampus argenteus*

Chinese silver pomfret *Pampus chinensis*

Pampus cinereus

Family: *Anabantidae* (Climbing gouramies)

Climbing perch *Anabas testudineus*

Family: *Osphronemidae* (Gouramies)

Banded gourami *Colisa fasciata*

Spiketail paradise fish *Pseudosphromenus cupanus* Dwarf gourami *Colisa lalia*

Frail gourami *Ctenops nobilis*

Honey gourami *Trichogaster chuna*

Family: *Channidae* (Snakeheads)

Barca snakehead

Channa barca

Channa gachua

Great snakehead *Channa marulius.*

Walking snakehead *Channa orientalis*

Spotted snakehead *Channa punctata*

Snakehead murrel *Channa striata*

22 **Order: Pleuronectiformes (Flatfishes)**

Family: *Psettodidae* (Psettodids)

Indian spiny turbot *Psettodes erumei*

Family: *Paralichthyidae* (Largetooth flounders)

Largetooth flounder *Pseudorhombus arsius*

Deep flounder *Pseudorhombus elevatus*

Malayan flounder *Pseudorhombus malayanus*

Three spotted flounder *Pseudorhombus triocellatus*

Family: *Citharidae* (Chitarids)

Yellow-dabbled flounder *Brachypleura novaezeelandiae*

Family: *Soleidae* (Soles)

Oriental sole *Brachirus orientalis*

Pan sole *Brachirus pan*

Eyed sole *Heteromycteris oculus*

Kaup's sole *Synaptura albomaculata*

Highfin sole *Zebrias altipinnis*

Family: *Cynoglossidae* (Tongue fishes)

Largescale tongue-sole *Cynoglossus arel*

Bengal tongue-sole *Cynoglossus cynoglossus*

Long tongue-sole *Cynoglossus lingua*

Malabar tongue-sole *Cynoglossus macrostomus*

Speckled tongue-sole *Cynoglossus puncticeps*

Bengal tongue-sole *Cynoglossus semifasciatus*

Doublelined tongue-sole *Paraplagusia bilineata*

23. **Order: Tetradontiformes (Puffers and Filefishes)**

Family: *Triacanthidae* (Triplespines) Short-nosed tripodfish *Triacanthus biaculeatus*

Family: *Tetraodontidae* (Puffers)

Immaculate puffer *Arothron immaculatus*

Milkspotted puffer *Chelonodon patoca*

Green rough-backed puffer *Lagocephalus lunaris*

Lattice blaasop *Takifugu oblongus*

Ocellated pufferfish *Tetraodon cutcutia*

Green pufferfish *Tetraodon fluviatilis*

Annexure 2B: A Checklist of Members of Class Amphibia in Indian Sundarbans (*Source* Mitra 2020)

1. **Order: Anura**

Family: *Bufonidae*

Common Indian toad *Bufo melanostictus*

Family: *Microhylidae*

Ornate narrow-mouthed frog *Microhyla ornate*

Family: *Rhacophoridae*

Common tree frog *Polypedates maculatus*

Family: *Ranidae*

Skittering frog *Euphlyctis cyanophlyctis*

Indian pond frog *Euphlyctis hexadactylus*

Indian bull frog *Hoplobatrachus tigerinus*

Cricket frog *Limnonectes limnocharis*

Annexure 2C: A Checklist of Members of Class Reptilia in Indian Sundarbans (*Source* Mitra 2020)

1. **Order: Crocodilia**

Family: *Crocodylidae*

Mugger crocodile *Crocodylus palustris*

Saltwater crocodile *Crocodylus porosus*

2. **Order: Testudines or Chelonia or Testudinata**

Family: *Dermochelyidae*

Leatherback sea turtle *Dermochelys coriacea*

Family: *Cheloniidae*

Loggerhead sea turtle *Caretta caretta*

Green sea turtle *Chelonia mydas*

Hawksbill sea turtle *Eretmochelys imbricata*

Olive ridley sea turtle *Lepidochelys olivacea*

Family: *Bataguridae*

Indian roofed turtle *Kachuga tecta*

Red crowned roofed turtle *Kachuga kachuga*

River terrapin *Batagur baska*

Spotted pond turtle *Geoclemys hamiltonii*

Family: *Trionychidae*

Narrow-headed soft turtle *Chitra indica*

Asian giant softshell turtle *Pelochelys cantorii*

Dark softshell turtle *Aspideretes nigricans* (B)

Indian flapshell turtle *Lissemys punctata*

3. **Order: Squamata; Suborder: Lacertila**

Family: *Gekkonidae*

Tokay gecko *Gekko gecko*

Brook's house gecko *Hemidactylus brookii*

Yellow-green house gecko *Hemidactylus flaviviridis*

Family: *Agamidae*

Indian garden lizard *Calotes versicolor*

Family: *Chamaeleonidae*

Indian chameleon *Chamaeleozeylanicus*

Family: *Scincidae*

Keeled grass skink *Mabuyacarinata*

Family: *Varanidae*

Bengal monitor *Varanus bengalensis*

Water monitor *Varanus salvator*

Yellow monitor *Varanus flavescens*

4. **Order: Squamata; Suborder: Serpents**

Family: *Acrochordidae*

Wart snake or file snake *Acrochordus granulatus*

Family: *Boidae*

Common sand boa *Gongylophis conicus*

Indian rock python *Python molurus*

Family: *Colubridae*

Common vine snake *Ahaetulla nasuta*

Striped keelback *Amphiesma stolatum*

Dog-faced watersnake *Cerberus rynchops*

Ornate flying snake *Chrysopelea ornata*

Common bronzeback tree snake *Dendrelaphis tristis*

Common smooth water snake *Enhydris enhydris*

White-bellied mangrove snake *Fordonia leucobalia*

Glossy marsh snake *Gerarda prevostiana*

Common wolf snake *Lycodon aulicus*

Banded kukri snake *Oligodon arnensis*

Indian rat snake *Ptyas mucosa*

Checkered keelback *Xenochrophis piscator*

Family: *Elapidae*

Common Indian krait *Bungarus caeruleus*

Banded krait *Bungarus fasciatus*

Monocled cobra *Naja kaouthia*

King cobra *Ophiophagus hannah*

Family: *Hydrophiidae*

Hook-nosed sea snake *Enhydrina schistosa*

Blue sea snake *Hydrophis caerulescens*

Annulated sea snake *Hydrophis cyanocinctus*

Black-banded sea snake *Hydrophis nigrocinctus*

Estuarine sea snake *Hydrophis obscurus*

Ornate sea snake *Hydrophis ornatus*

Malabar sea snake *Lapemis curtus*

Banded laticauda *Laticauda laticaudata*

Family: *Viperidae*

Russell's viper *Daboia russelii*

Spot-tailed pit viper *Trimeresurus erythrurus*

Annexure 2D: A Checklist of Members of Class Aves in Indian Sundarbans (*Source* Mitra 2020)

Vernacular name	Scientific name
Resident	
Little cormorant	*Phalacrocoraxniger*
Great cormorant	*P. carbo*
Grey heron	*Ardea cinerea*
Purple heron	*A. purpurea*
Indian pond-heron	*Ardeola grayi*
Cattle egret	*Bulbucus ibis*
Open bill stork	*Anastomus oscitans*
Great egret	*Egretta alba*
Intermediate egret	*E. intermedia*
Little egret	*E. garzetta*

(continued)

(continued)

Vernacular name	Scientific name
Black-crowned night heron	*Nycticorax nycticorax*
Greater adjutant	*Leptoptilos dubius*
Black-necked stork	*Xenorhynchus asiaticus*
Black-headed ibis	*Threskiornis melanocephala*
White-breasted water hen	*Amaurornis phoenicurus*
Bronze-winged jacana	*Metopidius indicus*
Pheasant -tailed jacana	*Hydrophasianus chirurgus*
Black-capped kingfisher	*Halcyon pileata*
Collared kingfisher	*H. chloris*
Brown-winged kingfisher	*Pelargopsis amauroptera*
Migratory species	
Spot-billed pelican	*Pelecanus philippensis*
Northern pintail	*Anas acuta*
Common teal	*A. crecca*
Tufted duck	*Aythya fuligula*
Common pochard	*A. ferina*
Ruddy shelduck	*Tadorna ferruginea*
Whimbrel	*Numenius phaeopus*
Black-tailed godwit	*Limosa limosa*
Little stint	*Calidris minuta*
Great knot	*C. tenuirostris*
In shallow water	
Eastern golden plover	*Pluvialis dominica*
Kentish plover	*Charadrius alexandrinus*
Eurasian curlew	*Numenius arquata*
Wood sandpiper	*Tringa glareola*
Terek sandpiper	*Xenus cinereus*
Reclaimed area	
Indian darter	*Anhinga melanogaster*
Chestnut bittern	*Ixobrychus cinnanoneus*
Black bittern	*Butoridea stellaris*
Spoonbill	*Plataleateu corodia*
Red-wattled lapwing	*Vanellus indicus*
Grey-headed lapwing	*V. cinereus*
Herring gull	*Larus argentatus* (rare visitor)
Great crested tern	*Sterna bergii* (rare visitor)
Lesser crested tern	*S. bengalensis* (rare visitor)

(continued)

(continued)

Vernacular name	Scientific name
Sooty tern	*S. fuscata* (rare visitor)
Red jungle fowl	*Gallus gallus (found in forest areas only)*
Swamp partridge	*Francolinus gularia*
Greater coucal	*Centropus sinensis*
Occasional visitors to the estuary	
Osprey	*Pandion haliaetus*
Black kite	*Milvus migrans*
Brahminy kite	*Haliastur indus*
White-bellied sea-eagle	*Haliaeetus leucogaster*
Crested serpent eagle	*Spilornis cheela*
Spotted dove	*Streptopelia chinensis*
Yellow-footed pigeon	*Treron phoenicoptera*
Large Indian parakeet	*Psittacula eupatria*
Rose-ringed parakeet	*P. krameri*
Barn owl	*Tyto alba*
Brown fish-owl	*Bubo zeylonensis*
Spotted owlet	*Athene brama*
Magpie robin	*Copsychus saularis*
Asian paradise flycatcher	*Terpsiphone paradise*
Black drongo	*Dicrurus adsimilis*
Treepie	*Crypsirina vagabonda*
Grey shrike	*Lanius excubitor*
Common swallow	*Hirundo rustica*
Indian roller	*Coracias bengalensis*

Annexure 2E: A Checklist of Members of Class Mammalia in Indian Sundarbans (*Source* Mitra 2020)

1. **Order: Primates**

Family: *Cercopithecidae*

Rhesus macaque *Macaca mulatta*

2. **Order: Artiodactyla**

Family: *Cervidae*

Indian muntjac *Muntiacus muntjak* (B)

Spotted deer *Axis axis*

Family: *Suidae*

Wild boar *Sus scrofa*

3. **Order: Carnivora**

Family: *Canidae*

Jackal *Canis aureus*

Indian fox *Vulpes bengalensis*

Family: *Felidae*

Tiger *Panthera tigris*

Jungle cat *Felis chaus*

Leopard cat *Prionailurus bengalensis*

Fishing cat *Prionailurus viverrinus*

Family: *Mustelidae*

Smooth-coated otter *Lutrogale perspicillata*

Small-clawed otter *Amblonyx cinereus*

Family: *Viverridae*

Small Indian civet *Viverricula indica*
Large Indian civet *Viverra zibetha*

Common palm civet *Paradoxurus hermaphroditus*

Family: *Herpestidae*

Grey mongoose *Herpestes edwardsii*

Small Indian mongoose *Herpestes javanicus*

4. **Order: Insectivora**

Family: *Soricidae.*

House shrew *Suncus murinus*

5. **Order: Rodentia**

Family: *Hystricidae*

Himalayan crestless porcupine *Hystrix brachyura* (B)

Family: *Sciuridae*

Five-striped palm squirrel *Funambulus pennantii*

Family: *Muridae*

Large bandicoot-rat *Bandicota indica*

Lesser Bandicoot-rat *Bandicota bengalensis*

House rat *Rattus rattus*

Long-tailed tree mouse *Vandeleuria oleracea*

Little Indian field mouse *Mus booduga*

6. **Order: Chiroptera**

Family: *Pteropodidae*

Indian flying fox *Pteropus giganteus*

Short-nosed fruit bat *Cynopterus sphinx*

Family: *Rhinopomatidae*

Lesser mouse-tailed bat *Rhinopoma hardwickii*

Family: *Emballonuridae*

Long-winged tomb bat *Taphozous longimanus*

Family: *Hipposideridae*

Tail-less leaf-nosed bat *Coelops frithii*

Family: *Megadermatidae*

Greater false vampire *Megaderma lyra*

Family: *Vespertilionidae*

Asiatic greater yellow house bat *Scotophilus heathii*

Lesser Asiatic yellow house bat *Scotophilus kuhlii*

Indian pipistrelle *Pipistrellus coromandra*

Indian pygmy bat *Pipistrellus tenuis*

7. **Order: Cetacea**

Family: *Platanistidae*

Ganges river dolphin *Platanista gangetica*

Family: *Phocoenidae*

Finless porpoise *Neophocaena phocaenoides*

Family: *Delphinidae*

Irrawaddy dolphin *Orcaella brevirostris*

Indo-Pacific bottlenose dolphin *Tursiops aduncus*

Common bottlenose dolphin *Tursiops truncates*

Short-beaked common dolphin *Delphinus delphis*

Annexure 2F: Feedback Questionnaire on the Fisheries and Aquaculture Sector of Blue Economy

1. **Name**

2. **Profession**

   Tick all that apply.

   - [ ] Scientist
   - [ ] Teacher / Faculty
   - [ ] Student / Research Scholar
   - [ ] Fisherman
   - [ ] Farmer
   - [ ] Aquaculturist
   - [ ] Others

3. **If others, please mention**

4. **Have you got any idea of Blue Economy in your area?**

   Mark only one oval.

   - ◯ Yes
   - ◯ No
   - ◯ Maybe

5. **Is coastal aquaculture harmful to mangrove ecosystem?**

   Tick all that apply.

   - [ ] Yes
   - [ ] No
   - [ ] Maybe

6. **In which sector of Indian Sundarbans you stay?**

   Tick all that apply.

   - [ ] Western sector
   - [ ] Central sector
   - [ ] Eastern sector

7. Do you think that fishing in the estuaries and seas strengthen the local level economy?

Mark only one oval.

	1	2	3	4	5	
Least role	◯	◯	◯	◯	◯	Highest role

8. Do you believe that institutional support plays a role in the fishery sector?

Mark only one oval.

	1	2	3	4	5	
Least role	◯	◯	◯	◯	◯	Highest role

9. How do you see the role of aquaculture in strengthening the local economy?

Mark only one oval.

	1	2	3	4	5	
Least role	◯	◯	◯	◯	◯	Highest role

10. Do you believe that Government intervention/role is needed to combat the threats on fishery sector?

Mark only one oval.

	1	2	3	4	5	
Least role	◯	◯	◯	◯	◯	HIghest role

11. Have you got any idea on innovation in the aquaculture sector in context to the present geographical locale? If so, can you point out few options?

Tick all that apply.

☐ Yes
☐ No
☐ Maybe

12. What are the major threats to fishery and aquaculture sector in your area?

Annexure 2G: Feedback Questionnaire on the Shipping and Transportation

1. **Name**

2. **Profession**
Tick all that apply.
- [ ] Teacher / Faculty
- [ ] Student / Research Scholar
- [ ] Hotel Owner
- [ ] Farmer
- [ ] Fisherman
- [ ] Staff
- [ ] Others

3. **If others, please mention**

4. **Do you think that estuaries in Sundarbans have sufficient depth for shipping purpose?**

Mark only one oval.
- ◯ Yes
- ◯ No
- ◯ Maybe

5. **For what purposes the navigation channels in Sundarbans are mostly used?**
Tick all that apply.
- [ ] Local transportation
- [ ] Tourism
- [ ] Transport of goods
- [ ] Defense

6. **What are the hindrances to smooth navigation in the Sundarban estuaries?**
Tick all that apply.
- [ ] Siltation
- [ ] Depth
- [ ] Submerged shoals
- [ ] Fishing nets
- [ ] Natural disasters

7. **How do you relate shipping with the ambient water quality?**
Mark only one oval.

	1	2	3	4	5	
Poor	◯	◯	◯	◯	◯	Excellent

8. **Do you think that ballast water brings microbes and other pollutants in the ports around Indian Sundarbans (like Haldia)?**

Mark only one oval.

◯ Yes
◯ No
◯ Maybe

9. **How the formation of sand bar in the midst of the estuary causes problem to shipping?**
Mark only one oval.

	1	2	3	4	5	
Mil d	◯	◯	◯	◯	◯	Severe

10. **How the navigational problem in Sundarban estuaries can be solved?**
Mark only one oval.

	1	2	3	4	5	
Poor	◯	◯	◯	◯	◯	Excellent

11. **Why antifouling paints are used for conditioning fishing vessels and trawlers?**

References

Agarwal S, Chakrabarty SP, Zaman S, Pramanick P, Biswas P, Mukhopadhyay N, Dutta J, Mitra A (2021) Assessment of carbon storage potential of dominant tree species in an educational campus, Kolkata. West Bengal. Ela J Forest Wildlife 10(2):869–878

Banerjee K, Sengupta K, Raha AK, Mitra A (2013) Salinity based allometric equations for biomass estimation of Sundarban mangroves. Biomass Bioenergy (ELSEVIER) 56:382–391

Biswas S, Pal N, Zaman S, Mitra A (2019) Samosa and Kachuri from Mangrove associate species: an innovative utilization of coastal flora. Int J Res Appl Sci Eng Technol 7(1):73–79

Chaudhuri AB, Choudhuri A (1994) Mangroves of the Sundarbans, vol 1. India. IUCN, Bangkok, Thailand.

Kim JK, Duston J, Corey P, Garbary DJ (2013) Marine finfish effluent bioremediation: effects of stocking density and temperature on nitrogen removal capacity of Chondrus crispus and Palmaria palmata (Rhodophyta). Aquac 414–415:210–216

Mandal AK, Nandi NC (1989) Fauna of sundarban mangrove ecosystem, West Bengal. Fauna Conserv Areas, Zool Surv India, India

Mitra A (2000) The North West Coast of the Bay of Bengal and Deltaic Sundarbans. Seas at the millennium: an environmental evaluation, U.K. II:145–160

Mitra A (2013) Sensitivity of Mangrove ecosystem to changing climate. Published by Springer New Delhi Heidelberg New York Dordrecht London, ISBN-10: 8132215087; ISBN-13: 978-8132215080. ISBN 978-81-322-1509-7 (eBook)

Mitra A (2020) Mangrove forests in India exploring ecosystem services. Springer, Chem, eBook ISBN 978-3-030-20595-9, XV, pp 361. https://doi.org/10.1007/978-3-030-20595-9

Mitra A, Biswas S (2021) Mangroves of Indian Sundarbans as sources of alternative livelihoods. New Millenium Graphics Publication, ISBN: 978-81-949433-9-6, pp 157

Mitra A, Gati RC (2015) Carbon census in the mangrove ecosystem of lower Gangetic delta. Economol J V:11–27

Mitra A, Sengupta K, Banerjee K (2011) Standing biomass and carbon storage of above-ground structures in dominant mangrove trees in the Sundarbans. For Ecol Manage. https://doi.org/10.1016/j.foreco.2011.01.012 261(7):1325–1335

Mitra A, Zaman A (2020) Environmental science—a ground zero observation on the Indian subcontinent. Springer, Chem, eBook ISBN 978-3-030-49131-4, XIV, pp 478. https://doi.org/10.1007/978-3-030-49131-4

Mitra A, Zaman S (2014) Carbon sequestration by coastal floral community, published by The Energy and Resources Institute (TERI) TERI Press, India. ISBN 978-81-7993-551-4

Mitra A, Zaman S (2015) Blue carbon reservoir of the blue planet published by Springer, ISBN 978-81-322-2106-7, XII, pp 299. https://doi.org/10.1007/978-81-322-2107-4

Mitra A, Zaman S (2016) Basics of marine and estuarine ecology; Publisher Springer India, ISBN 978-81-322-2707-6, pp 1–481

Mitra A, Zaman S (2021) Estuarine acidification: exploring the situation of mangrove dominated Indian Sundarban estuaries++. Springer, Chem, e-Book ISBN 978-3-030-84792-0, XII, pp 402. https://doi.org/10.1007/978-3-030-84792-0

Pramanick P, Zaman S, Mitra A (2014) Processing of fruits with special reference to S. apetala fruit jelly preparation. Int J Univ Pharm Bio Sci 3(5):36–49

Pramanick P, Zaman S, Rudra T, Guha A, Mitra A (2016) Comparative account of heavy metal level in two species of Chlorophyceae inhabiting Indian Sundarbans. Int J Res Chem Environ 6(2):65–68

Raha AK, Bhattacharyya SB, Zaman S, Banerjee K, Sengupta K, Sinha S, Sett S, Chakraborty S, Datta S, Dasgupta S, Roy Chowdhury M, Ghosh R, Mondal K, Pramanick P, Mitra A (2013) Carbon census in dominant mangroves of Indian Sundarbans. J Energy Environ Sci 127:345–354

Sengupta K, Roy Chowdhury M, Bhattacharyya SB, Raha A, Zaman S, Mitra A (2013) Spatial variation of stored carbon in *Avicennia alba* of Indian Sundarbans. Discov Nat (ISSN: 2319–5703) 3(8):19–24

Song SH, Lee C, Lee S (2013) Analysis of microflora profile in Korean traditional nuruk. J Microbiol Biotechnol 23:40–46

Chapter 3
Blue Economy and Livelihoods in Indian Sundarbans

3.1 Fisheries and Aquaculture Based Livelihoods

According to the World Resources Institute, about 39% of the World's population (which is around 7.9 billion in 2022), is approximately 3.0 billion people. This considerable chunk of global population lives on or within 100 km (60 miles) of the seashore and therefore may be considered as coastal population. Their lives and livelihoods are totally dependent on the resources available from the marine and estuarine ecosystems. However, very few regional policies have been adopted to safe guard and sustain the livelihoods of this considerable percentage of population from the cruel clutches of sea level rise and natural disasters (like tsunami, cyclone etc.).

Livelihood is a broad term encompassing ecological, social, economic assets of a Nation being driven by human resource (capital and capabilities). In other words, it may be stated that the livelihood comprises of human efficiencies, assets (including natural, material and social resources) and activities required for means of living.

A livelihood may be regarded as sustainable when it has the ability to cope with and recover from stresses and shocks and maintain or enhance its capabilities and assets both for present and future generations, without posing an adverse impact on the natural resource base. Livelihood assets may be of different categories like human, social, physical, financial, and natural resources. These assets have the potential in maintaining a satisfactory (which again is a relative term) living standard. Lack or insufficient quantity of any one or more of these assets would give rise to a situation where, if left uncompensated, would pose a substantial constraint to attain livelihood goals.

A sustainable livelihood can be achieved if there is maximum utilization of available resources with minimum wastage. This can be expressed as:

$$\mathbf{X} = \mathbf{R_u}/\mathbf{R_a}$$

where, X = livelihood standard, R_a = resources available to the society/community/individual and R_u = resources utilized.

A. Mitra et al., *Blue Economy in Indian Sundarbans*,
https://doi.org/10.1007/978-3-031-07908-5_3

In most cases the available resources remain unutilized due to lack of perfect and local-specific technology, management, policy etc. and sometimes social and political interventions and hence the value of X hardly approaches towards unity. It is important to note in this context that sustainable livelihood is always eco-friendly in nature and over-exploitation of resources to get short term benefit does not fit in this domain. Hence focus must be given on the term available resources (R_a) and not the total resources. The total resources need to be conserved through proper policy, appropriate technology and harnessing renewable energy sources.

The Indian part of Sundarban mangrove forest is comprised of some 105 islands; of which 54 islands are inhabited by humans sustaining about 4.4 million populations. The Indian region is demarcated by the River Hooghly on the west, the Bay of Bengal on the south, the Ichamati-Kalindi-Raimongal Rivers on the east and the Dampier-Hodges line on the north (Fig. 3.1).

The Indian Sundarban is comprised of 19 Community Development blocks of which 6 are in North and 13 are in South 24-Parganas districts that comprise of 190 g Panchayats and 1,064 villages in total. According to census report 2001 and 2011, the region spreads across 4,118.51 sq. km. It covers 64.46 and 31.81% area of South 24 Parganas and North 24 Parganas districts respectively. Total population of this region (as per 2011 census) is 44,26,259. According to the census report, the population of Indian Sundarbans increased by 354% in the last 50 years, which clearly speaks on imparting major thrust on alternative livelihoods for the local

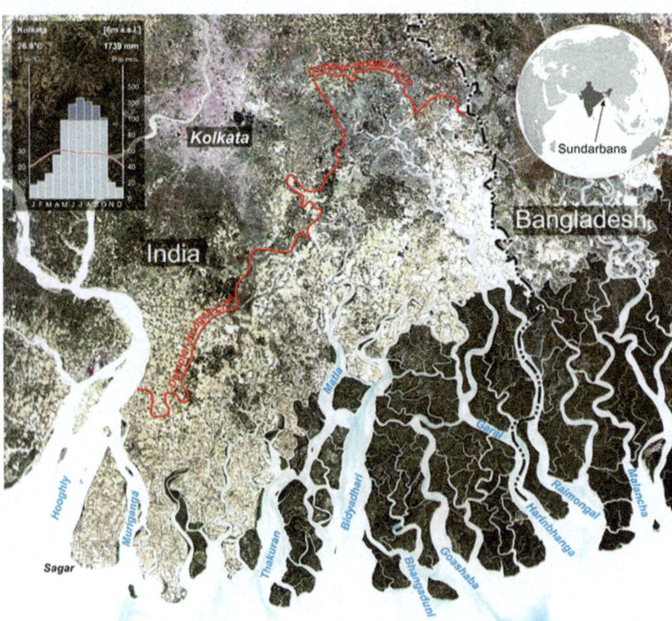

Fig. 3.1 Map of Indian Sundarbans showing the northern boundary (Dampier-Hodges line) with red line

Table 3.1 General information about Indian Sundarbans

Subject	District		Total
	South 24 Parganas	North 24 Parganas	
CD blocks	13	6	19
Area	3,054.15	1,064.36	4,118.51
% of area covered	64.46	31.81	50.94
Total population	33,03,526	11,16,733	44,26,259
Male	16,92,424	5,71,709	22,64,133
Female	16,17,102	5,45,024	21,62,126
Rural	30,83,950	10,88,298	41,72,248
Urban	2,25,576	28,435	2,54,011
Sex ratio	955	953	954

population. The population growth rate, population density and sex ratio are 15.11%, 1,074.72 persons/km^2 and 954 female/1,000 male respectively. Total male and female populations in the region are 22,62,126 and 21,62,126 respectively. Total rural and urban populations are 41,72,248 and 2,54,011 respectively (Table 3.1).

It has been witnessed that poverty stricken people (preferably the poor island dwellers) are more susceptible to adverse impact of climate change. They have no proper shelter, no fund for resettlement and no insurances against their lives, health and properties. Institutional help hardly reaches them. Hence reduction of poverty through development of sustainable livelihoods is an important component in fighting against temperature rise, sea water intrusion, and disease outbreaks, which are the clutches of climate change that frequently, scratch the economics of the region.

The Indian Sundarbans Delta (ISD) is a hot spot of West Bengal coastal fishery. The rivers are the live matrix of deltaic complex, on which the unique spectrum of biological diversity is embedded. In Indian Sundarbans, approximately 2,069 sq. km. area is occupied by tidal river system or estuaries which finally end up in the Bay of Bengal. The aquatic phase is highly congenial for breeding, growth and survival of a large variety of fishes and hence aquaculture has flourished in this region as the primary backbone of livelihoods.

In this estuarine region, there are large number of poor coastal fishermen who primarily depend on the Hooghly-Matla estuarine complex and the adjacent sea for their livelihoods. A wide variety of commercially important fishes are available in and around the region preferably in the seas, bays, estuaries, creeks and inlets (Table 3.2), but several risk/cost factors are involved in the sector of capture fisheries like (i) man-animal conflict, (ii) natural calamities like cyclones, (iii) wave actions due to high wind in the deep sea, (iv) excessive rise of fossil fuel cost in recent times etc.

The economic benefit is much greater in aquaculture practice (particularly in the verticals of shrimp culture and crab fattening) than traditional fishing and hence there is a general tendency of fisherman population to shift from capture fishery

Table 3.2 List of fish species commonly caught by the fisherman of Indian Sundarbans

Sl. No.	Order	Scientific name	Common name
1	Beloniformes	*Oryzias dancena*	Rice fish
2	Perciformes	*Ambassis kopsii*	Freckled hawkfish
3	Perciformes	*Ambassis nalua*	Scalloped perchlet
4	Perciformes	*Chanda nama*	Elongate glass-perchlet
5	Perciformes	*Parambassis baculis*	Himalayan glassy perchlet
6	Perciformes	*Parambassis ranga*	Indian glassy fish
7	Perciformes	*Anabas cobojius*	Gangetic Koi
8	Perciformes	*Anabas testdineus*	Climbing perch
9	Anguilliformes	*Anguilla bengalensis*	Indian mottled eel
10	Anguilliformes	*Anguilla bicolor*	Indonesian shortfin eel
11	Anguilliformes	*Anguilla nebulosa*	Mottled eel
12	Cyprinodontiformes	*Aplocheilus panchax*	Blue panchax
13	Siluriformes	*Arius arius*	Threadfin sea catfish
14	Siluriformes	*Arius gagora*	Gagora catfish
15	Siluriformes	*Arius jella*	Blackfin sea catfish
16	Siluriformes	*Batrachocephalus mino*	Beardless sea catfish
17	Siluriformes	*Cephalocassis jatia*	River catfish
18	Siluriformes	*Hemiarius sona*	Sona sea catfish
19	Siluriformes	*Hexanematichthys sagor*	Sagor catfish
20	Siluriformes	*Nemapteryx nenga*	Engraved catfish
21	Siluriformes	*Netuma thalassina*	Giant sea catfish
22	Siluriformes	*Osteogeneiosus militaris*	Soldier catfish
23	Siluriformes	*Picofollis platystomus*	Flatmouth sea catfish
24	Perciformes	*Badis badis*	Badis
25	Siluriformes	*Mystus cavasius*	Gangetic mystus
26	Siluriformes	*Mystus gulio*	Long wiskers catfish
27	Siluriformes	*Mystus tengara*	Tengara mystus
28	Siluriformes	*Mystus vittatus*	Striped dwarf catfish
29	Siluriformes	*Sperata aor*	Long whiskered catfish
30	Siluriformes	*Sperata seenghala*	Giant river catfish
31	Batrachoidiformes	*Allenbatrachus grunniens*	Grunting toadfish
32	Beloniformes	*Strongylura leiura*	Banded needlefish
33	Beloniformes	*Strongylura strongylura*	Spottail needlefish

(continued)

Table 3.2 (continued)

Sl. No.	Order	Scientific name	Common name
34	Beloniformes	*Tylosuruscrocodylus crocodylus*	Hound needlefish
35	Beloniformes	*Xenentodon cancila*	Freshwater garfish
36	Gadiformes	*Bregmaceros mcclellandi*	Spotted codlet
37	Perciformes	*Callionymus fluviatilis*	River dragonet
38	Perciformes	*Callionymus fluviatilis*	River dragonet
39	Perciformes	*Callionymus megastomas*	Indian megamouth dragonet
40	Perciformes	*Callionymus sagitta*	Arrow dragonet
41	Perciformes	*Alectis indicus*	Indian threadfish
42	Perciformes	*Atule mate*	Yellowtail scad
43	Perciformes	*Carangoides chrysophrys*	Longnose trevally
44	Perciformes	*Carangoides malabaricus*	Malabar trevally
45	Perciformes	*Caranx ignobilis*	Giant trevally
46	Perciformes	*Caranx sexfasciatus*	Bigeye trevally
47	Perciformes	*Megalaspis cordyla*	Torpedo scad
48	Perciformes	*Parastromateus niger*	Black pomfret
49	Perciformes	*Scomberoides tala*	Barred queenfish
50	Perciformes	*Scomberoidis lysan*	Doublespotted queenfish
51	Perciformes	*Trachinotus blochii*	Snubnose pompano
52	Carcharhiniformes	*Carcharhinus hemiodon*	Pondicherry shark
53	Carcharhiniformes	*Carcharhinus leucus*	Bull shark
54	Carcharhiniformes	*Carcharhinus limbatus*	Blacktip shark
55	Carcharhiniformes	*Carcharhinus melanopterus*	Blacktip reef shark
56	Carcharhiniformes	*Carcharhinus sorrah*	Spot tail shark
57	Carcharhiniformes	*Galeocerdo cuvier*	Tiger shark
58	Carcharhiniformes	*Glyphis gangeticus*	Ganges shark
59	Carcharhiniformes	*Lamiopsis temminckii*	Broad fin Shark
60	Carcharhiniformes	*Rhizoprionodon acutus*	Milk shark
61	Carcharhiniformes	*Scoliodon laticaudus*	Spadenose shark
62	Gonorhynchiformes	*Chanos chanos*	Milkfish
63	Perciformes	*Channa gachua*	Dwarf snakehead
64	Perciformes	*Channa marulius*	Great snakehead
65	Perciformes	*Channa punctata*	Spotted snakehead

(continued)

Table 3.2 (continued)

Sl. No.	Order	Scientific name	Common name
66	Perciformes	*Channa striata*	Stripted snakehead
67	Clupeiformes	*Chirocentrus dorab*	Dorab wolfherring
68	Siluriformes	*Clarias magur*	Magur
69	Carcharhiniformes	*Eusphyra blochii*	Winghead shark
70	Clupeiformes	*Andontostoma thailandiae*	Thai gizzard shad
71	Clupeiformes	*Anodontostoma chacunda*	Chacunda gizzard shad
72	Clupeiformes	*Corica soborna*	Ganges river spart
73	Clupeiformes	*Escualosa thoracata*	White sardine
74	Clupeiformes	*Gonoalosa manmina*	Ganges river gizzard shad
75	Clupeiformes	*Gudusia chapra*	Indian river shad
76	Clupeiformes	*Hilsa kelee*	Kelee shad
77	Clupeiformes	*Nematalosa galatheae*	Galathea gizzard shad
78	Clupeiformes	*Nematalosa nasus*	Bloch's gizzard shad
79	Clupeiformes	*Tenualosa toli*	Toli Shad
80	Clupeiformes	*Tenualosa ilisha*	Hilsa shad, river shad
81	Cypriniformes	*Lepidocephalichthys guntea*	Guntea loach
82	Perciformes	*Coryphaena hippurus*	Common dolphin fish
83	Pleuronectiformes	*Cynoglossus arel*	Largescale tongue-sole
84	Pleuronectiformes	*Cynoglossus cynoglossus*	Bengal tongue-sole
85	Pleuronectiformes	*Cynoglossus lingua*	Long tongue-sole
86	Pleuronectiformes	*Cynoglossus macrostomus*	Malabar tonguesole
87	Pleuronectiformes	*Cynoglossus puncticeps*	Speckled tongue-sole
88	Pleuronectiformes	*Paraplagusia bilineata*	Doublelined tongue-sole
89	Cypriniformes	*Amblypharyngodon mola*	Mola carplet
90	Cypriniformes	*Chela laubuca/Laubuka laubuka*	Indian glass barb
91	Cypriniformes	*Danio rerio*	Zebra danio
92	Cypriniformes	*Esomus danrica*	Flying barb
93	Cypriniformes	*Pethia phutunio*	Spotted sail barb
94	Cypriniformes	*Puntius chola*	Swamp barb
95	Cypriniformes	*Puntius conchonius*	Rosy barb

(continued)

Table 3.2 (continued)

Sl. No.	Order	Scientific name	Common name
96	Cypriniformes	*Puntius sarana/Systomus sarana*	Olive barb
97	Cypriniformes	*Puntius sophore*	Pool barb
98	Cypriniformes	*Puntius terio*	Onespot barb
99	Cypriniformes	*Puntius ticto*	Ticto barb
100	Cypriniformes	*Rasbora daniconius*	Slender rasbora
101	Cypriniformes	*Salmostoma bacaila*	Large razorbelly minnow
102	Cypriniformes	*Pethia gelius*	Golden barb
103	Myliobatiformes	*Dasyatis zugei*	Pale-edged stingray
104	Myliobatiformes	*Himamtura imbricata*	Scaly whipray
105	Myliobatiformes	*Himantura bleekeri*	Bleeker's whipray
106	Myliobatiformes	*Himantura fava*	honeycomb whipray
107	Myliobatiformes	*Himantura gerrardi*	Sharpnose stingray
108	Myliobatiformes	*Himantura marginatus*	Blackedge whipray
109	Myliobatiformes	*Himantura walga*	Dwarf whipray
110	Myliobatiformes	*Himantura uarnak*	Honeycomb stingrey
111	Myliobatiformes	*Pastinachus sephen*	Cowtail stingray
112	Perciformes	*Datnioides polota*	Four-banded tigerfish
113	Perciformes	*Drepane longimana*	Concertina fish
114	Perciformes	*Drepane punctata*	Spotted sicklefish
115	Perciformes	*Butis butis*	Duckbill sleeper
116	Perciformes	*Butis humeralis*	Blackspotted gudgeon
117	Perciformes	*Eleotris fusca*	Dusky sleeper
118	Perciformes	*Eleotris melanosoma*	Broadhead sleeper
119	Perciformes	*Giuris margaritacea*	Snalehead gudgeon
120	Perciformes	*Odonteleotris macrodon*	Gangetic sleeper
121	Perciformes	*Ophiocara ophicephalus*	Northern mud gudgeon
122	Elopiformes	*Elops machnta*	Ladyfish
123	Clupeiformes	*Coilia dussumieri*	Goldspotted grenadier anchovy
124	Clupeiformes	*Coilia neglecta*	Neglected grenadier anchovy
125	Clupeiformes	*Coilia ramcarati*	Ramcarat grenadier anchovy
126	Clupeiformes	*Coilia reynaldi*	Reynald's grenadier anchovy
127	Clupeiformes	*Setipinna tenuifilis*	Common hairfin anchovy
128	Clupeiformes	*Setipina Taty*	Scaly hairfin anchovy
129	Clupeiformes	*Setipinna phasa*	Gangetic hairfin anchovy

(continued)

Table 3.2 (continued)

Sl. No.	Order	Scientific name	Common name
130	Clupeiformes	*Stolephorus baganensis*	Bagan anchovy
131	Clupeiformes	*Stolephorus commersonnii*	Commerson's anchovy
132	Clupeiformes	*Stolephorus indicus*	Indian anchovy
133	Clupeiformes	*Thryssa dussumieri*	Dussumier's thryssa
134	Clupeiformes	*Thryssa gautamiensis*	Gautama thryssa
135	Clupeiformes	*Thryssa hamiltonii*	Hamilton's thryssa
136	Clupeiformes	*Thryssa malabarica*	Malabar thryssa
137	Clupeiformes	*Thryssa purava*	Oblique jaw thryssa
138	Clupeiformes	*Thryssa stenosoma*	Slender thryssa
139	Clupeiformes	*Setipinna brevifils*	Short hairfin anchovy
140	Perciformes	*Gerres filamentosus*	Whipfin silver-biddy
141	Perciformes	*Gerres oyena*	Common silver-biddy
142	Perciformes	*Gerres phaiya*	Strong spine silver-biddy
143	Perciformes	*Gerres setifer*	Small Bengal silver-biddy
144	Perciformes	*Acentrogo biuscaninus*	Tropical sand goby
145	Perciformes	*Acentrogo biusviridpunctatus*	Spotted green goby
146	Perciformes	*Amblyeleotris gymnocephala*	Masked shrimp goby
147	Perciformes	*Amblyotrypauchen arctocephalus*	Armoureel goby
148	Perciformes	*Apocryptes bato*	Flat-toothed goby
149	Perciformes	*Apocryptobon madurensis*	Madura goby
150	Perciformes	*Awaouichthys mennoni*	Menon's goby
151	Perciformes	*Bathygobius fuscus*	Frill goby
152	Perciformes	*Boleophthalmus boddarti*	Boddart's goggle-eyed goby
153	Perciformes	*BoleophthaSllmus dussumieri*	Dussumier's mudskipper
154	Perciformes	*Brachygobius nunus*	Bumblebee goby
155	Perciformes	*Caragobius urolepis*	Sumatra eel-goby
156	Perciformes	*Drombus globiceps*	Bighead goby
157	Perciformes	*Glossogobius giuris*	Tank goby
158	Perciformes	*Gnatholepis cauerensis*	Eyebar goby
159	Perciformes	*Gobiopsis macrostoma*	Longjaw goby
160	Perciformes	*Gobiopterus chuno*	Glass goby

(continued)

Table 3.2 (continued)

Sl. No.	Order	Scientific name	Common name
161	Perciformes	*Hemigobius hoevenii*	Banded mulletgoby
162	Perciformes	*Istigobius ornatus*	Ornate goby
163	Perciformes	*Odontamblyopus rubicundus*	Rubicanduseelgoby
164	Perciformes	*Oligolepis acutipennis*	Sharptail goby
165	Perciformes	*Oxuderces dentatus*	Chinese sharp-toothed goby
166	Perciformes	*Parapocryptes serperaster*	Largescaled flat-toothed goby
167	Perciformes	*Paratrypauchen microcephalus*	Comb goby
168	Perciformes	*Periophthalmodon schlosseri*	Giant mudskipper
169	Perciformes	*Periophthalmodon septemradiatus*	Hamilton's mudskipper
170	Perciformes	*Periophthalmus Argentilineatus*	Barred mudskipper
171	Perciformes	*Periophthalmus barbarus*	Atlantic mudskipper
172	Perciformes	*Periophthalmus kalolo*	Common mudskipper
173	Perciformes	*Periophthalmus novemradiatus*	Pearse's mudskipper
174	Perciformes	*Periophthalmus variabilis*	Variable mudskipper
175	Perciformes	*Pseudapocryptes elongatus*	Lanceolate false flat-teethed goby
176	Perciformes	*Pseudotrypauchen Multiradiatus*	
177	Perciformes	*Scartelaos histophorus*	Walking goby
178	Perciformes	*Taenioides anguillaris*	Eel worm goby
179	Perciformes	*Taenioides buchanani*	Burmese goby eel
180	Perciformes	*Taenioides cirratus*	Bearded worm goby
181	Perciformes	*Trypauchen vagina*	Burrowing goby
182	Perciformes	*Trypauchenichthys sumatrensis*	Indonesian eel goby
183	Myliobatiformes	*Gymnura japonica*	Japanese butterfly ray
184	Myliobatiformes	*Gymnura poecilura*	Longtail butterfly ray
185	Perciformes	*Plectorhinchus gibbosus*	Harry hotlip
186	Perciformes	*Pomadasys argenteus*	Silver grunt

(continued)

Table 3.2 (continued)

Sl. No.	Order	Scientific name	Common name
187	Perciformes	*Pomadasys kaakan*	Javelin grunt
188	Perciformes	*Pomadasys maculatus*	Saddle grunt
189	Beloniformes	*Hemiramphus far*	Black barred halfbeak
190	Beloniformes	*Hyporhamphus limbatus*	Congaturi halfbeak
191	Beloniformes	*Rhynchorhamphus georgii*	Long billed halfbeak
192	Orectolobiformes	*Chiloscyllium griseum*	Grey bambooshark
193	Siluriformes	*Heteropneustes fossilis*	Stinging catfish
194	Siluriformes	*Pachypterus atherinoides*	Indian potasi
195	Perciformes	*Kurtus indicus*	Indian hump head
196	Perciformes	*Lates calcarifer*	Barramundi, giant seaperch
197	Perciformes	*Aurigequula fasciata*	Sriped ponyfish
198	Perciformes	*Eubleekeria splendens*	Splendid ponyfish
199	Perciformes	*Gazza minuta*	Toothed pony
200	Perciformes	*Karalla dussumieri*	Dussumier's ponyfish
201	Perciformes	*Leiognathus equula*	Common ponyfish
202	Perciformes	*Nuchequula blochii*	Twoblotch ponyfish
203	Perciformes	*Nuchequula gerreoides*	Decorated ponyfish
204	Perciformes	*Photopectoralis bindus*	Orangefin ponyfish
205	Perciformes	*Sceutor ruconius*	Pugnose ponyfish
206	Perciformes	*Secutor insidiator*	Deep pugnose ponyfish
207	Perciformes	*Lujanus indicus*	Indian snapper
208	Perciformes	*Lutijanus argentimaculatus*	Mangrove red snapper
209	Perciformes	*Lutjanus fulvus*	Blacktail snapper
210	Perciformes	*Lutjanus johnii*	John's snapper
211	Synbranchiformes	*Macrognathus aral*	One-stripe spiny eel
212	Synbranchiformes	*Macrognathus pancalus*	Barred spiny eel
213	Synbranchiformes	*Mastacembelus armatus*	Zig-zag eel
214	Elopiformes	*Megalops cyprinoides*	Indo-Pacific tarpon
215	Anguilliformes	*Moringua guthriana*	Bengal spaghetti-eel
216	Anguilliformes	*Moringuar aitaborua*	Purple spaghetti eel
217	Mugiliformes	*Ellochelon vaigiensis*	Squaretail mullet
218	Mugiliformes	*Liza macrolepis*	Largescale mullet
219	Mugiliformes	*Liza melanoptera*	Otomebora mullet

(continued)

Table 3.2 (continued)

Sl. No.	Order	Scientific name	Common name
220	Mugiliformes	*Liza parsia*	Goldspot mullet
221	Mugiliformes	*Liza subviridis*	Greenback mullet
222	Mugiliformes	*Liza tade*	Tade gray mullet
223	Mugiliformes	*Mugil cephalus*	Flathead gray mullet
224	Mugiliformes	*Rhinomugil corsula*	Corsula
225	Mugiliformes	*Valamugil buchanani*	Bluetail mullet
226	Mugiliformes	*Valamugil cunnesius*	Longarm mullet
227	Mugiliformes	*Valamugil seheli*	Bluespot mullet
228	Mugiliformes	*Valamugil speigleri*	Speigler's mullet
229	Anguilliformes	*Congresox talabon*	Yellow pike conger
230	Anguilliformes	*Congresox talabonoides*	Indian pike conger
231	Anguilliformes	*Muraenesox bagio*	Common pike conger
232	Anguilliformes	*Muraenesox cinereus*	Daggertooth pike conger
233	Anguilliformes	*Gymnothorax tile*	Indian mud moray
234	Anguilliformes	*Strophidon sathete*	Slender giant moray
235	Myliobatiformes	*Aetobatus ocellatus*	Ocellated eagle ray
236	Perciformes	*Nnadus nandus*	Gangetic leaf fish
237	Torpediniformes	*Narcine brunnea*	Brown numbfish
238	Osteoglossiformes	*Notopterus notopterus*	Bronze featherback
239	Anguilliformes	*Caecula pterygera*	Finny snake eels
240	Anguilliformes	*Lamnostoma orientalis*	Oriental worm-eel
241	Anguilliformes	*Ophichthus apicalis*	Bluntnose snake-eel
242	Anguilliformes	*Pisodonophis boro*	Rice paddy eels
243	Anguilliformes	*Uroconger lepturus*	Slender conger
244	Perciformes	*Colisa fasciata*	Banded gourami
245	Perciformes	*Trichogaster chuna*	Honey gourami
246	Perciformes	*Trichogaster lalius*	Dwarf gourami
247	Siluriformes	*Pangasius pangasius*	Pangas catfish
248	Pleuronectiformes	*Pseudorhombus elevatus*	Deep flounder
249	Pleuronectiformes	*Pseudorhombus javanicus*	Javan flounder
250	Pleuronectiformes	*Pseudorhomus arsius*	Largetooth flounder
251	Scorpaeniformes	*Grammoplites scaber*	Rough flathead
252	Scorpaeniformes	*Platycepphalus indicus*	Bartail flathead
253	Siluriformes	*Plotosus canius*	Grey eel catfish
254	Siluriformes	*Plotosus lineatus*	Striped eel catfish

(continued)

Table 3.2 (continued)

Sl. No.	Order	Scientific name	Common name
255	Perciformes	*Eleutheronema tetradactylum*	Fourfinger threadfin
256	Perciformes	*Leptomelanosoma indicum*	Indian threadfin
257	Perciformes	*Polynemus paradiseus*	Paradise threadfin
258	Perciformes	*Polydactylus plebeius*	Striped threadfin
259	Perciformes	*Polydactylus sextarius*	Blackspot threadfin
260	Pristiformes	*Anoxypristis cuspidata*	Knifetooth sawfish
261	Pristiformes	*Pristis clavata*	Dwarf sawfish
262	Pristiformes	*Pristis pristis*	Large tooth sawfish
263	Clupeiformes	*Ilisha alongata*	Elongate ilisha
264	Clupeiformes	*Ilisha kampeni*	Kampen's ilisha
265	Clupeiformes	*Ilisha megaloptera*	Bigeye ilisha
266	Clupeiformes	*Ilisha melastoma*	Indian ilisha
267	Clupeiformes	*Ilisha sirishai*	Lobejawilisha
268	Clupeiformes	*Opisthopterus tardoore*	Tardoore
269	Clupeiformes	*Pellona ditchela*	Indian pellona
270	Clupeiformes	*Raconda russeliana*	Raconda
271	Rajiformes	*Glaucostegus granulatus*	Sharpnose guitarfish
272	Rajiformes	*Glaucostegus obtusus*	Widenose guitarfish
273	Rajiformes	*Glaucostegus typus*	Giant shovelnose ray
274	Rajiformes	*Rhinobatos annandalei*	Annandale's guiterfish
275	Rajiformes	*Rhina ancylostoma*	Bowmouth guitarfish
276	Rajiformes	*Rhynchobatus laevis*	Smooth nose wedgefish
277	Rajiformes	*Rhinobatos lionotus*	Smootback guiterfish
278	Perciformes	*Scatophagus argus*	Spotted scat
279	Siluriformes	*Ailia coila*	Gangetic ailia
280	Siluriformes	*Silonia silondia*	Silond catfish
281	Perciformes	*Bahaba chaptis*	Chaptisbahaba
282	Perciformes	*Chrysochir aureus*	Reeve's croaker
283	Perciformes	*Dasyciaena albida*	Bengal corvina
284	Perciformes	*Dendrophysa russelii*	Goatee croaker
285	Perciformes	*Johnieops borneensis*	Sharpnose hammer croaker
286	Perciformes	*Johnius gangeticus*	Ganges croaker
287	Perciformes	*Macrospinosa cuja*	Largespine croaker
288	Perciformes	*Nibea coibor*	Coibor croaker

(continued)

Table 3.2 (continued)

Sl. No.	Order	Scientific name	Common name
289	Perciformes	*Nibea soldado*	Soldier croaker
290	Perciformes	*Otolithoides biauritus*	Bronze croaker
291	Perciformes	*Otolithoides pama*	Pama croaker
292	Perciformes	*Panna microdon*	Panna croaker
293	Perciformes	*Protonibea diacanthus*	Blackspotted croaker
294	Perciformes	*Pterotolithus maculatus*	Blotched tiger-tooth croaker
295	Perciformes	*Epinephelus coioides*	Orange-spotted grouper
296	Perciformes	*Epinephelus lanceolatus*	Giant grouper
297	Perciformes	*Epinephelus malabaricus*	Malabar grouper
298	Perciformes	*Siganus canaliculatus*	Whitespotted Spinefoot
299	Perciformes	*Siganus javas*	Streaked spinefoot
300	Perciformes	*Sillagenopsis domina*	Flathead sillago
301	Perciformes	*Sillago sihama*	Silver sillago
302	Siluriformes	*Ompok bimaculatus*	Butter catfish
303	Siluriformes	*Ompok pabda*	Pabdah catfish
304	Siluriformes	*Wallago attu*	Wallago
305	Siluriformes	*Bagarius bagarius*	Goonch
306	Pleuronectiformes	*Brachirus pan*	Pan sole
307	Pleuronectiformes	*Synaptura albomaculata*	Kaup's sole
308	Pleuronectiformes	*Synaptura commersonnii*	Commerson's sole
309	Perciformes	*Acanthopagrus berda*	Goldsilk seabream
310	Perciformes	*Acanthopagrus longispinnis*	Longfin seabream
311	Perciformes	*Rhabdosargus sarba*	Goldlined seabream
312	Perciformes	*Sparidentex datnia*	Yellowfin seabream
313	Perciformes	*Sphyraena obtusata*	Obtuse barracuda
314	Orectolobiformes	*Stegostoma fasciatum*	Zebra shark
315	Perciformes	*Pampus argenteus*	Silver pomfret
316	Perciformes	*Pampus chinensis*	Chinese silver pomfret
317	Perciformes	*Scomberomorus guttatus*	Indo-Pacific king mackerel
318	Synbranchiformes	*Monopterus cuchia*	Cuchia
319	Synbranchiformes	*Ophisternon bengalense*	Bengals wampeel
320	Synbranchiformes	*Hippichthys spicifer*	Bellybarred pipefish

(continued)

Table 3.2 (continued)

Sl. No.	Order	Scientific name	Common name
321	Synbranchiformes	*Ichthyocampus carce*	Freshwater pipefish
322	Synbranchiformes	*Microphis cuncalus*	Crocodiletooth pipefish
323	Aulopiformes	*Harpadon nehereus*	Bombay duck
324	Perciformes	*Terapon jarbua*	Jarbuaterapon
325	Perciformes	*Terapon puta*	Small-scaledterapon
326	Perciformes	*Terapon theraps*	Large-scaledterapon
327	Tetradontiformes	*Chelonodon patoca*	Milkspotted puffer
328	Tetradontiformes	*Lagocephalus lunaris*	Greenrough-backed puffer
329	Tetradontiformes	*Takifuga oblongus*	Lattice blaasop
330	Tetradontiformes	*Tetraodon cutcutia*	Ocellated pufferfish
331	Tetradontiformes	*Tetraodon fluviatilis*	Green pufferfish
332	Perciformes	*Toxotes chatareus*	Largescale archerfish
333	Perciformes	*Toxotes jaculatrix*	Banded archerfish
334	Tetradontiformes	*Pseudotriacanthus strigilifer*	Long-spined tripodfish
335	Tetradontiformes	*Triacanthus biaculeatus*	Short-nosed tripodfish
336	Perciformes	*Eupleurogram musmuticus*	Smallhead hairtail
337	Perciformes	*Lepturacanthus pantului*	Coromandel hairtail
338	Perciformes	*Lepturacanthus savala*	Savalani hairtail
339	Perciformes	*Trichiurus gangeticus*	Gangetic hairtail
340	Perciformes	*Trrichiurus lepturus*	Largehead hairtail
341	Beloniformes	*Zenarchopterus buffonis*	Buffon's river-garfish
342	Beloniformes	*Zenarchopterus dispar*	Feathered river-garfish
343	Beloniformes	*Zenarchopterus ectuntio*	Ectuntio halfbeak
344	Beloniformes	*Zenarchopterus striga*	Hooghly halfbeak

to aquaculture (Bhattacharyya et al. 2010). Fishermen along with other dwellers of Indian Sundarbans have adopted brackishwater farming with remarkable success, which has become an important tool for livelihood development in such a vulnerable eco-region. Classical example of traditional brackishwater aquaculture is in practice in West Bengal in the vicinity of Indian Sundarbans for centuries and these fisheries are locally called as 'Bheri' (Fig. 3.2). In this method, low lying areas near the banks of saline water rivers and creeks are encircled by peripheral dyke and tidal water is allowed to enter in the impoundment along with natural seeds of various species of shrimps, crabs and fishes. Water is retained with periodical exchanges during lunar cycles and the species are allowed to grow. After 3–5 months, harvesting is done partially during lunar cycles. Productivity in this system ranged between 500 and

Fig. 3.2 A traditionally managed brackish water farm (Bheri) in Indian Sundarban

900 kg ha^{-1} of which about 30% is occupied by prawns/shrimps and 70% by mullets (Saha et al. 2001).

The age old traditional farming methods of fin fish and shrimp with some modifications still exist because of their simplicity and low cost. These are being gradually replaced by modern improved farming practices. Commercial shrimp farming started gaining roots only during the mid-eighties. The boom period of commercial-scale shrimp farming in West Bengal started in 1990 and the bust came in 1995–1996, with the outbreak of viral disease (Mehta 2009). Later with the advent of bio-secured closed culture technology using better management practices, semi-intensive and intensive shrimp farming started to regain its lost glory during early years of this century (Fig. 3.3). In this method apart from providing scientific feed, aerators are used (Fig. 3.4) to control the optimum level of dissolved oxygen (DO), which is around 5.5 ppm.

The semi-intensive method of shrimp culture is profitable with production at the range of 4–6 tonnes ha^{-1} (Sahu et al. 2012), but there are so many drawbacks in this practice particularly in Indian Sundarbans region e.g., due to absence of shrimp hatcheries in the state, people of Sundarbans (preferably the women) screen the estuarine water for seeds of tiger prawn. This causes a lot of hazard in terms of health and socio-economic stability of the prawn seed catchers.

We carried out an assessment on the threats and problems faced by the women population (preferably tiger prawn seed collectors) of Indian Sundarbans during 2019 by scaling the threats posed on women population engaged in aquaculture sector (through Composite Threat Scale—CTS),

Fig. 3.3 Semi-intensive method of shrimp culture backed up with scientific management

Fig. 3.4 Scientifically managed semi-intensive shrimp farm with aerators

The various steps for this approach are listed here.

1. Assess the threats inflicted on women population engaged in aquaculture sector through stakeholders meet and interactions with local people.
2. Scaling the threats with score and develop CTS for individual threats.

The threats on women prawn seed collectors of Indian Sundarbans basically crop up from various issues as listed here.

(1) Employment problem (low involvement of the local women in male dominated occupations/professions like deep sea fishing, driving fishing vessels and trawlers, timber and wood collection from the deep forest, honey, bee wax collection etc.)
(2) Natural hazards (tidal waves, cyclones, erosion, siltation, flooding etc.)
(3) Health related problems (like dermatological problems, gastric problem, vision problem etc.)
(4) Problems in education and skill development (preferably for young girls)
(5) Victims of animal attack i.e., tiger, crocodile, shark etc.

The methodology encompasses three stages (i) identification of respondents (policy maker, researcher, fisherman, agriculturist and local inhabitant), (ii) identification of major threats (8 in number) and (iii) evaluation of the respondent's response to construct the threats scale through ranking and voting. Although threats can be of various types, the present list captures the major threats operating on the woman prawn seed catchers in three sectors of Indian Sundarbans and was ranked in terms of their importance by building a Threat Assessment Matrix (TAM). However, as there is high probability of variation of this ranking with the category of respondents, therefore the views of the respondents were also considered (by inclusion of the % of voting along with their respective ranking factor) and finally Composite Threat Scale (CTS) was constructed as per the expression:

$$\mathrm{CTS} = \mathrm{TS1} + \mathrm{TS2} + \mathrm{TS3} + \mathrm{TS4} + \mathrm{TS5}$$

where, TS = Threat Rank (TR) of five categories of respondents × % of Vote.

It is to be noted in this context that the sample size of respondents are variable e.g., for policy maker it is 38, but for other five groups of respondents, n = 1,545 for each group.

The deltaic region of Indian Sundarbans exhibits marked differentiation in terms of salinity, geomorphological features and vegetation and hence three distinct sectors have been marked viz., the western, central and eastern sectors (Barua et al. 2011; Mitra et al. 2015; Chakraborty et al. 2016; Pal et al. 2016; Mukherjee et al. 2017; Pal et al. 2017; Islam et al. 2018; Trivedi et al. 2018; Saha et al. 2019; Dutta et al. 2020). We therefore attempted to segregate the CTS for these three different sectors of Indian Sundarbans and observed highly significant spatial variation.

The results generated after the present exercise revealed few interesting findings as listed here (Tables 3.3, 3.4 and 3.5; Figs. 3.5, 3.6, 3.7, 3.8, 3.9, 3.10, 3.11, 3.12,

Table 3.3 Threats to women prawn seed collectors with scoring in the western Indian Sundarbans

Threat	Policy maker (Respondent type 1)		
	TR	% of vote	TS 1
Education related problem	8	11.1	88.8
Political problem	2	2.1	4.2
Religious problem	1	1.1	1.1
Water quality related problem	3	3.0	9
Animal attack and natural hazard	4	12.7	50.8
Gastric problem	7	17.3	121.1
Vision problem	6	25.8	154.8
Dermatological problem	8	26.9	215.2
Threat	Researcher (Respondent type 2)		
	TR	% of vote	TS 2
Education related problem	7	9.8	68.6
Political problem	1	1.9	1.9
Religious problem	2	1.7	3.4
Water quality related problem	3	4.4	13.2
Animal attack and natural hazard	5	13.5	67.5
Gastric problem	8	16.9	135.2
Vision problem	7	25.4	177.8
Dermatological problem	8	26.4	211.2
Threat	Fisherman (Respondent type 3)		
	TR	% of vote	TS 3
Education related problem	7	10.5	73.5
Political problem	2	1.8	3.6
Religious problem	1	2.3	2.3
Water quality related problem	3	4.2	12.6
Animal attack and natural hazard	4	13.2	52.8
Gastric problem	6	15.6	93.6
Vision problem	5	25.1	125.5
Dermatological problem	8	27.3	218.4
Threat	Agriculturist (Respondent type 4)		
	TR	% of vote	TS 4
Education related problem	7	9.2	64.4
Political problem	2	1.9	3.8
Religious problem	1	2.5	2.5
Water quality related problem	3	3.9	11.7
Animal attack and natural hazard	5	13.8	69

(continued)

Table 3.3 (continued)

Threat	Policy maker (Respondent type 1)		
	TR	% of vote	TS 1
Gastric problem	6	16.2	97.2
Vision problem	7	25.7	179.9
Dermatological problem	8	26.8	214.4
Threat	Local inhabitant (Respondent type 5)		
	SR	% of vote	TS 5
Education related problem	8	10.8	86.4
Political problem	1	2.4	2.4
Religious problem	1	1.9	1.9
Water quality related problem	4	4.3	17.2
Animal attack and natural hazard	5	13.1	65.5
Gastric problem	7	16.2	113.4
Vision problem	7	25.1	175.7
Dermatological problem	8	26.2	209.6

3.13, 3.14, 3.15, 3.16, 3.17, 3.18, 3.19, 3.20, 3.21, 3.22, 3.23, 3.24, 3.25, 3.26, 3.27, 3.28 and 3.29).

In western Indian Sundarbans the threats inflicted on the women tiger prawn seed collectors varied as per the order dermatological problem (213.76) > vision problem (162.74) > gastric problem (108.86) > education related problem (76.34) > animal attack and natural hazard (61.12) > water quality related problem (12.74) > political problem (3.18) > religious problem (2.24).

In central Indian Sundarbans the threats on the women tiger prawn seed collectors varied as per the order dermatological problem (212.96) > vision problem (145.12) > gastric problem (103.5) > education related problem (81.92) > animal attack and natural hazard (63.26) > water quality related problem (12.42) > political problem (3.38) > religious problem (1.98).

In eastern Indian Sundarbans the threats on the women tiger prawn seed collectors varied as per the order dermatological problem (212.8) > vision problem (145.6) > gastric problem (109.16) > education related problem (84.8) > animal attack and natural hazard (50.66) > water quality related problem (11.9) > political problem (3.98) > religious problem (1.70).

The region-wise CTS values are highlighted in Fig. 3.29.

In addition to the above socio-economic and health related problems, the intensification of aquaculture, particularly shrimp farming has raised a wide range of environmental concerns such as effluent discharge (Mitra and Bhattacharyya 2003), absence of bio-treatment pond, accumulation of ammonia and organic load on the pond bottom and dependence on commercial feed (Granada et al. 2016). Viewing the risks associated with shrimp farming and decreasing profitability, farmers have felt the need of diversification to secure income and sustainability in very recent

Table 3.4 Threats to women prawn seed collectors with scoring in the central Indian Sundarbans

Threat	Policy maker (Respondent type 1)		
	TR	% of vote	TS 1
Education related problem	7	10.2	71.4
Political problem	2	1.6	3.2
Religious problem	1	2.8	2.8
Water quality related problem	4	3.3	13.2
Animal attack and natural hazard	4	12.2	48.8
Gastric problem	6	16.7	100.2
Vision problem	5	25.4	127
Dermatological problem	8	27.6	220.8
Threat	Researcher (Respondent type 2)		
	TR	% of vote	TS 2
Education related problem	7	11.7	81.9
Political problem	1	2.4	2.4
Religious problem	1	1.3	1.3
Water quality related problem	3	2.7	8.1
Animal attack and natural hazard	5	13.6	68
Gastric problem	6	17.9	107.4
Vision problem	7	24.9	174.3
Dermatological problem	8	25.5	204
Threat	Fisherman (Respondent type 3)		
	TR	% of vote	TS 3
Education related problem	7	10.1	70.7
Political problem	2	1.5	3
Religious problem	1	2.7	2.7
Water quality related problem	4	3.4	13.6
Animal attack and natural hazard	5	13.1	65.5
Gastric problem	7	16.5	115.5
Vision problem	5	25.1	125.5
Dermatological problem	8	27.6	220.8
Threat	Agriculturist (Respondent type 4)		
	TR	% of vote	TS 4
Education related problem	8	11.3	90.4
Political problem	1	2.7	2.7
Religious problem	1	1.7	1.7
Water quality related problem	4	3.2	12.8
Animal attack and natural hazard	5	13.6	68

(continued)

Table 3.4 (continued)

Threat	Policy maker (Respondent type 1)		
	TR	% of vote	TS 1
Gastric problem	6	16.1	96.6
Vision problem	6	25.4	152.4
Dermatological problem	8	26.0	208
Threat	Local inhabitant (Respondent type 5)		
	TR	% of vote	TS 5
Education related problem	8	11.9	95.2
Political problem	2	2.8	5.6
Religious problem	1	1.4	1.4
Water quality related problem	4	3.6	14.4
Animal attack and natural hazard	5	13.2	66
Gastric problem	6	16.3	97.8
Vision problem	6	24.4	146.4
Dermatological problem	8	26.4	211.2

years. The diversification encompasses either switching off the shrimp farming and stepping towards other species and/or polyculture practices.

The aquatic phase around Sundarban mangroves is rich in nutrients and is extremely favourable in terms of aquaculture. We carried out a pilot project during 2018 and 2019 on tilapia (*Oreochromis niloticus*) culture after selecting two ponds with different salinities at Bali Island (22°04′35.17″ N; 88°44′55.70″ E). One of the ponds was stocked with estuarine water having 10.0 psu salinity and the other pond was diluted with stocked rain water to lower down the salinity to 4.0 psu.

The results obtained in the present pilot scale programme are presented here in points.

Growth parameters

It is observed that the growth is better in the pond with 4.0 psu salinity (hyposaline pond) compared to the pond having salinity of 10.0 psu (hypersaline pond), as the b values are 3.0715 and 3.0086 in 2018 and 2019 respectively in hyposaline pond indicating a uniform growth of the cultured species. On contrary, the b values were around 2.0 in the hypersaline pond (Figs. 3.30, 3.31, 3.32 and 3.33).

Proximate composition

Protein, carbohydrate and lipid exhibited unique seasonal variations in both 2018 and 2019. The level of protein was highest in monsoon followed by postmonsoon and premonsoon. Carbohydrate exhibited highest value in premonsoon followed by postmonsoon and monsoon. In case of lipid highest value was observed in monsoon followed by postmonsoon and premonsoon. This seasonal trend of biochemical

Table 3.5 Threats to women prawn seed collectors with scoring in the eastern Indian Sundarbans

Threat	Policy maker (Respondent type 1)		
	TR	% of vote	TS 1
Education related problem	8	10.1	80.8
Political problem	2	1.3	2.6
Religious problem	1	2.2	2.2
Water quality related problem	4	3.6	14.4
Animal attack and natural hazard	3	13.1	39.3
Gastric problem	6	16.2	97.2
Vision problem	5	26.2	131
Dermatological problem	8	27.3	218.4
Threat	Researcher (Respondent type 2)		
	TR	% of vote	TS 2
Education related problem	8	9.3	74.4
Political problem	1	2.9	2.9
Religious problem	1	1.7	1.7
Water quality related problem	4	3.8	15.2
Animal attack and natural hazard	4	13.7	54.8
Gastric problem	6	17.9	107.4
Vision problem	6	24.1	144.6
Dermatological problem	8	26.6	212.8
Threat	Fisherman (Respondent type 3)		
	TR	% of vote	TS 3
Education related problem	8	11.9	95.2
Political problem	2	2.5	5
Religious problem	2	1.7	3.4
Water quality related problem	3	3.3	9.9
Animal attack and natural hazard	4	13.1	52.4
Gastric problem	7	16.3	114.1
Vision problem	6	25.0	150
Dermatological problem	8	26.2	209.6
Threat	Agriculturist (Respondent type 4)		
	TR	% of vote	TS 4
Education related problem	8	10.4	83.2
Political problem	2	2.2	4.4
Religious problem	1	1.6	1.6
Water quality related problem	3	3.2	9.6
Animal attack and natural hazard	4	13.5	54

(continued)

Table 3.5 (continued)

Threat	Policy maker (Respondent type 1)		
	TR	% of vote	TS 1
Gastric problem	7	17.7	123.9
Vision problem	6	25.1	150.6
Dermatological problem	8	26.3	210.4
Threat	Local inhabitant (Respondent type 5)		
	TR	% of vote	TS 5
Education related problem	8	11.3	90.4
Political problem	3	2.5	7.5
Religious problem	1	1.3	1.3
Water quality related problem	4	2.6	10.4
Animal attack and natural hazard	4	13.2	52.8
Gastric problem	6	17.2	103.2
Vision problem	6	25.3	151.8
Dermatological problem	8	26.6	212.8

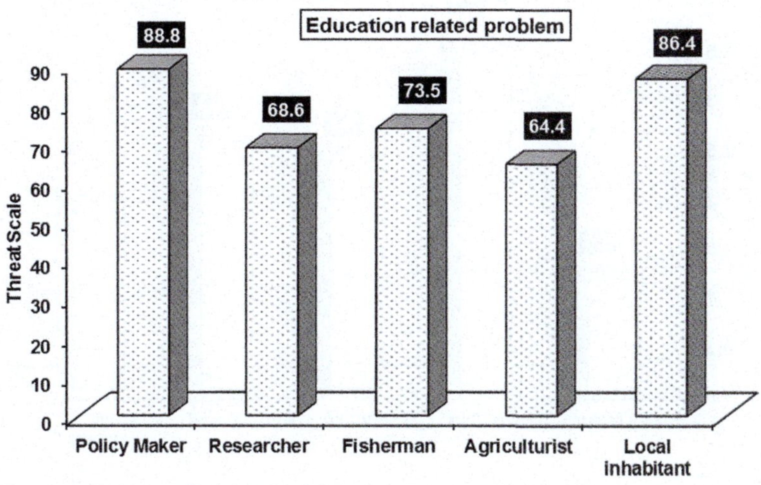

Fig. 3.5 Threat scores of education related problem of different respondents in western Indian Sundarbans

composition in the fish flesh was observed both in the hyposaline and hypersaline ponds (Figs. 3.34 and 3.35).

The cost and quality of fish depend on the microbial load in different organs, and hence we also carried out a study on the total and faecal coliform loads in the ambient media and different organs of the cultured species.

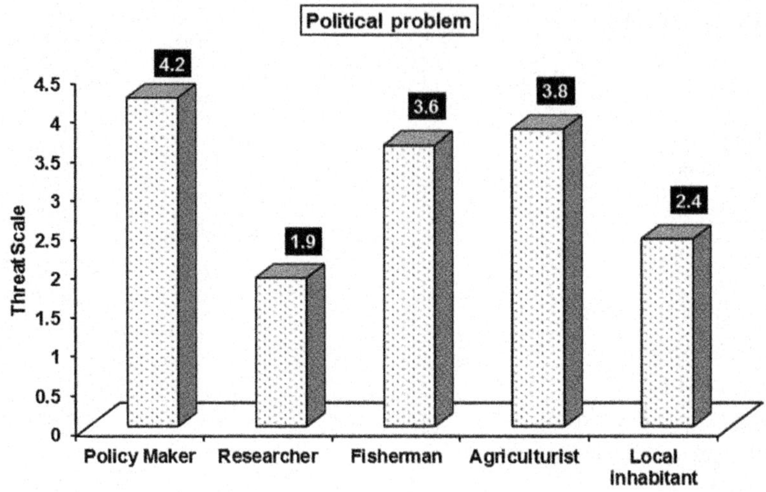

Fig. 3.6 Threat scores of political problem of different respondents in western Indian Sundarbans

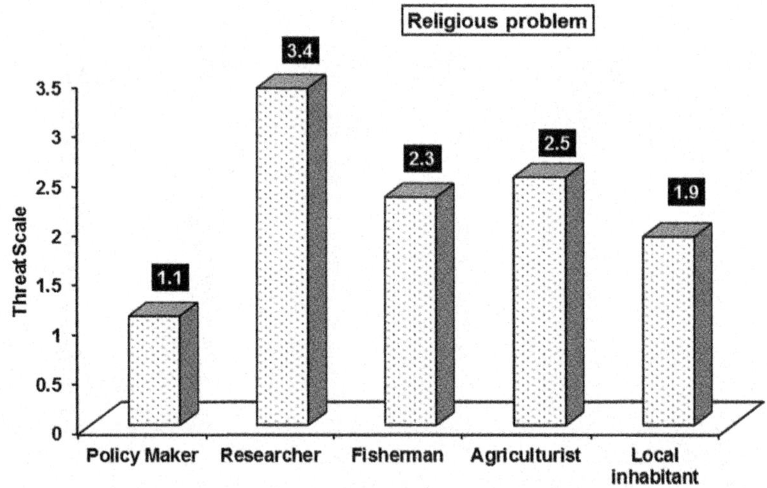

Fig. 3.7 Threat scores of religious problem of different respondents in western Indian Sundarbans

Total coliform in water

Total coliform in water exhibited highest value in monsoon followed by postmonsoon and premonsoon. This trend was observed both in hypersaline pond and hyposaline pond (Figs. 3.36 and 3.37).

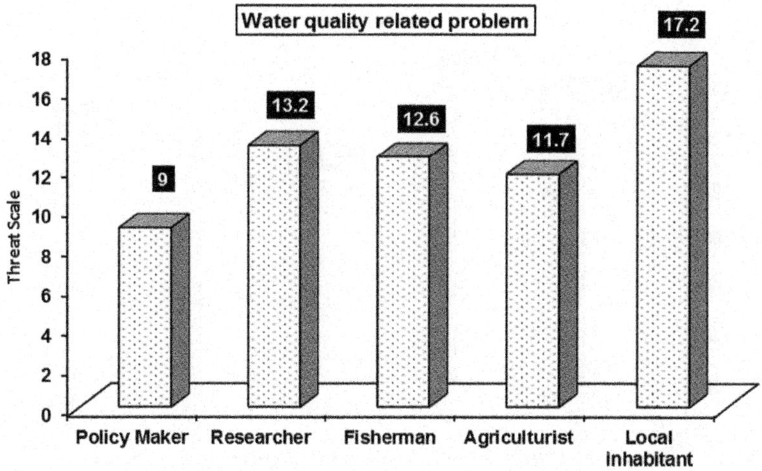

Fig. 3.8 Threat scores of water quality related problem of different respondents in western Indian Sundarbans

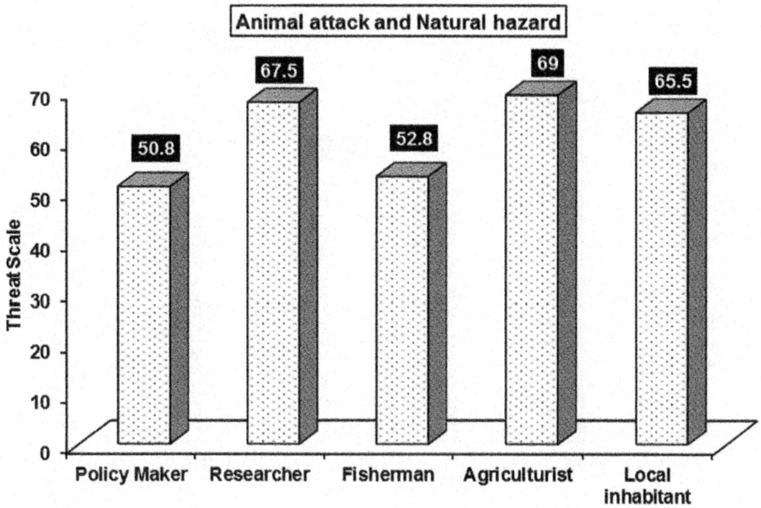

Fig. 3.9 Threat scores of animal attack and natural hazard of different respondents in western Indian Sundarbans

Total coliform in sediment

Total coliform in sediment exhibited highest value in monsoon followed by post-monsoon and premonsoon. This trend was observed both in hypersaline pond and hyposaline pond (Figs. 3.38 and 3.39).

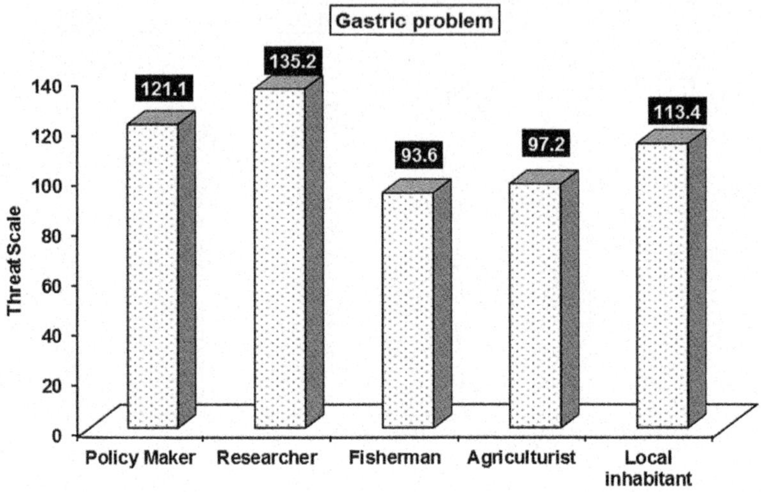

Fig. 3.10 Threat scores of gastric problem of different respondents in western Indian Sundarbans

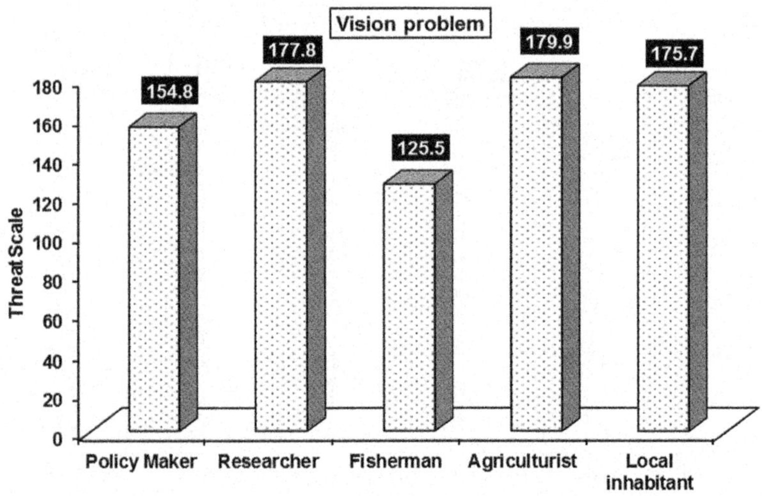

Fig. 3.11 Threat scores of Vision problem of different respondents in western Indian Sundarbans

Total coliform in fish muscle

Total coliform in fish muscle exhibited highest value in monsoon followed by post-monsoon and premonsoon. This trend was observed both in hypersaline pond and hyposaline pond (Figs. 3.40 and 3.41).

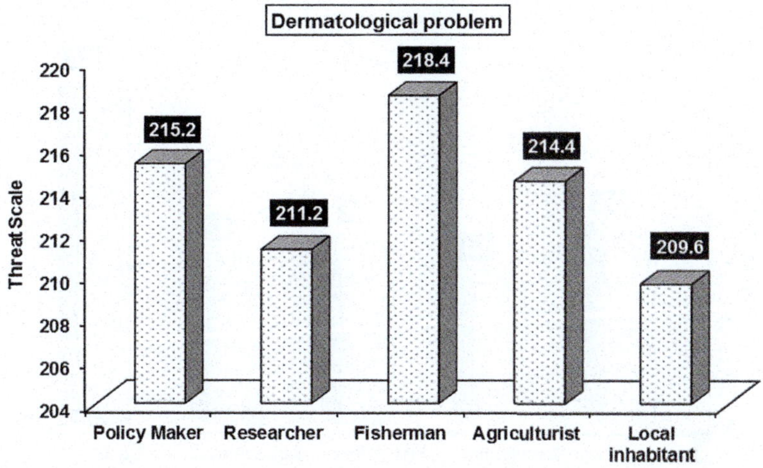

Fig. 3.12 Threat scores of dermatological problem of different respondents in western Indian Sundarbans

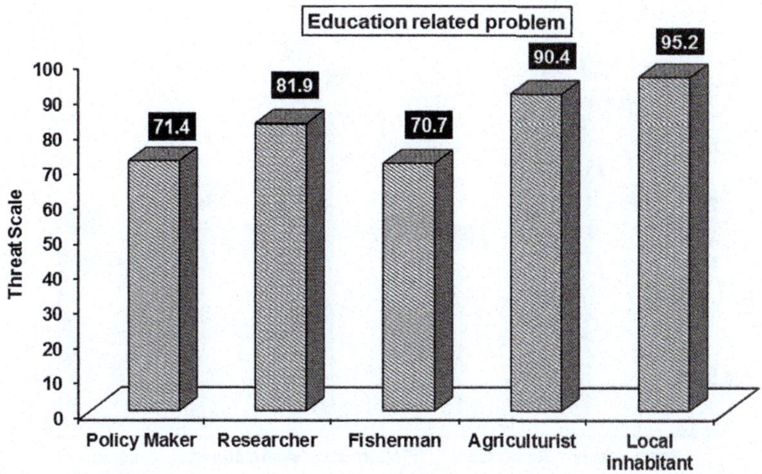

Fig. 3.13 Threat scores of education related problem of different respondents in central Indian Sundarbans

Total coliform in fish gills

Total coliform in fish gill exhibited highest value in monsoon followed by post-monsoon and premonsoon. This trend was observed both in hypersaline pond and hyposaline pond (Figs. 3.42 and 3.43).

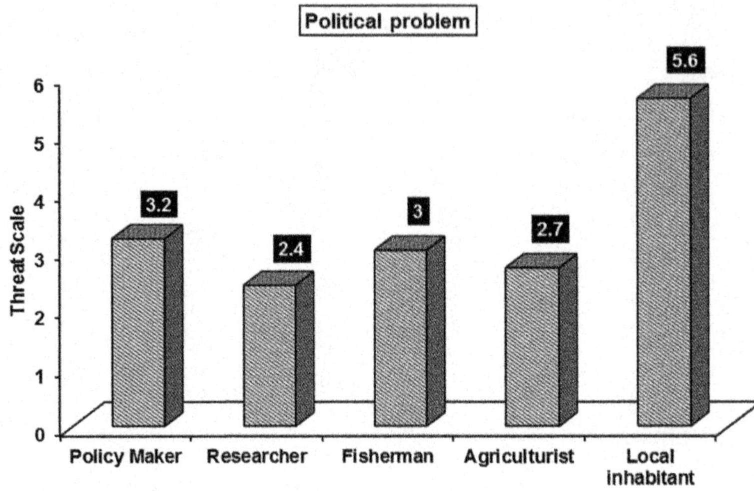

Fig. 3.14 Threat scores of political problem of different respondents in central Indian Sundarbans

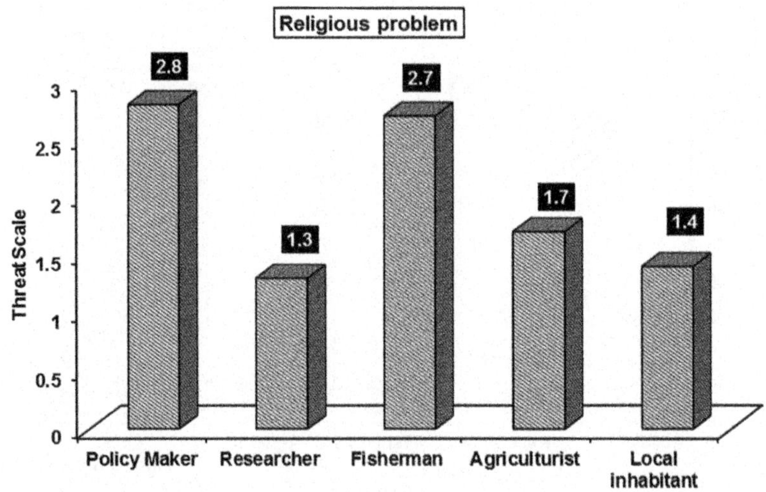

Fig. 3.15 Threat scores of religious problem of different respondents in central Indian Sundarbans

Total coliform in fish hepatopancreas

Total coliform in hepatopancreas exhibited highest value in monsoon followed by postmonsoon and premonsoon. This trend was observed both in hypersaline pond and hyposaline pond (Figs. 3.44 and 3.45).

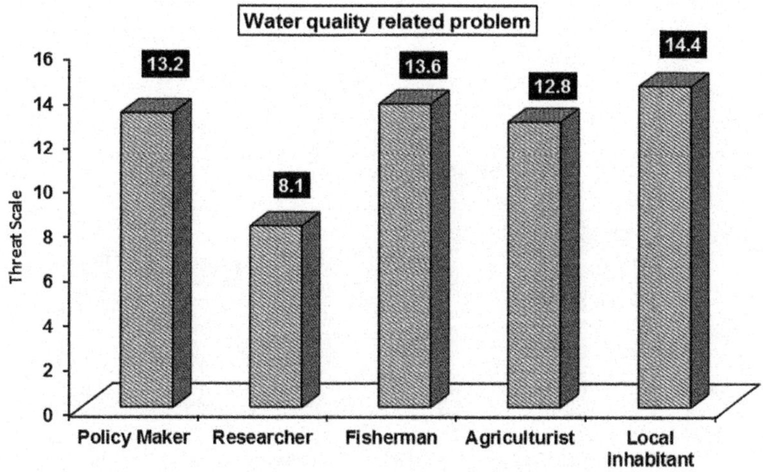

Fig. 3.16 Threat scores of water quality related problem of different respondents in central Indian Sundarbans

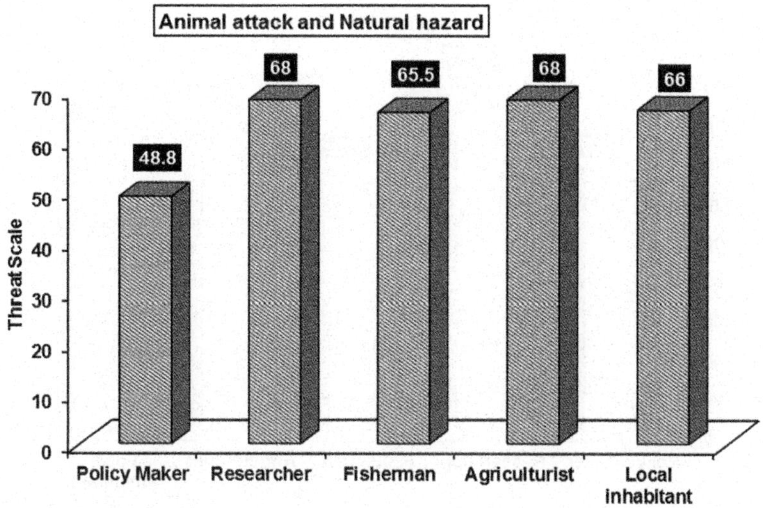

Fig. 3.17 Threat scores of animal attack and Natural hazard of different respondents in central Indian Sundarbans

Faecal coliform in culture pond water

Faecal coliform in water exhibited highest value in monsoon followed by post-monsoon and premonsoon. This trend was observed both in hypersaline pond and hyposaline pond (Figs. 3.46 and 3.47).

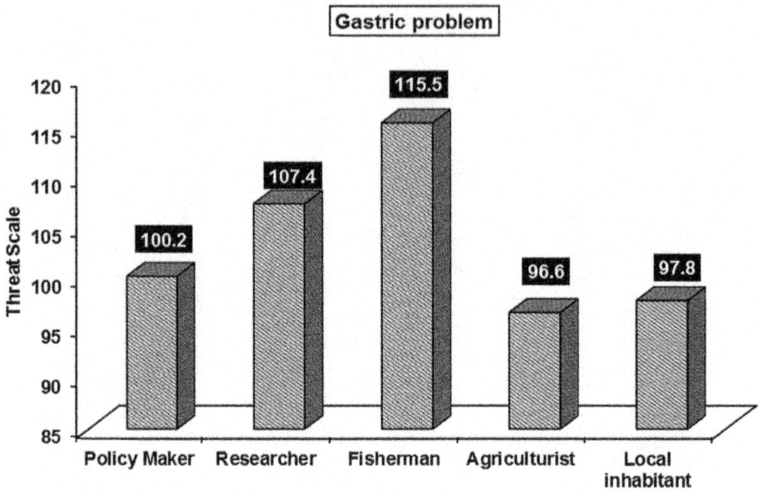

Fig. 3.18 Threat scores of gastric problem of different respondents in central Indian Sundarbans

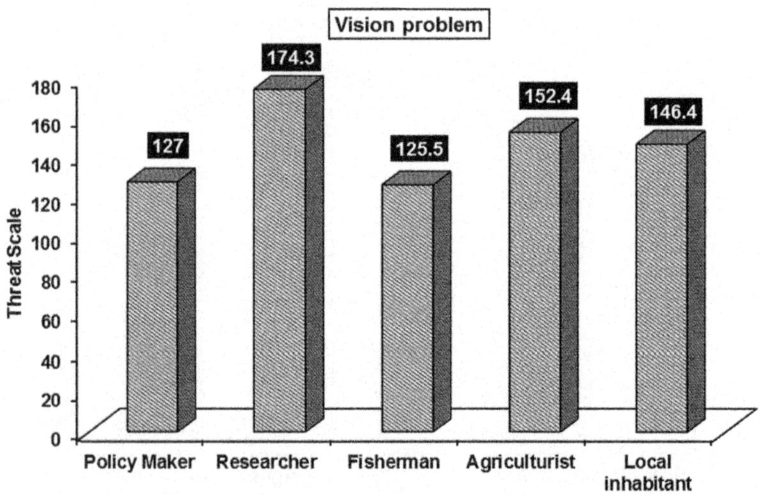

Fig. 3.19 Threat scores of vision problem of different respondents in central Indian Sundarbans

Faecal coliform in sediment

Faecal coliform in sediment exhibited highest value in monsoon followed by postmonsoon and premonsoon. This trend was observed both in hypersaline pond and hyposaline pond (Figs. 3.48 and 3.49).

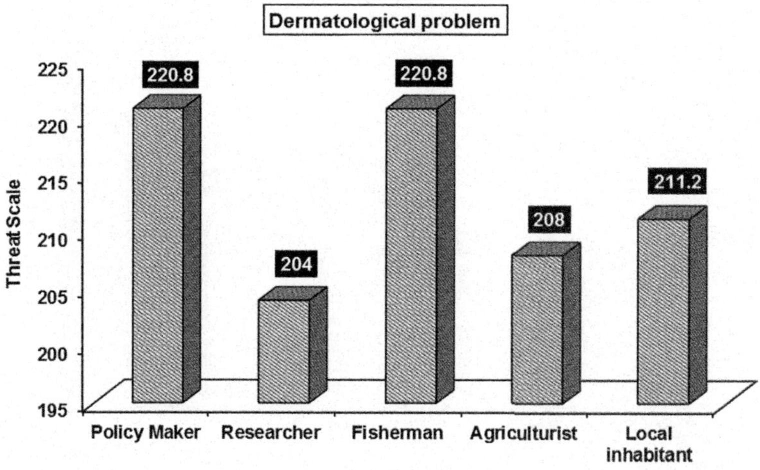

Fig. 3.20 Threat scores of dermatological problem of different respondents in central Indian Sundarbans

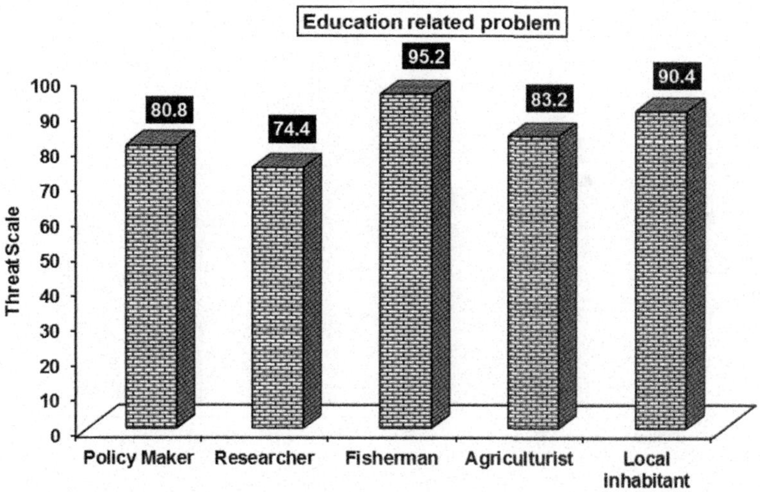

Fig. 3.21 Threat scores of education related problem of different respondents in eastern Indian Sundarbans

Faecal coliform in muscle

Faecal coliform in fish muscle exhibited highest value in monsoon followed by postmonsoon and premonsoon. This trend was observed both in hypersaline pond and hyposaline pond (Figs. 3.50 and 3.51).

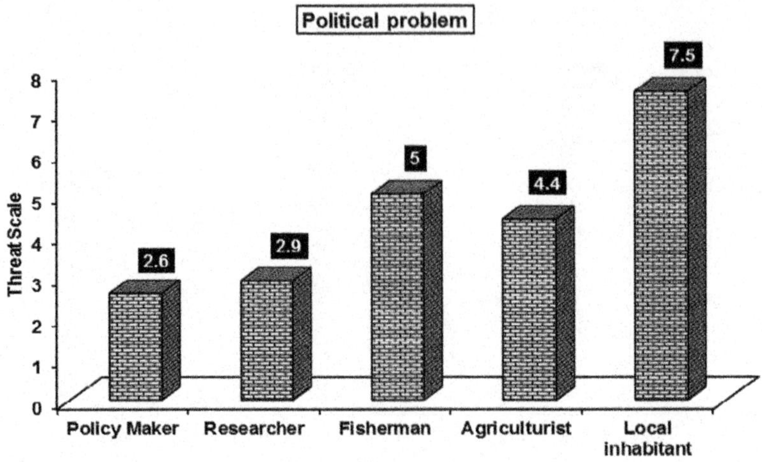

Fig. 3.22 Threat scores of political problem of different respondents in eastern Indian Sundarbans

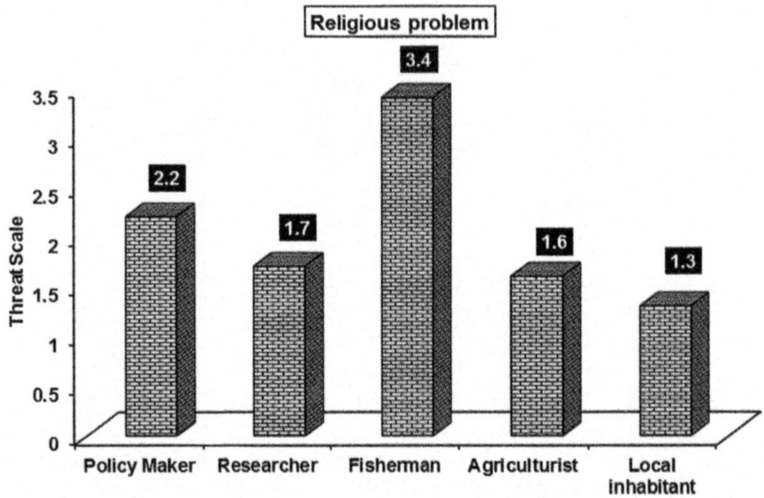

Fig. 3.23 Threat scores of religious problem of different respondents in eastern Indian Sundarbans

Faecal coliform in gill

Faecal coliform in fish gill exhibited highest value in monsoon followed by post-monsoon and premonsoon. This trend was observed both in hypersaline pond and hyposaline pond (Figs. 3.52 and 3.53).

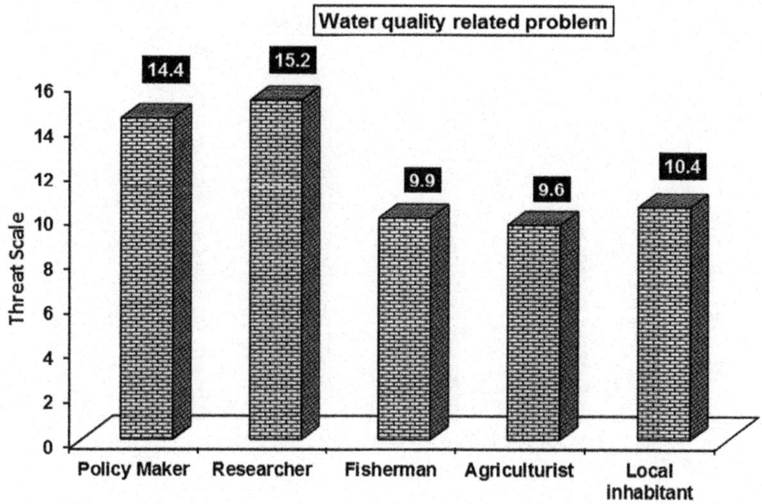

Fig. 3.24 Threat scores of water quality related problem of different respondents in eastern Indian Sundarbans

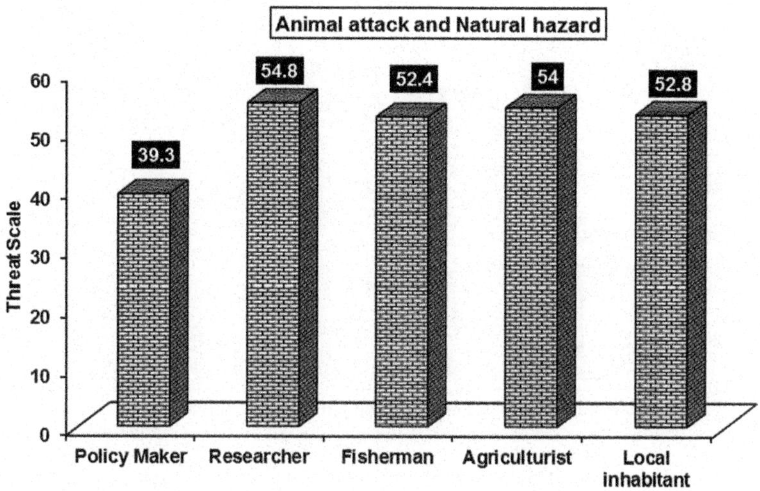

Fig. 3.25 Threat scores of animal attack and natural hazard of different respondents in eastern Indian Sundarbans

Faecal coliform in hepatopancreas

Faecal coliform in fish hepatopancreas exhibited highest value in monsoon followed by postmonsoon and premonsoon. This trend was observed both in hypersaline pond and hyposaline pond (Figs. 3.54 and 3.55).

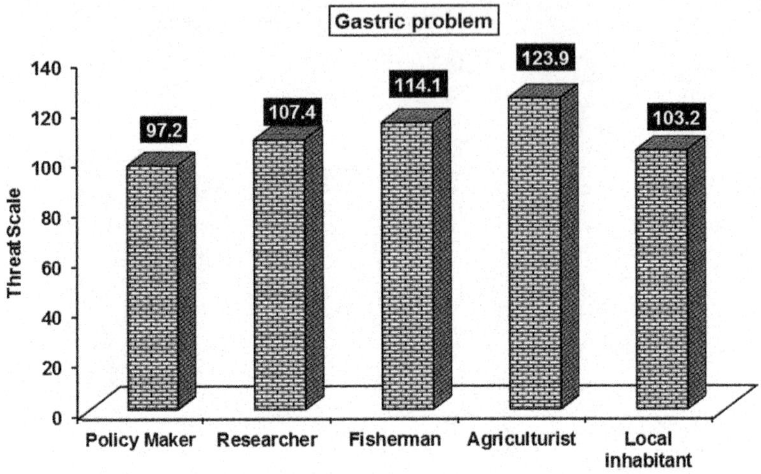

Fig. 3.26 Threat scores of gastric problem of different respondents in eastern Indian Sundarbans

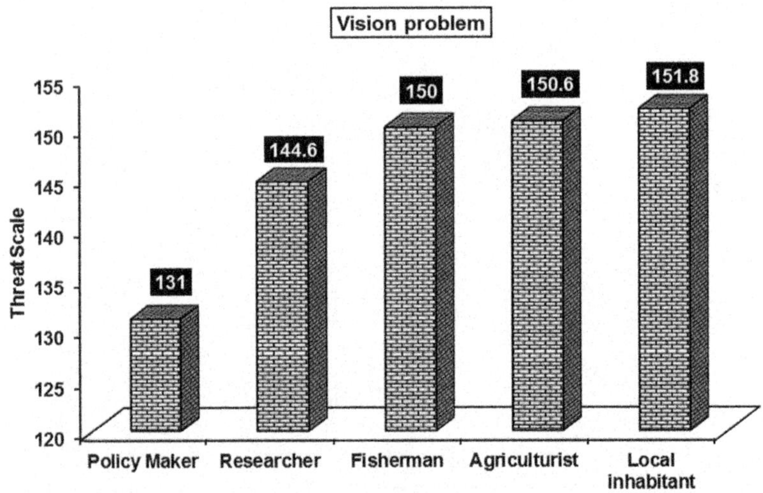

Fig. 3.27 Threat scores of vision problem of different respondents in eastern Indian Sundarbans

The length–weight relationship as highlighted in Figs. 3.31 and 3.33 and Figs. 3.30 and 3.32 for hyposaline pond and hypersaline pond respectively clearly indicates better condition of *O. niloticus* cultured in hyposaline pond compared to hypersaline pond. When the value of b is equal to 3.0, growth is isometric and when it is less or greater than 3.0, it is allometric (Enin 1994). More specifically, growth is positive allometric when organism weight increases more than length ($b > 3$). This is observed in case of hyposaline pond where the value of b is 3.0715 and 3.0086 in 2018 and 2019 respectively. The situation becomes negative allometric when the weight of

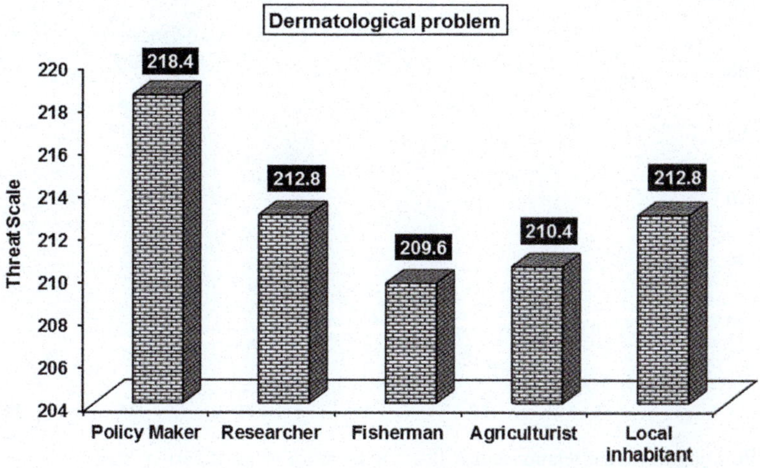

Fig. 3.28 Threat scores of dermatological problem of different respondents in eastern Indian Sundarbans

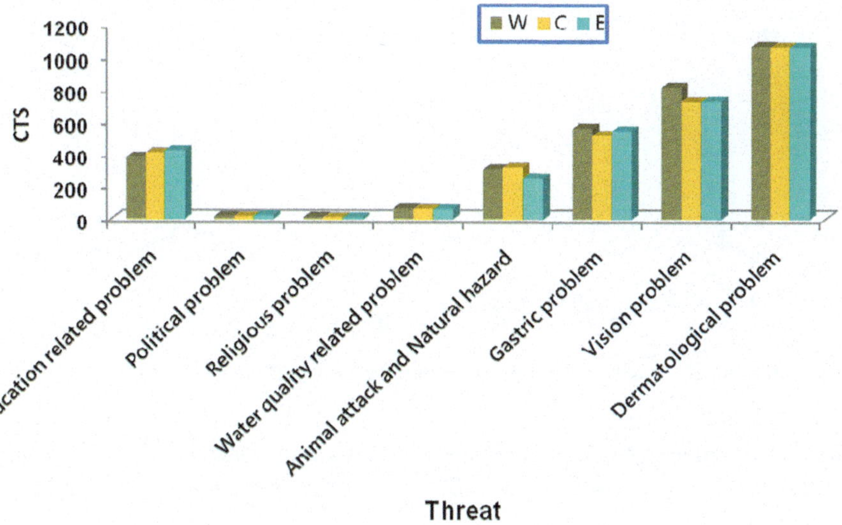

Fig. 3.29 CTS values in western, central and eastern Indian Sundarbans

cultured species decreases compared to the expected length b < 3. This situation is observed in hypersaline pond where the value of b is 2.6013 and 2.0878 in 2018 and 2019 respectively.

The FCR values of the cultured fishes in hyposaline pond were 1.02 and 1.17 during 2018 and 2019 respectively and in hypersaline pond the values were 2.54 and 2.96 during 2018 and 2019 respectively (Table 3.6). These values also point towards better growth of the tilapia species in hyposaline pond compared to hypersaline pond.

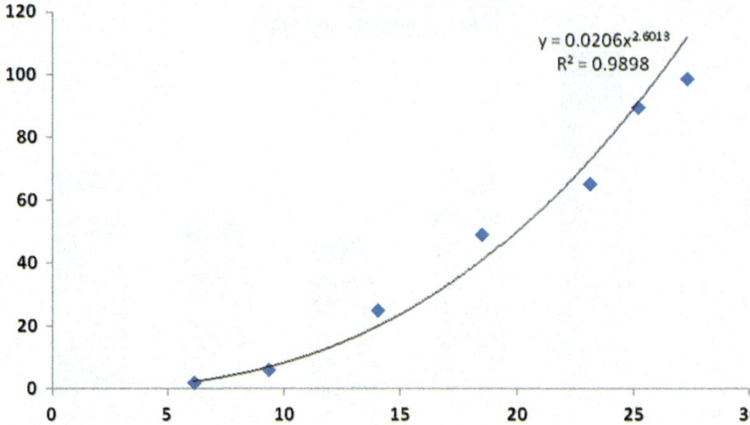

Fig. 3.30 Length–weight relationship of fishes in hypersaline pond during 2018

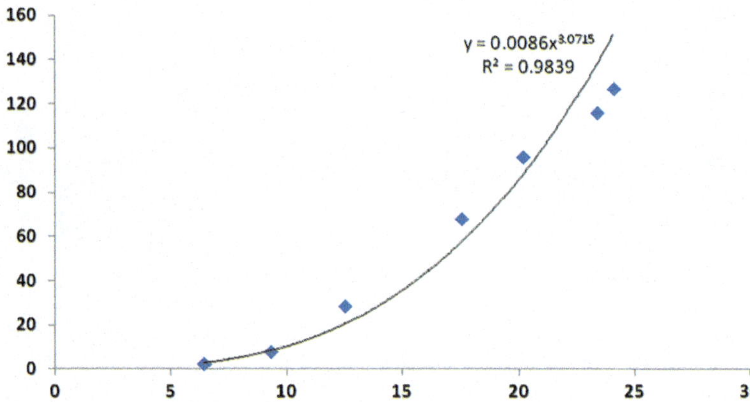

Fig. 3.31 Length–weight relationship of fishes in hyposaline pond during 2018

Lower the value of FCR, better is the growth as it indicates better food intake and conversion into biomass.

Proximate composition

Protein, carbohydrate and lipid exhibited unique seasonal variations in both 2018 and 2019. The level of protein was highest in monsoon followed by postmonsoon and premonsoon. Carbohydrate exhibited a different seasonal trend with highest value in premonsoon followed by postmonsoon and monsoon. More nitrates in the aquatic phase during monsoon may be the possible cause of highest protein value in the species during monsoon. In case of lipid, highest value was observed in monsoon followed by postmonsoon and premonsoon. This seasonal trend of

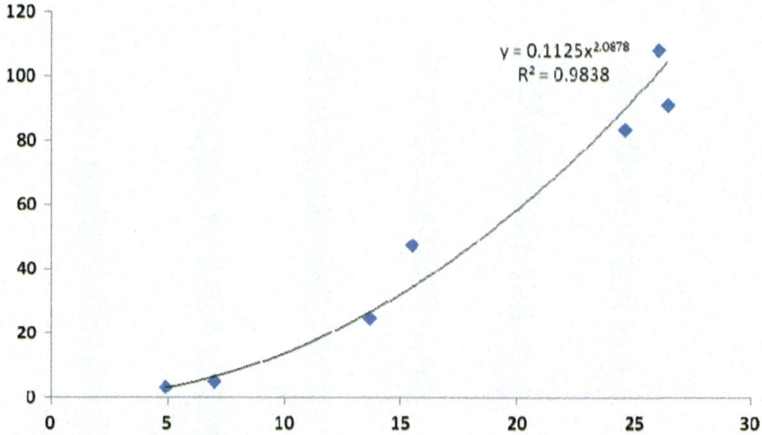

Fig. 3.32 Length–weight relationship of fishes in hypersaline pond during 2019

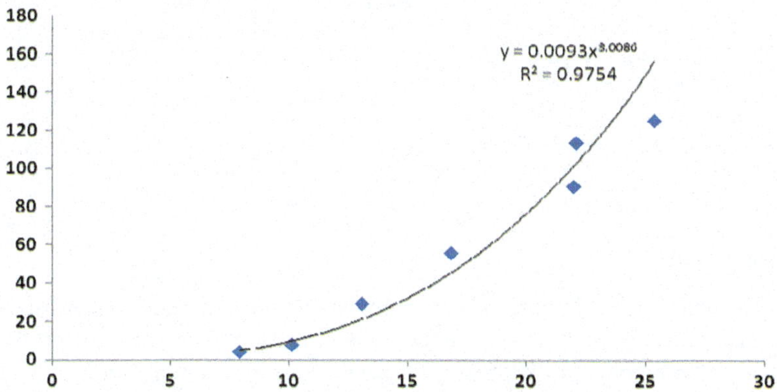

Fig. 3.33 Length–weight relationship of fishes in hyposaline pond during 2019

biochemical composition in the fish flesh was observed both in the hypersaline pond and hyposaline pond.

Hydrological parameters

The hydrological parameters were better in the hyposaline pond (particularly if DO is considered) compared to hypersaline pond, which may be attributed to more diffusion of atmospheric oxygen in the water of hyposaline pond having more surface area and phytoplankton compared to hypersaline pond.

Significant seasonal variations were observed in the values of selected hydrological parameters that may be considered as impact of monsoon and subsequent run-off from surrounding landscapes.

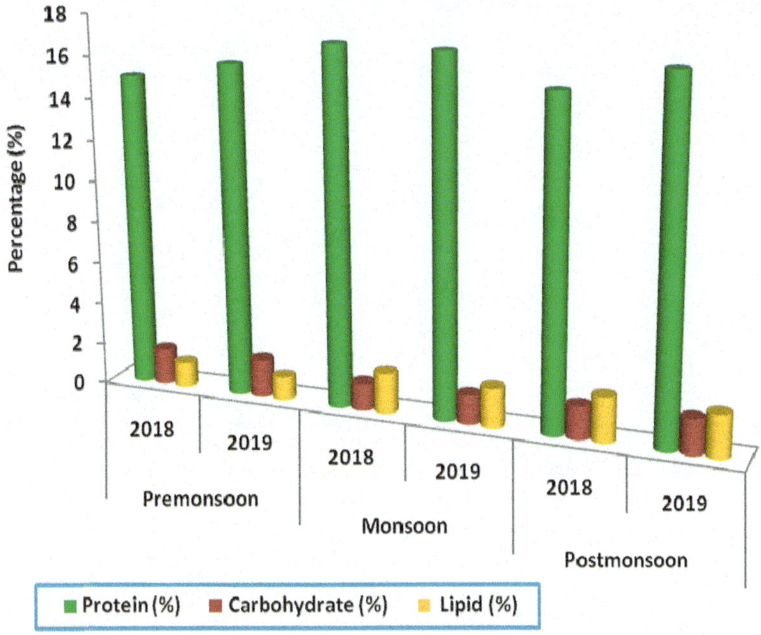

Fig. 3.34 Protein, carbohydrate and lipid percentages in the muscle of *O. niloticus* in hypersaline pond

Microbial load

The microbial load (TC and FC) in ambient media and body tissues exhibited contrasting seasonal picture with highest load during monsoon followed by post-monsoon and premonsoon. The run-off from the adjacent landmasses rich in organic detritus and faecal matter of grazing animals (in both types of ponds) may be one of the causes for this contrasting seasonal picture.

Tilapia has been considered as the Food Fish of the twenty-first century and is popularly known as aquatic chicken. The World aquaculture production of tilapia is 4.0 million tonnes with an estimated value of around $3 billion. FAO reports indicate that the Global tilapia production is expected to reach 7.3 million tonnes by 2030. In countries like China, Egypt, Philippines, Brazil, Thailand and Bangladesh tilapia contribute substantially to the Food Security Programmes. Therefore, it was felt at the National level that tilapia would be one of the candidate species which could be cultured to address the food security in general and cheap protein requirement of the growing population in particular. However, the market price of tilapia is not very appreciable (~INR 150–200/kg) and thus lacking the commercial momentum in the pisciculture practice in Indian Sundarbans.

Monoculture of shrimp is practiced in most of the states of India for its high productivity. *Penaeus monodon*, which is highly susceptible to diseases is replaced by *Penaeus vannamei* (Tables 3.7 and 3.8), but we think from our long term experience

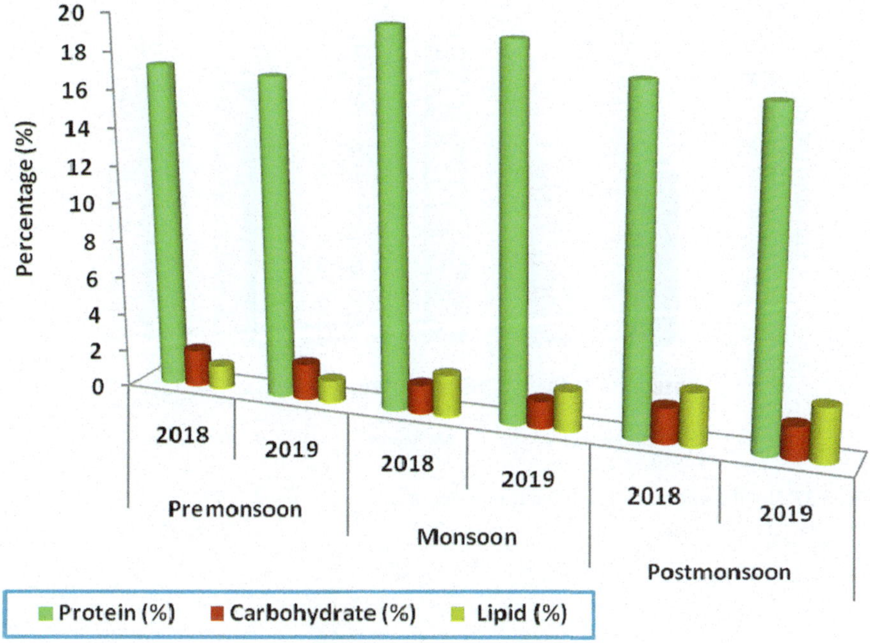

Fig. 3.35 Protein, carbohydrate and lipid percentages in the muscle of *O. niloticus* in hyposaline pond

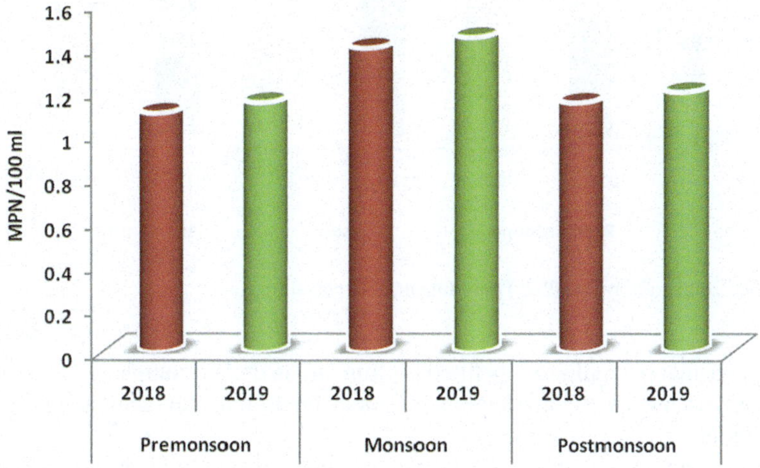

Fig. 3.36 Total coliform (MPN/100 ml) in water of hypersaline pond

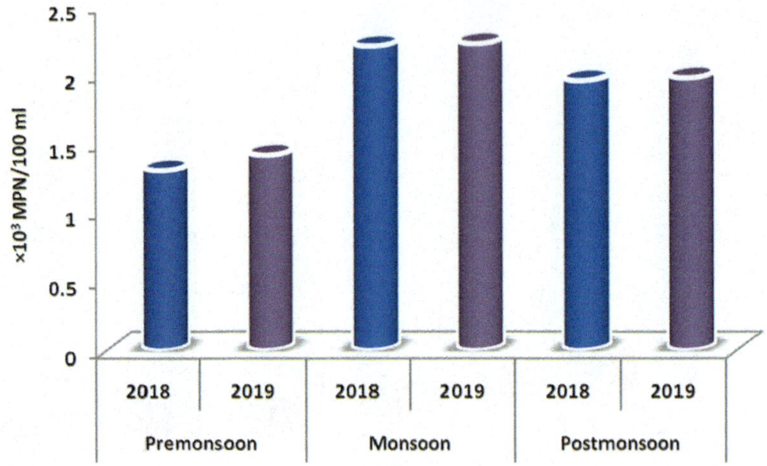

Fig. 3.37 Total coliform (MPN/100 ml) in water of hyposaline pond

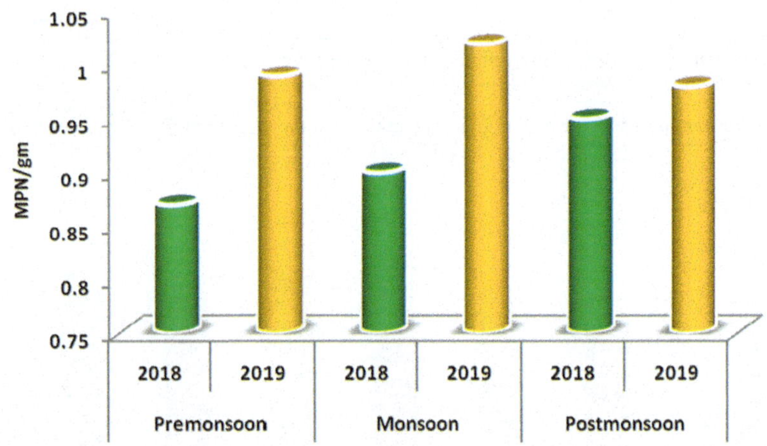

Fig. 3.38 Total coliform (MPN/g) in sediment of hypersaline pond

of 30 years that polyculture is befitted for Sundarban deltaic complex because of low risk from diseases, better production, minimum environmental issues and year round production.

Polyculture is potentially more sustainable than monoculture, due to reutilization of waste products of one species by another (Petrell et al. 1993; Chopin and Yarish 1999). Shrimp polyculture represents an important alternative to solving and/or minimizing some of the problems (environmental pollution, diseases and decreasing prices) that shrimp aquaculture has faced in the past two decades. Several species from diverse trophic levels have the potential to be co-cultured with shrimps. A good knowledge of the species that are candidates for polyculture and an adequately

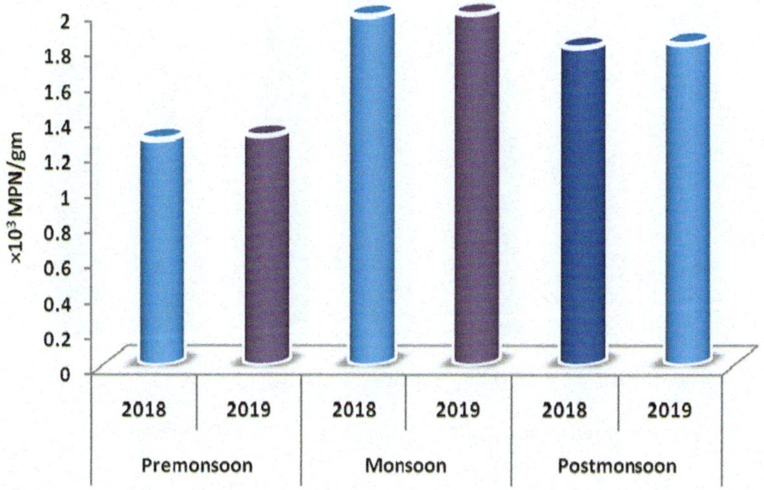

Fig. 3.39 Total coliform (MPN/g) in sediment of hyposaline pond

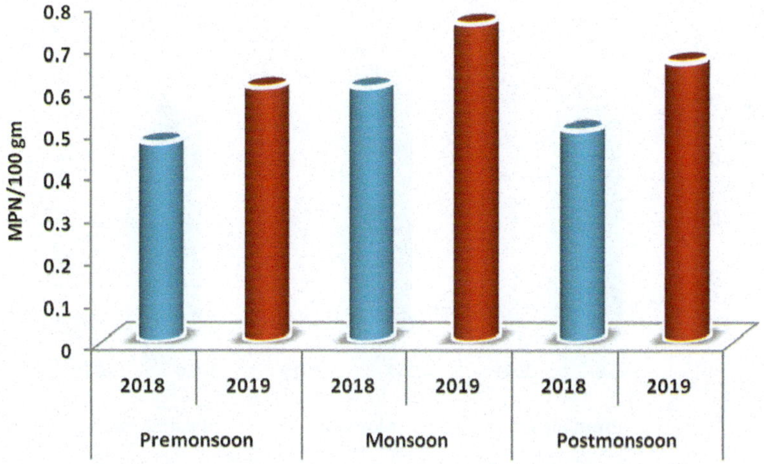

Fig. 3.40 Total coliform (MPN/100 g) in fish muscle in the hypersaline pond

designed culture system are the most important points to consider when co-culturing shrimp with other species (Martínez-Porchas et al. 2010). Modification of age old traditional brackishwater polyculture is a logical choice towards sustainable intensification. Selection of compatible species for better production and profitability is the first step. Improved polyculture of shrimp with non-carnivorous fishes is a viable alternative to shrimp monoculture as a low cost low risk and high profit venture. Improved farming practices like pond preparation, selective stocking, and rational use

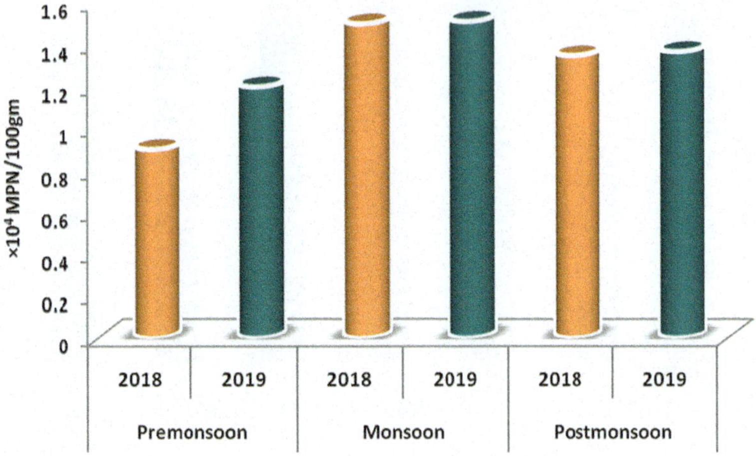

Fig. 3.41 Total coliform (MPN/100 g) in fish muscle in the hyposaline pond

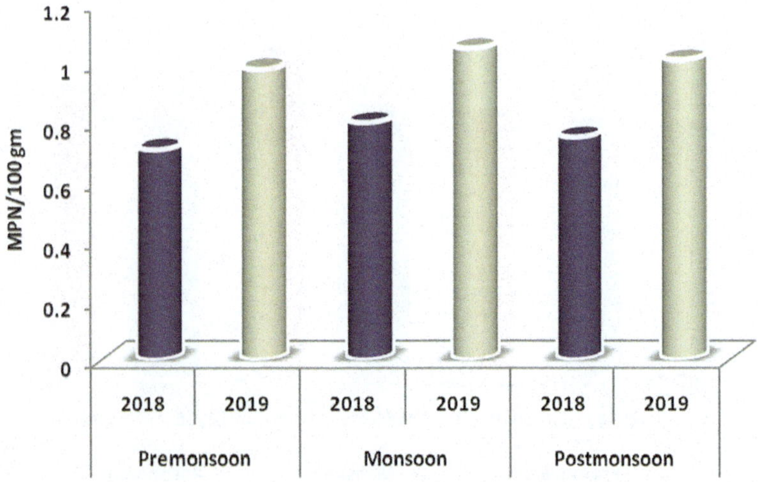

Fig. 3.42 Total coliform (MPN/100 g) in the fish gills in the hypersaline pond

of formulated feed and proper pond management are the steps involved in improvement of productivity in polyculture. Species combination and their compatibility are of great concern to achieve higher productivity and profitability in polyculture, but information in this context is scanty.

We carried out a pilot project at Jharkhali (22°05′52.82″ N; 88°41′47.25″ E) during 2019 by stocking tiger shrimp, parsia, grey mullet and milkfish (Table 3.9) with different species combination viz. shrimp-parsia-grey mullet (T1) shrimp-parsia-milkfish (T2), shrimp-grey mullet-milkfish (T3) and shrimp-parsia-grey mullet-milkfish (T4) (Table 3.10).

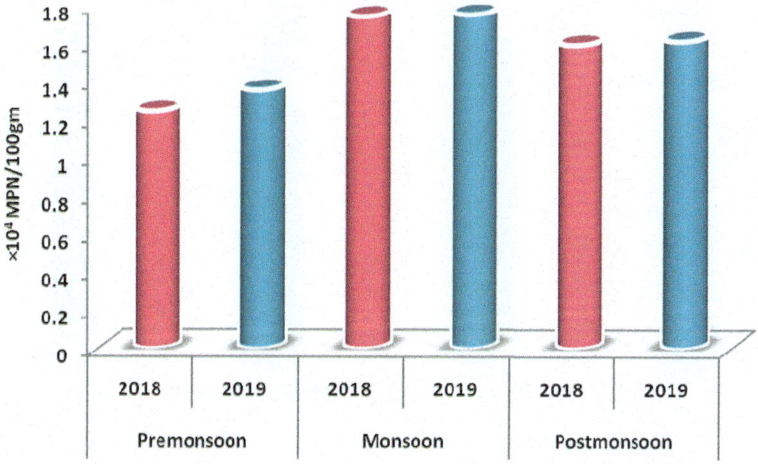

Fig. 3.43 Total coliform (MPN/100 g) in the fish gills in the hyposaline pond

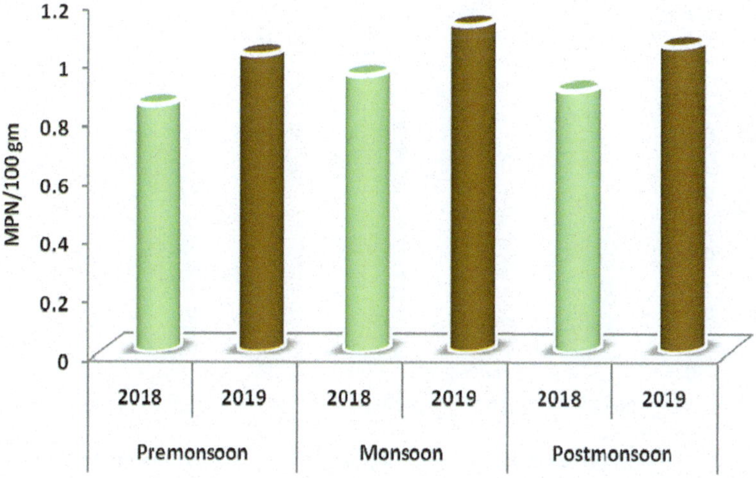

Fig. 3.44 Total coliform (MPN/100 g) in the fish hepatopancreas in the hypersaline pond

We observed highly satisfactory body weight and almost ideal length–weight relationships in almost all combinations (Fig. 3.56, 3.57, 3.58, 3.59, 3.60, 3.61, 3.62 and 3.63), which suggest the potential of polyculture practice in the framework of Indian Sundarbans.

Apart from the edible fishes discussed so far, ornamental fishes available in the estuaries, creeks and inlets of Indian Sundarbans have high potential to sustain livelihoods of the island dwellers. The demand of marine ornamental fishes is increasing remarkably due to their important role in the World trade for fish production. Singapore is the major supplier of ornamental fishes in the international market with an

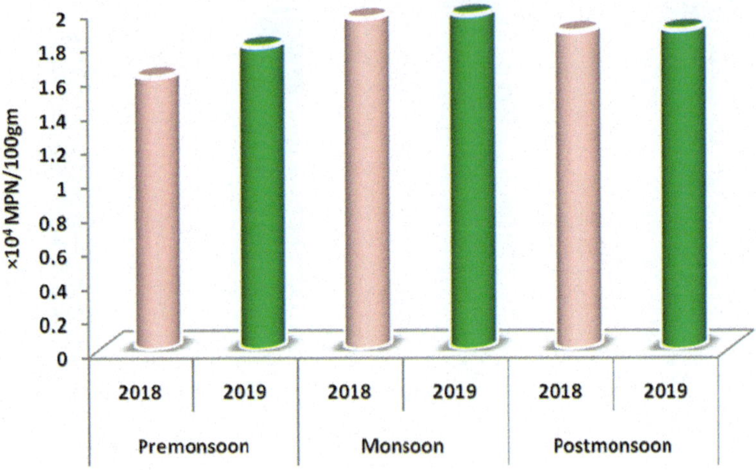

Fig. 3.45 Total coliform (MPN/100 g) in the fish hepatopancreas in the hyposaline pond

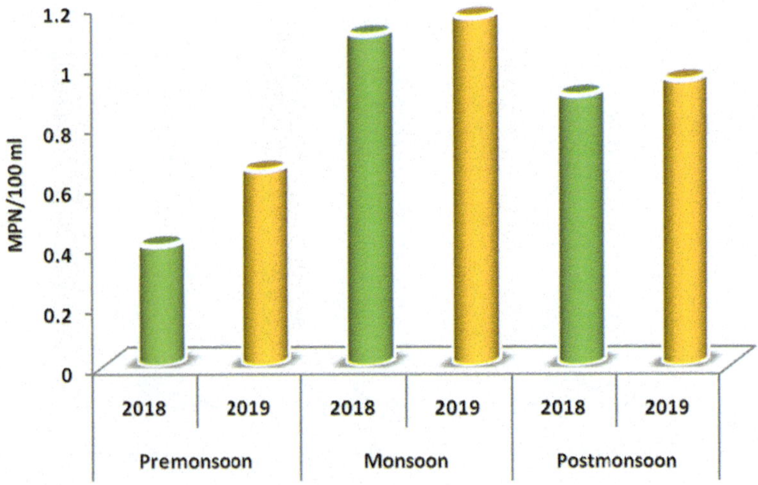

Fig. 3.46 Faecal coliform (MPN/100 ml) in the water of hypersaline pond

annual export of about US$80.0 million. However, the export of ornamental fishes from India is worth only US$0.20 million, which is not an appreciable figure with regard to the rising demand of these colorful faunal species in the World ornamental fish market. This may be attributed to the lack of involvement of skilled and knowledgeable people in ornamental fish farming. India has a rich biodiversity of marine ornamental fishes in the mangrove-dominated Indian Sundarbans in coastal West Bengal, along with the lagoons, and coral reefs of Andaman and Nicobar Islands, Palk Bay, Gulf of Mannar, Coast of Kerala, Lakshadweep, and Gulf of Kutch. About

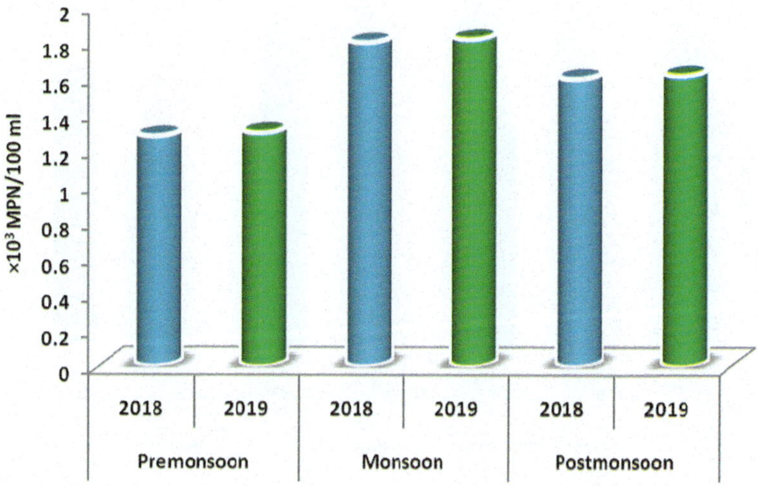

Fig. 3.47 Faecal coliform (MPN/100 ml) in the water of hyposaline pond

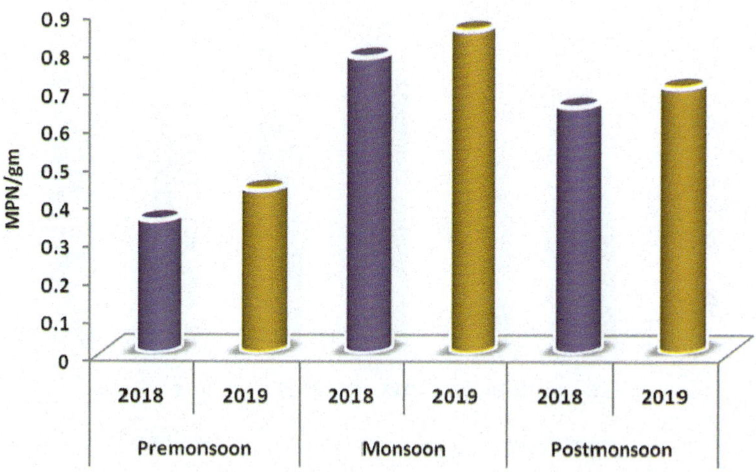

Fig. 3.48 Faecal coliform (MPN/g) in sediment of hypersaline pond

400 indigenous species/varieties of ornamental fishes have been reported from the marine ecosystems of India spreading over the Bay of Bengal, Laccadive Sea in the Indian Ocean, and the Arabian Sea. Therefore, India has tremendous potential to take a giant leap in the domain of marine/brackishwater ornamental fishery industry owing to the vast repertoire of diverse ornamental fish species encompassing both hyposaline and hypersaline regions, congenial tropical and sub-tropical climatic conditions, and availability of cheap/non-skilled labor. This in turn will open numerous job opportunities and alternative livelihood options for the poor inland and island dwellers alike.

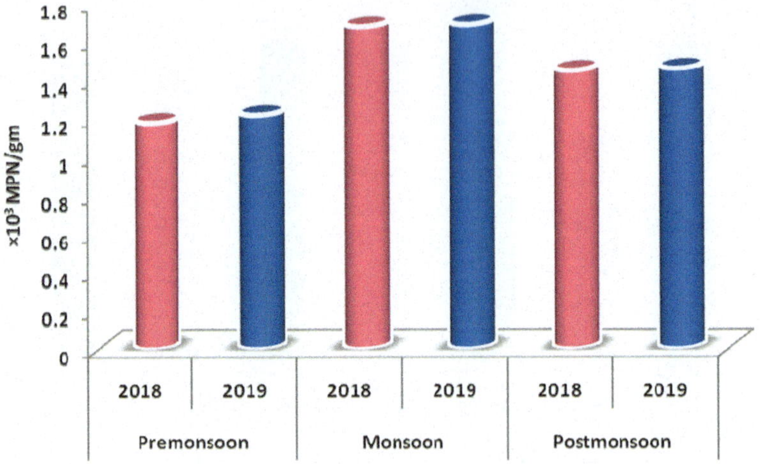

Fig. 3.49 Faecal coliform (MPN/g) in sediment of hyposaline pond

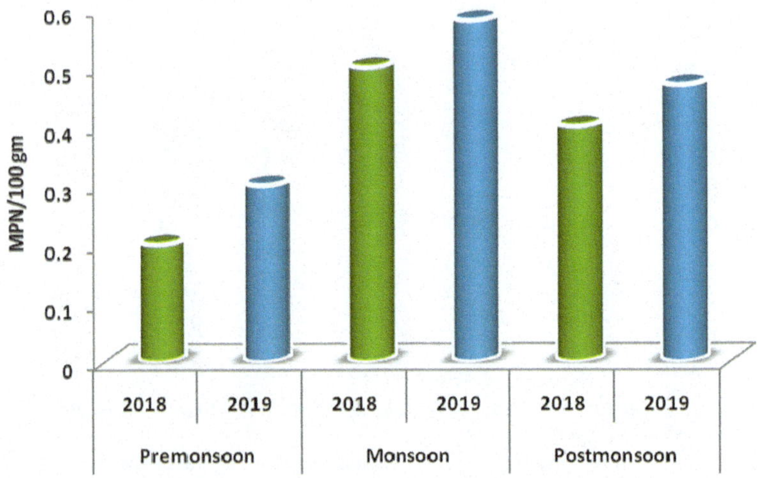

Fig. 3.50 Faecal coliform (MPN/100 g) in fish muscle in hypersaline pond

The estuaries of Indian Sundarbans are the hotspots of several categories of ornamental fishes that are primarily distributed as a function of salinity. We carried out a survey at 10 stations distributed in the western and central sectors of Indian Sundarbans (Table 3.11; Fig. 3.64) during November, 2019 to January, 2020. We recorded 19 ornamental fish species during our survey, which exhibited significant distribution pattern based on the salinity of the ambient aquatic phase. It is interesting to note that the stations in the western sector of Indian Sundarbans namely, Harinbari (Stn. 1), Chemaguri (Stn. 2), Sagar South (Stn. 3), Lothian island (Stn. 4) and Prentice island (Stn. 5) showed the presence of all types of fishes, where as in the estuarine

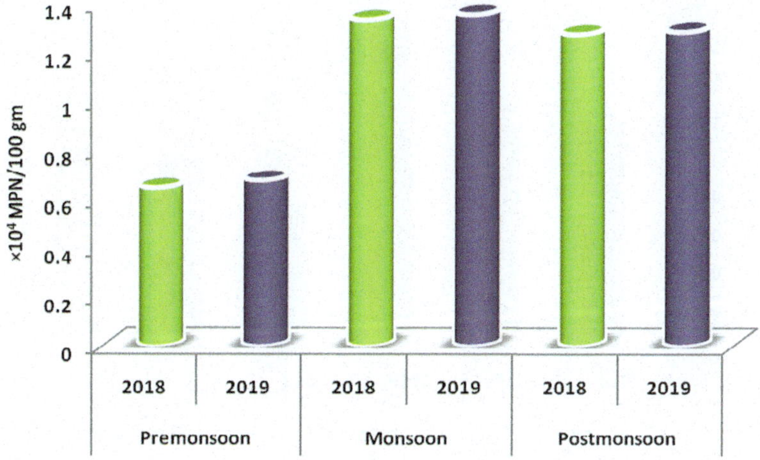

Fig. 3.51 Faecal coliform (MPN/100 g) in fish muscle in hyposaline pond

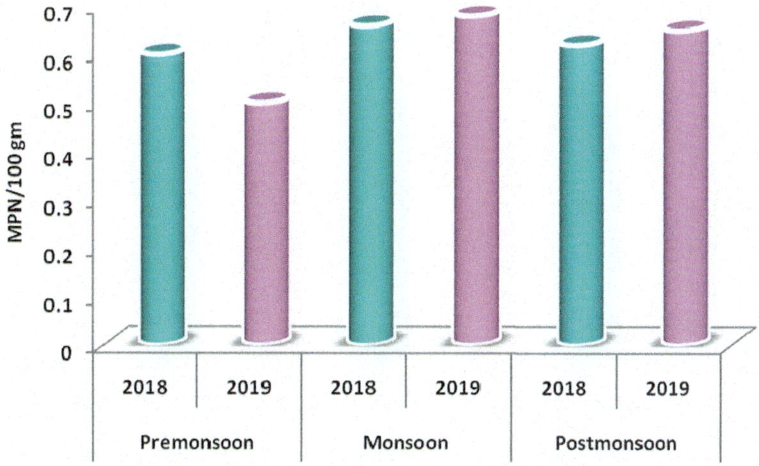

Fig. 3.52 Faecal coliform (MPN/100 g) in fish gill in hypersaline pond

water surrounding the stations of central Indian Sundarbans namely, Canning (Stn. 6), Sajnekhali (Stn.7), Chotomollakhali (Stn. 8), Satjelia (Stn. 9) and Pakhiralaya (Stn. 10) most of the species are absent (Table 3.12). The hypersalinity of central Indian Sundarbans might pose an adverse impact on the ornamental fish diversity in this sector of Indian Sundarbans.

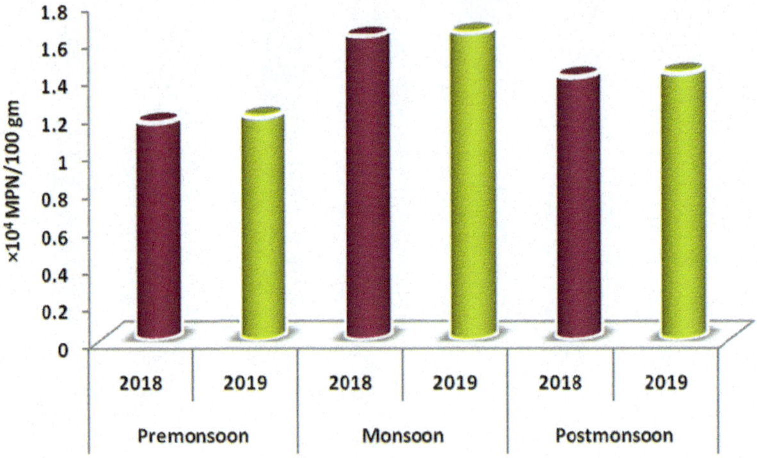

Fig. 3.53 Faecal coliform (MPN/100 g) in fish gill in hyposaline pond

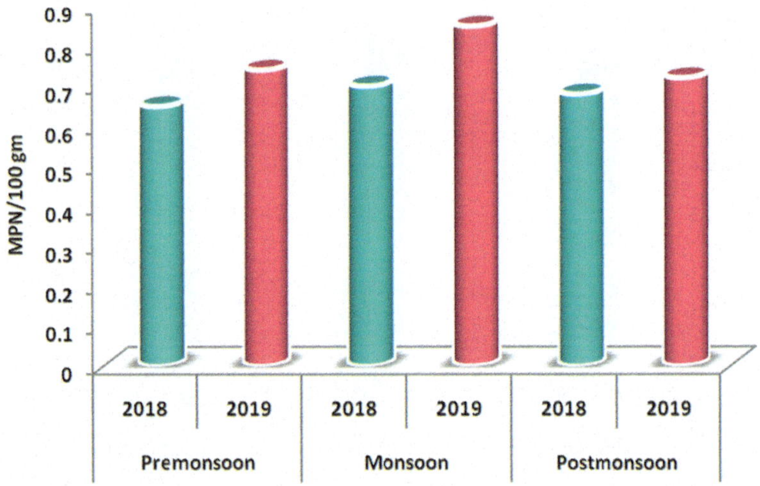

Fig. 3.54 Faecal coliform (MPN/100 g) in fish hepatopancreas in hypersaline pond

3.2 Mangrove Based Livelihoods

Mangroves constitute a small fragment of the total global forest cover, but still provide a wide range of livelihoods. The principal goods and services provided by mangroves may be very broad in dimension, but we will consider here only few innovative mangrove floral based livelihoods, not addressed so far in context to the frame work of Indian Sundarbans (Fig. 3.65).

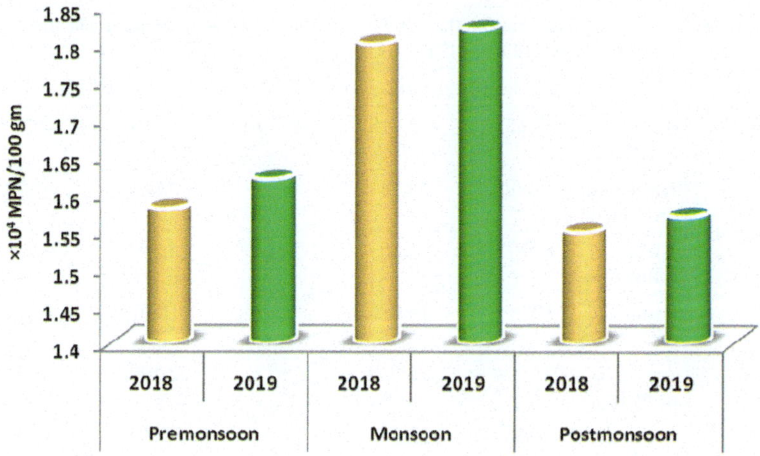

Fig. 3.55 Faecal coliform (MPN/100 g) in fish hepatopancreas in hyposaline pond

Table 3.6 Zoo-technical parameters related to condition of cultured species

Parameters	2018		2019	
	Hyposaline pond	Hypersaline pond	Hyposaline pond	Hypersaline pond
Survival rate (%)	63.9	51.5	65.6	49.9
FCR	1.02	2.54	1.17	2.96
SGR (% d^{-1})	1.88	1.79	1.58	1.42

3.2.1 *Preparation of Mangrove (*Sonneratia apetala*) Fruit Based Coconut Ladoo (Narkel Naru)*

Introduction

Fruits are known as protective food due to presence of vitamins (like vitamin A, vitamin B_1, vitamin B_2, vitamin B_3, vitamin B_6, and vitamin C), minerals (like sodium, potassium, zinc, calcium, magnesium, phosphorus etc.), antioxidants and phytochemicals. Regular consumption of fruits plays a *vital* role in nutrition as well as increase of life expectancy and some chronic diseases such as cardiovascular diseases, diabetes, cancer, hypertension, obesity, asthma and depression are cured. Fruits get easily spoiled due to contamination by different types of microorganisms during storage and transport. To avoid this spoilage, it is a common practice to preserve the fruit quality through making different by-products from the fruits. Fruit based desert {Coconut Ladoo (Narkel Naru)} is such type of by-product which was prepared in the present pilot scale alternative livelihood programmes from the fruits of a fresh water loving mangrove species, *Sonneratia apetala* during the Durga Puja festival of 2021. Ladoo is a ball shaped sweet desert which is very popular in Indian sub-continent. Flours, sugar and other ingredients are required to make ladoo (Naidu

Table 3.7 State wise area utilization (ha) and production (Tonnes) of tiger shrimp (*Penaeus monodon*) over last 5 years (MPEDA 2017)

State		2011–2012	2012–2013	2013–2014	2014–2015	2015–2016
West Bengal	Area utilization	48,558	48,410	48,730	48,848	50,593
	Estimated production	45,999	52,581	53,049	53,526	61,998
Orissa	Area utilization	8,597	6,256	5,523	5,392	4,552
	Estimated production	10,901	14,096	11,075	10,075	9,191
Andhra Pradesh	Area utilization	35,274	15,020	2,522	2,190	2,637
	Estimated production	51,081	25,948	2,883	2,962	3,739
Tamil Nadu	Area utilization	5,360	6,293	343	38	409
	Estimated production	12,097	17,220	916	73	1,103
Kerala	Area utilization	12,809	12,917	12,719	13,230	8,306
	Estimated production	8,138	5,175	3,360	3,643	3,490
Karnataka	Area utilization	650	240	94	689	1,948
	Estimated production	609	180	56	499	682
Goa	Area utilization	53	30	53	7	0
	Estimated production	51	48	14	16	0
Maharashtra	Area utilization	1,098	1,047	817	126	3
	Estimated production	1,721	2,010	1,083	178	6
Gujarat	Area utilization	1,971	1,992	1,375	881	398
	Estimated production	4,869	6,045	4,362	2,184	1,243
Total	Area utilization	114,370	93,110	72,177	71,400	68,846
	Estimated production	135,466	123,303	76,798	73,156	81,452

Table 3.8 State wise area utilization (ha) and production (Tonnes) of white leg shrimp (*Penaeus vannamei*) over last 5 years (MPEDA 2017)

State		2011–2012	2012–2013	2013–2014	2014–2015	2015–2016
West Bengal	Area utilization	0	0	130	326	1,387
	Estimated production	0	0	479	395	6,776
Orissa	Area utilization	25	46	485	2,340	4,439
	Estimated production	100	436	2,907	11,866	19,241
Andhra Pradesh	Area utilization	7,128	20,198	49,764	37,560	39,800
	Estimated production	75,385	133,135	210,639	276,077	295,332
Tamil Nadu	Area utilization	397	1,511	5,087	5,037	7,615
	Estimated production	2,863	8,595	26,281	32,688	44,453
Kerala	Area utilization	0	0	0	6	22
	Estimated production	0	0	0	12	74
Karnataka	Area utilization	72	154	157	125	333
	Estimated production	232	484	517	623	1,045
Goa	Area utilization	0	1	29	27	10
	Estimated production	0	15	67	88	33
Maharashtra	Area utilization	127	439	908	1,275	1,356
	Estimated production	941	1,503	3,291	4,901	6,118
Gujarat	Area utilization	88	367	707	3,545	4,154
	Estimated production	1,195	3,348	6,326	26,763	32,946
Total	Area utilization	7,837	22,716	57,267	50,241	59,116
	Estimated production	80,717	147,516	250,507	353,413	406,018

Table 3.9 Stocking density and initial body weight of each species used in combinations during 2019

Species	Stocking density (numbers ha^{-1})	Fish weight at stocking (g)
Tiger shrimp	20,000	0.02 ± 0.01
Parsia	5,000	0.39 ± 0.08
Grey mullet	5,000	3.29 ± 1.81
Milkfish	2,500	7.29 ± 1.41

Table 3.10 Growth parameters of tiger shrimp and fishes in Shrimp-parsia-grey mullet (T1) shrimp-parsia-milkfish (T2), shrimp-grey mullet-milkfish (T3) and shrimp-parsia-grey mullet-milkfish (T4) polyculture during 2019

Species	Combination	Final average body weight (g)	Daily weight gain (g day^{-1})	Specific growth rate (% day^{-1})	Fulton's condition factor (K)	Exponent of length weight relationship (b)
Tiger shrimp	T1	42.65 ± 1.25	0.18 ± 0.02	3.19 ± 1.13	0.90 ± 0.05	2.99
	T2	35.54 ± 1.95	0.15 ± 0.02	3.12 ± 1.30	0.71 ± 0.02	2.94
	T3	31.63 ± 1.42	0.13 ± 0.02	3.07 ± 1.23	0.63 ± 0.02	2.90
	T4	44.17 ± 1.53	0.20 ± 0.02	3.21 ± 1.37	1.15 ± 0.04	3.07
Parsia	T1	54.16 ± 6.34	0.22 ± 0.037	1.21 ± 0.15	1.13 ± 0.03	3.01
	T2	40.27 ± 4.98	0.16 ± 0.029	1.34 ± 0.28	0.96 ± 0.03	2.84
	T4	45.24 ± 5.58	0.19 ± 0.026	1.29 ± 0.26	0.96 ± 0.28	3.04
Grey mullet	T1	487.17 ± 20.42	1.97 ± 0.30	1.44 ± 0.15	0.74 ± 0.03	3.18
	T3	367.33 ± 13.55	1.48 ± 0.21	1.33 ± 0.16	0.58 ± 0.02	2.91
	T4	367.52 ± 14.84	1.43 ± 0.20	1.32 ± 0.14	0.56 ± 0.02	2.97
Milk fish	T2	382.91 ± 18.39	1.48 ± 0.17	1.18 ± 0.20	0.67 ± 0.02	3.10
	T3	366.35 ± 14.94	1.43 ± 0.20	1.17 ± 0.19	0.53 ± 0.01	3.05
	T4	311.43 ± 13.22	1.21 ± 0.25	1.10 ± 0.12	0.51 ± 0.01	2.99

et al. 2012). It is a good source of elements like fibre, iron, calcium, phosphorus, carotene, niacin and amino acids. Coconut ladoo (coconut is used as main ingredient) is mainly prepared during Durga Puja, one of the Hindu festivals and locally this sweet is known as "Narkel Naru". Durga Puja is one of the most famous festivals celebrated in West Bengal and particularly in Kolkata, for worshipping Goddess Durga during the period of Navaratri (Fig. 3.66). It is celebrated for 10 days. However starting from the sixth day upto the ninth day, the Pandals (structures where the Goddesses are kept) with grand idols of Goddess Durga is open for the visitors. The tenth day, also known as Dashami marks the Visarjan (immersion in water) of the idol with grand celebrations and processions. On the last day, distribution of this dessert, coconut ladoo takes place in the social gatherings and in 2021, we prepared this ladoo with the fruit extract of *Sonneratia apetala*.

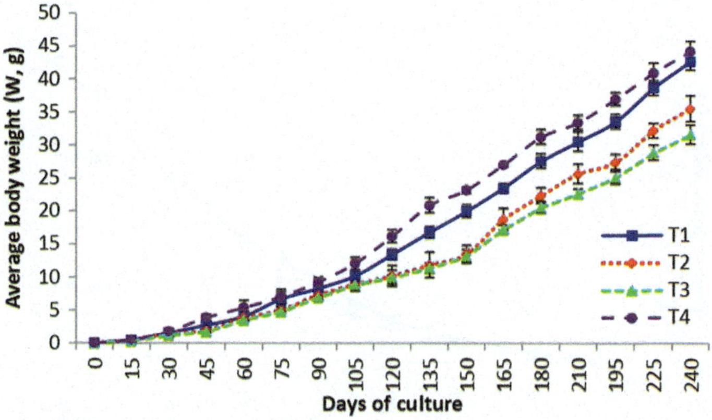

Fig. 3.56 Growth patterns of tiger shrimp in shrimp-parsia-grey mullet (T1), shrimp-parsia-milkfish (T2), shrimp-grey mullet-milkfish (T3) and shrimp-parsia-grey mullet-milkfish (T4) polyculture during 2019

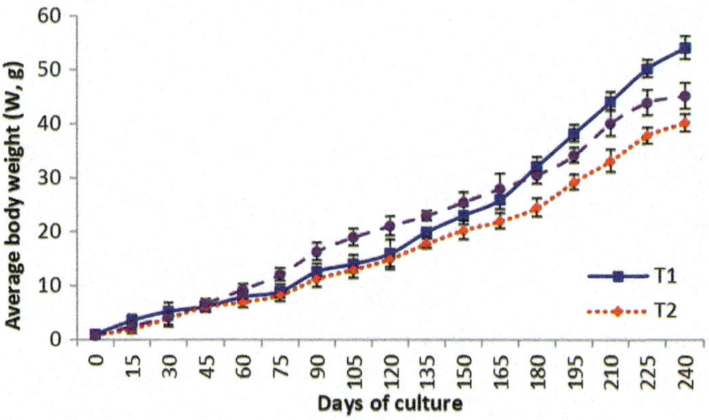

Fig. 3.57 Growth patterns of parsia in shrimp-parsia-grey mullet (T1), shrimp-parsia-milkfish (T2), and shrimp-parsia-grey mullet-milkfish (T4) polyculture during 2019

Sonneratia apetala, belongs to the family Lythraceae, and the local name of this species is Keora. The species is also found in some coastal regions of Bangladesh, Myanmar, Malaysia and Africa. *S. apetala* fruit is mainly available in rainy season (August to October). The island dwellers of Sundarbans consume this fruit in raw form and also eat by making chatni/curry. Some years back, *S. apetala* fruit based jelly was also prepared by a group of researchers of Techno India University, West Bengal (Pramanick et al. 2014a, b).

In the present study, we have modified the formulation of the coconut ladoo by mixing the fruit pulp of *S. apetala* with all other ingredients. The aim of the present pilot project was to evaluate the nutritional value of the *S. apetala* fruit pulp along with the prepared fruit based coconut ladoo in terms of protein, carbohydrate and

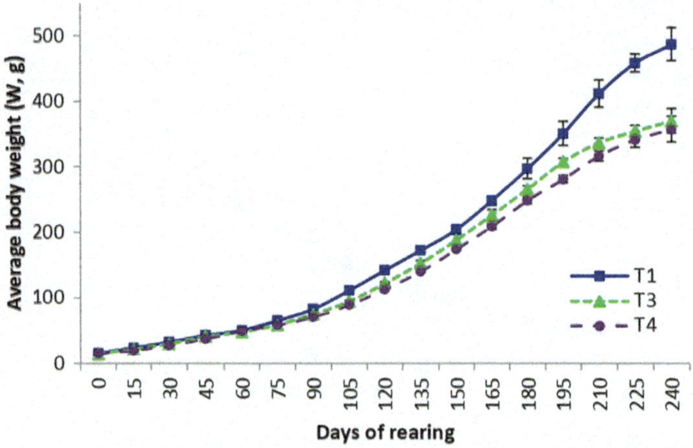

Fig. 3.58 Growth patterns of grey mullet in shrimp-parsia-grey mullet (T1), shrimp-grey mullet-milkfish (T3) and shrimp-parsia-grey mullet-milkfish (T4) polyculture during 2019

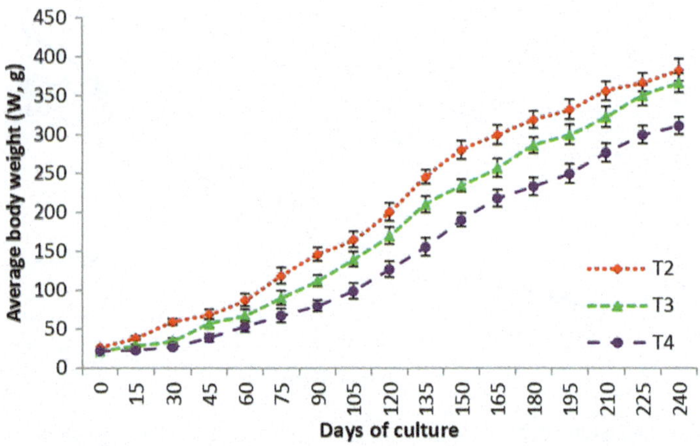

Fig. 3.59 Growth patterns of milkfish in shrimp-parsia-milkfish (T2), shrimp-grey mullet-milkfish (T3) and shrimp-parsia-grey mullet-milkfish (T4) polyculture during 2019

vitamin C. Microbial population was also monitored in terms of Total Coliform in both the fruit and prepared coconut ladoo samples.

Materials and Methods

Collection site

Sonneratia apetala is one of the true mangrove trees in Indian Sundarbans and mostly found in the hyposaline environment of western and eastern sectors of Indian Sundarbans (Mitra 2000, 2013; Mitra and Zaman 2015, 2016). For the present research

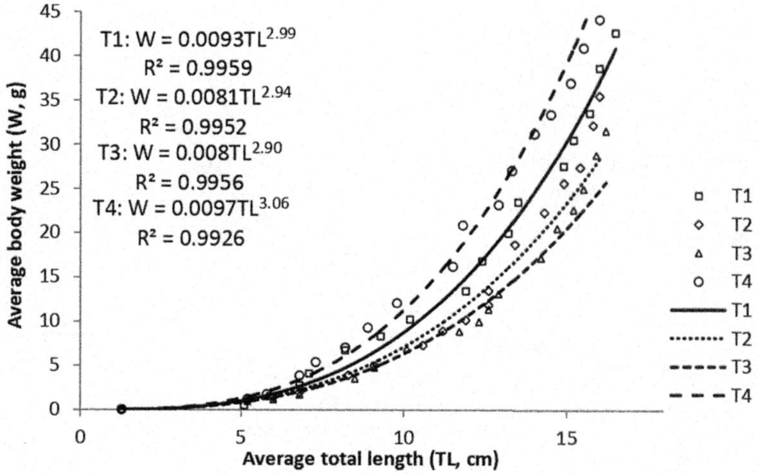

Fig. 3.60 Length–weight relationship of tiger shrimp in shrimp-parsia-grey mullet (T1), shrimp-parsia-milkfish (T2), shrimp-grey mullet-milkfish (T3) and shrimp-parsia-grey mullet-milkfish (T4) polyculture during 2019

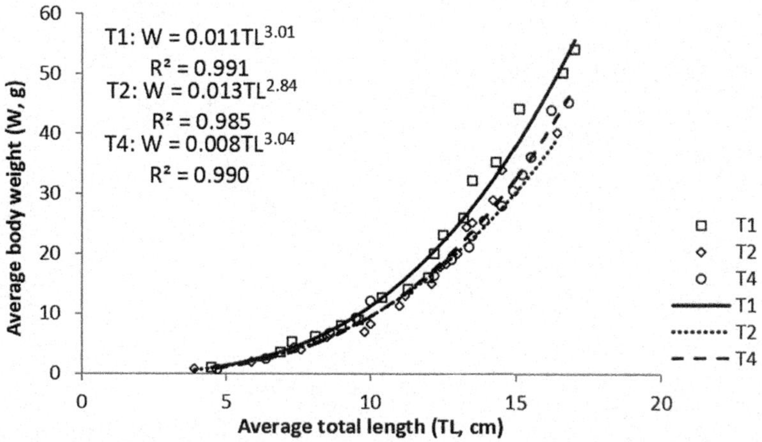

Fig. 3.61 Length–weight relationship of parsia in shrimp-parsia-grey mullet (T1), shrimp-parsia-milkfish (T2), and shrimp-parsia-grey mullet-milkfish (T4) polyculture in extensive system during 2019

work, we have collected *S. apetala* fruits from the Chemaguri (21°39′42.88″ N; 88°08′49.01″ E) in the western Indian Sundarbans.

Greenish coloured unripe *S. apetala* fruit becomes green-yellowish coloured after ripening in the month of August. During the last week of September 2021, fruits were collected from the inter-tidal mudflats of Chemaguri and washed them in normal water (Fig. 3.67).

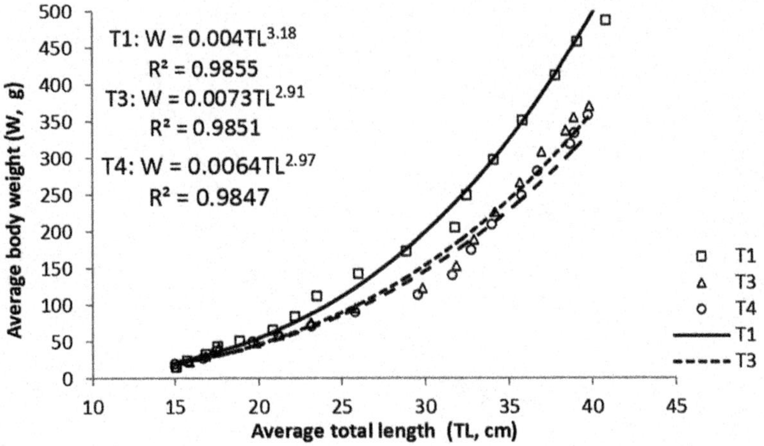

Fig. 3.62 Length–weight relationship of grey mullet in shrimp-parsia-grey mullet (T1), shrimp-grey mullet-milkfish (T3) and shrimp-parsia-grey mullet-milkfish (T4) polyculture in extensive system during 2019

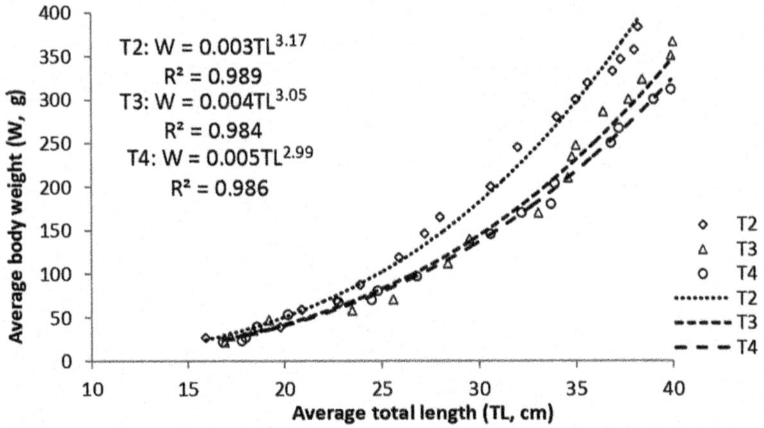

Fig. 3.63 Length–weight relationship of milkfish in shrimp-parsia-milkfish (T2), shrimp-grey mullet-milkfish (T3) and shrimp-parsia-grey mullet-milkfish (T4) polyculture in during 2019

Ladoo preparation

Coconut ladoo is a traditionally prepared sweet ball with fresh shredded coconut and other ingredients (Fig. 3.68). In the preparation process, at first the pulp was extracted from the collected fruits and then dried in hot-air oven and grinded in a mixer to make pulp powder. In the next phase, dehydrated coconut was grated and condensed milk with grated coconut were added in a frying pan and stirred until well mixed. Sugar and fruit pulp powder were added and continuously stirred. After

Table 3.11 Sampling stations with salient features

Study site	Latitude and longitude	Site description
Harinbari (Stn. 1)	21°46′53.07″ N; 88°04′22.88″ E	Situated in the western region of Indian Sundarbans almost in the middle of the Sagar Island; receives the water of the Hugli River
Chemaguri (Stn. 2)	21°39′42.88″ N; 88°08′49.01″ E	Situated on the south-eastern side of Sagar Island and receives the water of the Mooriganga River
Sagar South (Stn. 3)	21°37′49.90″ N; 88°04′ 0.51″ E	Situated on the south-western part of the Sagar Island at the confluence of the River Hugli and the Bay of Bengal. Anthropogenically stressed zone due to presence of passenger jetties, fishing activities and pilgrimage
Lothian island (Stn. 4)	21°39′08.04″ N; 88°19′8.47″ E	Situated east of Bakkhali island; a Wildlife sanctuary; faces the River Saptamukhi
Prentice island (Stn. 5)	21°42′43.31″ N; 88°17′3.62″ E	Situated north of Lothian island; receives the water of the Saptamukhi River
Canning (Stn. 6)	22°19′03.20″ N; 88°41′04.43″ E	Situated in the central part of the Indian Sundarbans and faces the mighty River Matla, a tide-fed river. Due to presence of fish landing stations, passenger jetties and busy market, the area is anthropogenically stressed
Sajnekhali (Stn. 7)	22°06′34.19″ N; 88°48′15.78″ E	A Wildlife Sanctuary and a part of Sundarban Tiger Reserve; adjacent to River Bidhya and Gomor. Tourism pressure is extremely high in this station particularly during postmonsoon
Chotomollakhali (Stn. 8)	22°10′40.00″ N; 88°54′26.71″ E	Situated in the upper portion of Central Indian Sundarban adjacent to Jhila forest; receives the water of Rangabelia and Korankhali Rivers
Satjelia (Stn. 9)	22°05′17.86″ N; 88°52′49.51″ E	Situated adjacent to River Duttar in the upper region of Central Indian Sundarban facing western part of the Jhilla forest
Pakhiralaya (Stn. 10)	22°07′07.23″ N; 88°48′29.00″ E	Situated adjacent to River Gomor; opposite to Sajnekhali Wild Life Sanctuary

cooling, the mixer was poured in another bowl and shaped into ball shaped ladoos (Fig. 3.69). The entire process is schematically diagramed in the Fig. 3.68.

Analysis

The collected *S. apetala* fruit and the prepared fruit based coconut ladoo were analyzed in the laboratory for their nutritional properties and microbial population in terms of Total Coliform.

Analysis of crude protein and total carbohydrate of the samples were done by Kjeldahl method (AOAC 1990) and phenol–sulphuric acid method (Dubois et al. 1956) respectively. Vitamin C of the fruit and ladoo were estimated by the standard protocol (AOAC 1995). The concentrations of protein and carbohydrate were expressed in percentage whereas vitamin C was expressed in mg/100 g.

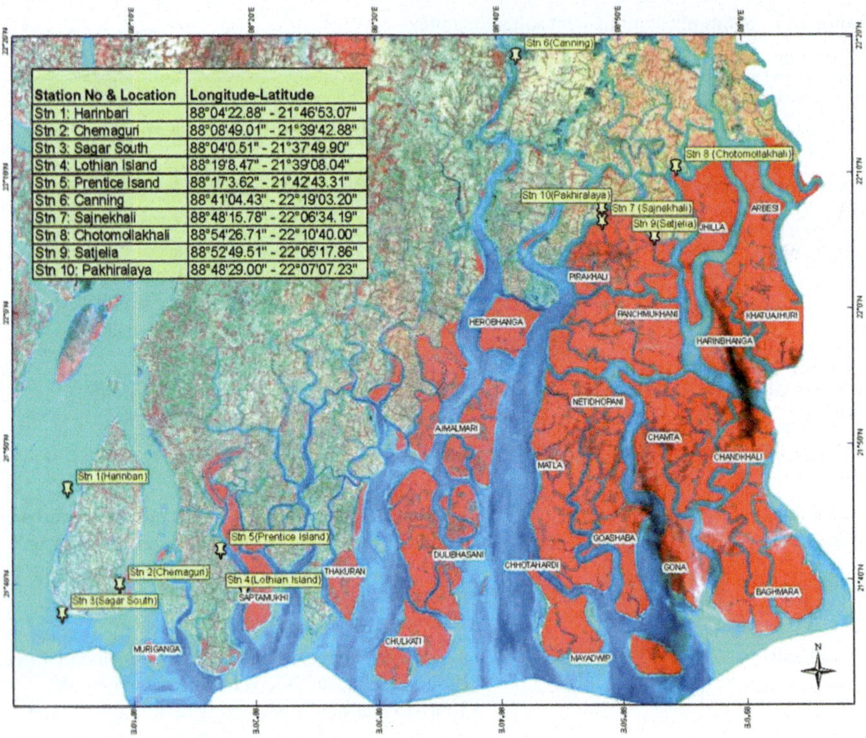

Fig. 3.64 Selected stations in Indian Sundarbans

Microbial populations (in terms of Total Coliform) in both the fruit and ladoo were analyzed by Standard MPN method using Lauryl Tryptose Broth (LTB) and Brilliant Green Bile Broth (BGLB) broths (FAO 1992) and the results were expressed in CFU/g.

Result

The concentrations of protein, carbohydrate and vitamin C of the fruit and prepared ladoo were presented in Figs. 3.70, 3.71 and 3.72 respectively and the order is fruit > ladoo for protein and vitamin C whereas in case of carbohydrate, the order is reverse. In case of Total Coliform, the value was less (27.23×10^4 CFU/g) in the prepared ladoo (Fig. 3.73) and hence may be considered safe from the human consumption point of view.

Discussion

In the human diet chart, fruits are known to boost immunity. The island dwellers, who do not have the facility to meet the physician frequently, take their dietary supplement from the mangrove fruits like *Sonneratia apetala* fruit. The protein content of this mangrove fruit is 0.41% that is similar to the apple but higher than the citrus fruits

Table 3.12 Distribution and availability of different types of ornamental fishes in Indian Sundarbans documented during our survey from November, 2019–January, 2020

Species	Stn.1	Stn.2	Stn.3	Stn.4	Stn.5	Stn.6	Stn.7	Stn.8	Stn.9	Stn.10
Colisa fasciata	Abundant	Moderately available	Rare	Abundant	Abundant	Absent	Abundant	Rare	Rare	Moderately available
Badis badis	Moderately available	Rare	Moderately available	Abundant	Abundant	Moderately available	Moderately available	Moderately available	Abundant	Abundant
Oryzias malastigma	Abundant	Abundant	Rare	Abundant	Abundant	Rare	Abundant	Absent	Rare	Moderately available
Brachygobius sua	Abundant	Moderately available	Abundant	Rare	Rare	Absent	Rare	Abundant	Moderately available	Abundant
Lepidocephalichthys guntea	Rare	Rare	Abundant	Abundant	Abundant	Absent	Abundant	Moderately available	Moderately available	Moderately available
Chanda ranga	Moderately available	Abundant	Moderately available	Rare	Rare	Rare	Abundant	Moderately available	Absent	Absent
Chanda nama	Rare	Moderately available	Moderately available	Abundant	Abundant	Absent	Rare	Absent	Moderately available	Moderately available
Brachydanio rerio	Moderately available	Moderately available	Rare	Abundant	Moderately available	Rare	Abundant	Rare	Moderately available	Abundant
Scatophagus argus	Abundant	Abundant	Abundant	Moderately available	Moderately available	Rare	Abundant	Absent	Absent	Moderately available
Ailia coila	Rare	Moderately available	Moderately available	Moderately available	Rare	Absent	Moderately available	Absent	Absent	Absent
Gobiopterus chun	Abundant	Moderately available	Abundant	Abundant	Abundant	Rare	Moderately available	Rare	Rare	Moderately available
Puntius ticto	Abundant	Rare	Rare	Rare	Moderately available	Absent	Moderately available	Absent	Absent	Rare
Puntius stigma	Moderately available	Moderately available	Rare	Moderately available	Abundant	Rare	Rare	Rare	Moderately available	Rare
Chela laubuca	Rare	Moderately available	Abundant	Moderately available	Moderately available	Rare	Moderately available	Moderately available	Absent	Rare
Notopterus notopterus	Moderately available	Rare	Rare	Rare	Rare	Moderately available	Absent	Absent	Absent	Absent
Nandus nandus	Moderately available	Abundant	Abundant	Moderately available	Moderately available	Rare	Rare	Absent	Moderately available	Moderately available
Aplocheilus panchax	Rare	Moderately available	Abundant	Abundant	Abundant	Absent	Absent	Rare	Moderately available	Rare
Mastacembelus pancalus	Rare	Moderately available	Rare	Rare	Moderately available	Rare	Moderately available	Moderately available	Rare	Absent
Puntius sarana	Abundant	Abundant	Moderately available	Rare	Rare	Absent	Moderately available	Absent	Abundant	Abundant

INDEX

ABSENT ABUNDANT MODERATELY AVAILABLE RARE

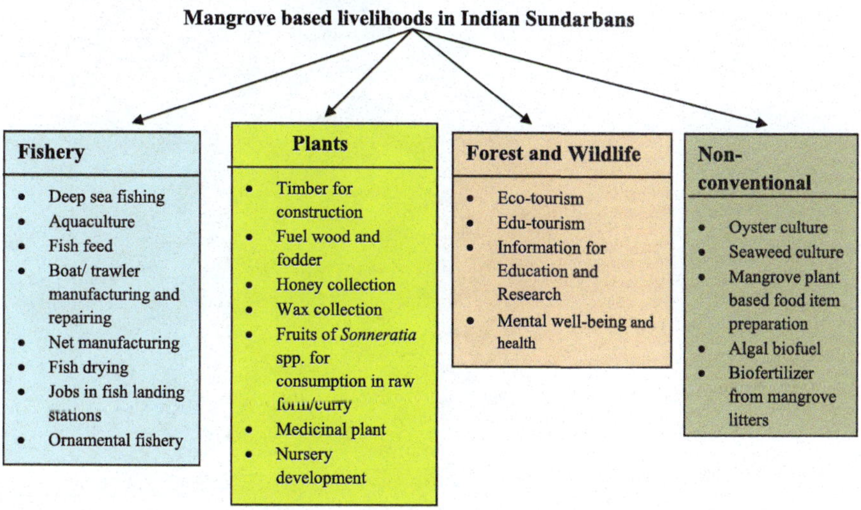

Fig. 3.65 Major mangrove based livelihoods in Indian Sundarbans

Fig. 3.66 Goddess Durga with 10 arms (Dashabhuja)

Fig. 3.67 *Sonneratia apetala* fruits after washing with normal water

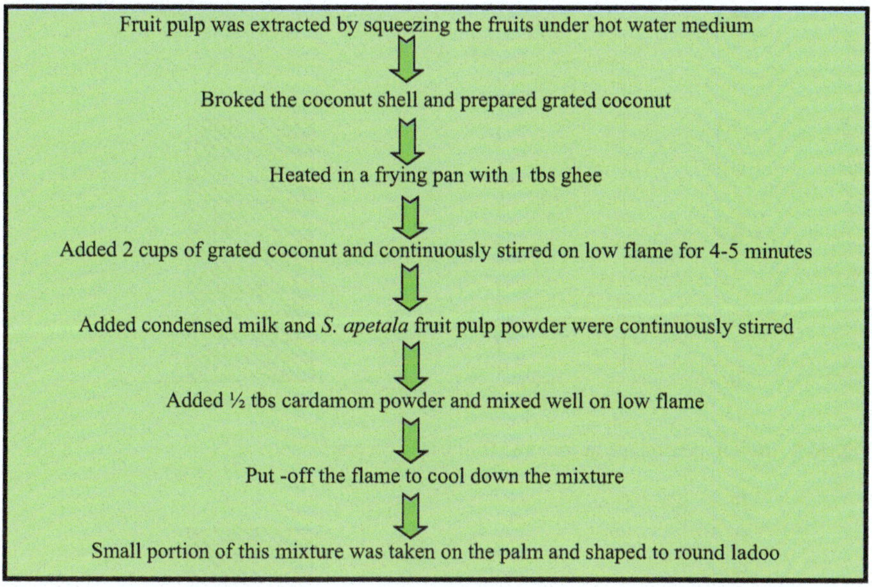

Fig. 3.68 Schematic diagram of the preparation of *S. apetala* fruit based coconut ladoo

Fig. 3.69 Preparation of *S. apetala* based coconut ladoo

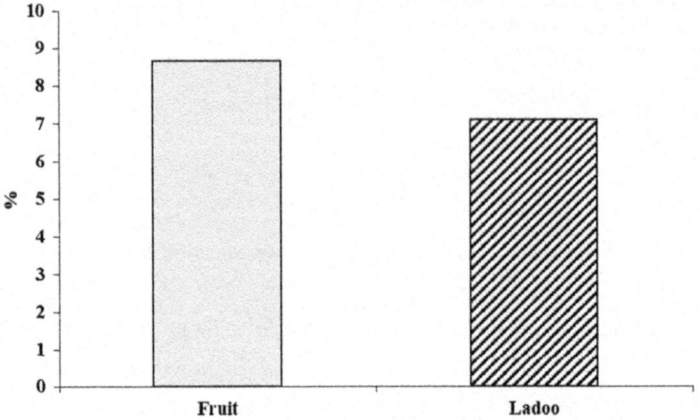

Fig. 3.70 Protein percentage in fruit and fruit based coconut ladoo

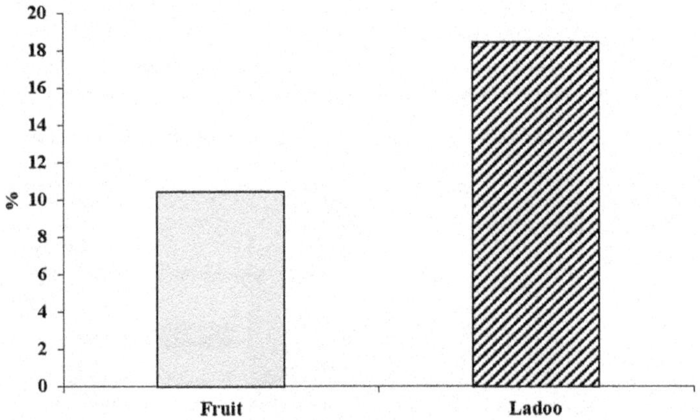

Fig. 3.71 Carbohydrate (total sugar) percentage in fruit and fruit based coconut ladoo

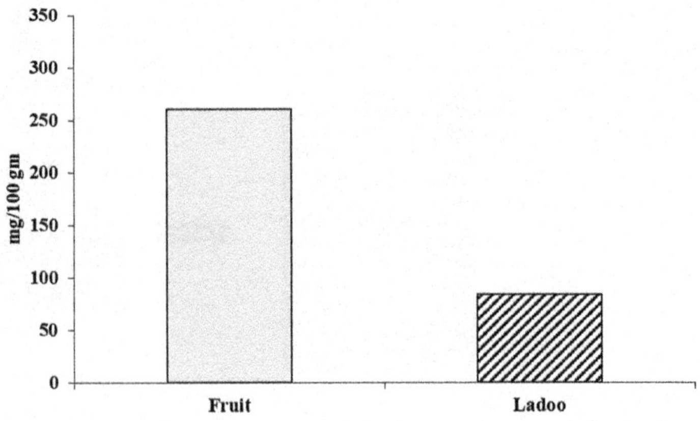

Fig. 3.72 Concentrations of Vitamin C (mg/100 g) in fruit and fruit based coconut ladoo

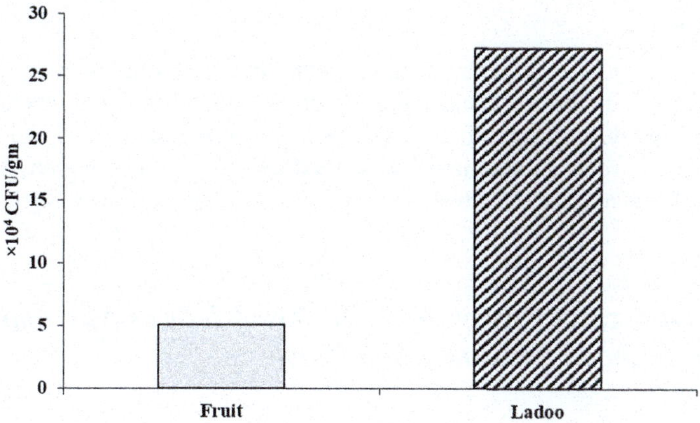

Fig. 3.73 Total Coliform (CFU/g) in *S. apetala* fruit and prepared ladoo

(Kazi et al. 2015). Protein is made up of amino acids' chains which are linked with peptide bonds. During preparation of coconut mixer on low flame, these peptide linkages were broken and therefore protein percentage was relatively less in the ladoo.

Carbohydrate is an important food component which acts as energy source of the different physiological processes. In the *S. apetala* fruit, sugar content is very low. But when the fruit based coconut ladoo is prepared, some extra sugar is added and therefore sugar content gets elevated. The main role of the sugar in the fruit products is to preserve the water content in the fruit and also to inhibit the growth of mould.

Another important food component is vitamin C which protects the body from cough and cold. Vitamin C is not synthesized in the human body, and hence it is taken from the external sources (like fruits) to meet the dietary requirement. *S. apetala* fruit is an example of the source of vitamin C. The concentrations of vitamin C in some fruits are 218 mg/100 g in guava, 150 mg/100 g in blackberry, 40 mg/100 g in lemon (Kazi et al. 2015) whereas in *S. apetala* fruit, the value is comparatively higher (261 mg/100 g). As vitamin C is very sensitive to heat and light, so the vitamin C level in the prepared ladoo is reduced to 84 mg/100 g, but still high in comparison to many other fruits.

During the harvesting and post-harvesting periods, fruits are contaminated by some microorganisms present in the ambient environment, which causes their spoilage. Fruit spoilage decreases the quality and shelf-life and in this project total coliform count is used as an indicator. We observed that ladoo prepared from the fruits exhibited higher level of Total Coliform compared to the fruit, which may be the impact of hand rolling during the process of ladoo preparation. Preparation of ladoo in aseptic condition using UV chamber may reduce the chance of contamination.

Conclusion

The present study reveals that both the *S. apetala* fruit and fruit based coconut ladoo are rich sources of nutrition. This unique desert has great demand in several festivals of India mainly during worship of Godesses like Durga, Laxmi, Kali and Saraswati and thus can serve as an alternative livelihood for the local population, provided a nursery of *S. apetala* can be developed for constant supply of raw materials.

3.2.2 Sonneratia Caseolaris *Fruit Based Jelly as Alternative Livelihood for Sundarban People*

Introduction

Fruits are good sources of fiber, minerals, vitamin C, organic acids, pigments and phytochemicals. They are known for boosting immunity potential and play significant roles in human nutrition and prevent some chronic non-communicable diseases. Vitamin C (ascorbic acid) has an important role in prevention of cough and cold. Doctors in the present times are also suggesting the intake of Vitamin C to combat the attack of COVID-19. Fruits are also the reservoir of several minerals that regulate various metabolic processes in human beings.

Sonneratia caseolaris is a true mangrove floral species in Indian Sundarbans, and belongs to Lythraceae family (Fig. 3.74). *S. caseolaris* is also known as ‘mangrove apple’ or ‘crab apple’ mangrove. It grows on the muddy area. The species prefer to grow in the low saline belt of Sundarbans. The fruit of this species is flattened with star like structure and is a rich source of pectin (Fig. 3.74). The fermented *S. caseolaris* fruit juice is used in the prevention of haemorrhage. The half-ripe fruit

Fig. 3.74 *Sonneratia caseolaris* with fruits in Indian Sundarbans

is used in the treatment of cough. For the treatment of worms, the old fruit walls are used. The unripe fruit is sour in taste and island dwellers consume it by making curries and chutnies (local term for a sweet sour jelly like food items with less viscosity). The ripe fruits are cheese like with sweetish sour taste and eaten as raw or processed. Jelly is one of the processed food items that are prepared from the fruits and save the food item from the microbial contamination. The present pilot project was undertaken during the COVID-19 phase of 2020 with the aim to prepare vitamin C rich jelly to provide natural vitamin C to the people by developing a sound supply chain.

Materials and Methods

Collection Site

Sonneratia caseolaris is mainly found in the hyposaline region where average salinity level is <10 psu. We collected the fruits from Chandkhali (21°51′13.59″ N; 89°00′44.68″ E) located in the eastern Indian Sundarbans.

Fruits of *S. caseolaris* become ripe during monsoon and early postmonsoon season (August–November). The greenish coloured fruits were collected from the selected site during October 2020 and then washed with normal water.

Jelly preparation

Fruit pulp extraction is the first step for jelly preparation (Fig. 3.75). For this extraction, the collected fruits were tied with a fresh cloth and boiled for 20 min and after cooling, they were squeezed to get the pulp portion. In the next step, we prepared the jelly (Fig. 3.76) through slight modification of the protocol as developed by Madhav

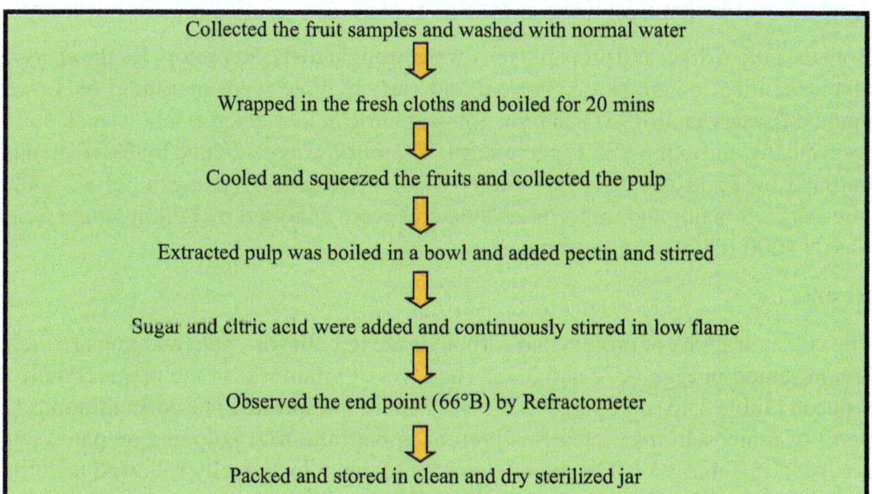

Fig. 3.75 Diagrammatic representation of Jelly preparation from *S. caseolaris* fruits

Fig. 3.76 Jelly prepared from *S. caseolaris* fruit

and Pushpalatha (2002). The entire flow chart for jelly preparation is highlighted in the Fig. 3.75.

Analysis of biochemical composition

Both the pulp extract and prepared jelly were brought to the laboratory for the analysis of biochemical composition. Protein and carbohydrate were measured by Lowry method (Lowry et al. 1951) and Phenol–sulphuric acid method (Dubois et al. 1956) respectively and expressed as percentage. Vitamin C was estimated by DCIP method outlined by Food Analysis Laboratory Manual Chapter 7 (Zvaigzne et al. 2009). The selected major and minor/trace elements were analysed by Perkin-Elmer Sciex ELAN 5000 ICP mass spectrometer.

Results

The concentrations of protein and carbohydrate in both fruit pulp and prepared jelly are presented in Figs. 3.77 and 3.78. The level of vitamin C in the prepared jelly is reduced (Table 3.13) compared to the raw fruit of the species. The concentrations (in ppm) of major and trace elements present in both the fruit pulp and prepared jelly are listed in Table 3.13. The major elements of the pulp and jelly followed the order K > Na > Ca > Mg. In case of trace elements, the order in the pulp is Zn > Cu > Co > Mo and in the prepared jelly it is Zn > Cu. Co and Mo was Below Detectable Level

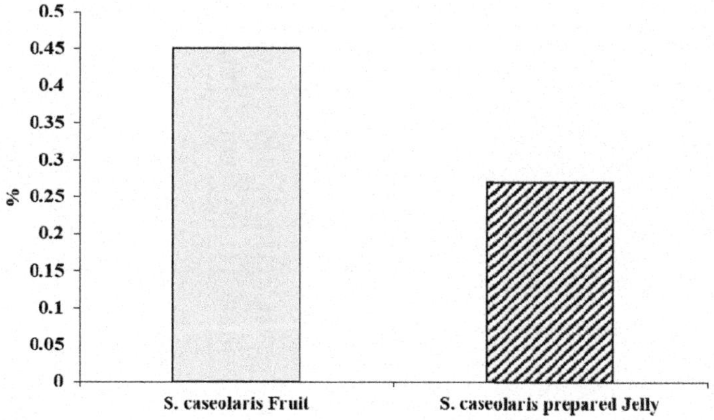

Fig. 3.77 Protein content in *S. caseolaris* fruit pulp and prepared jelly

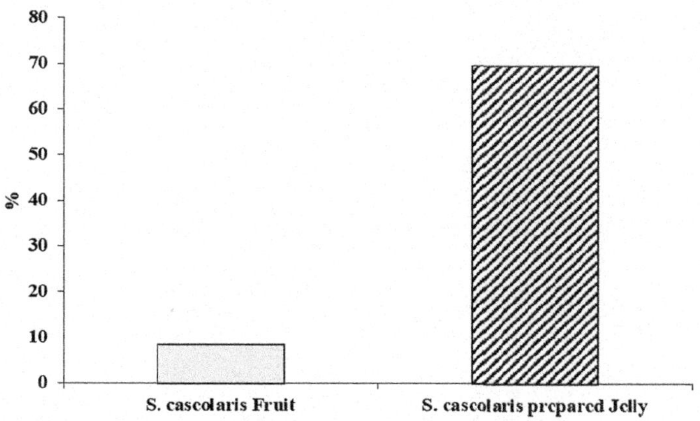

Fig. 3.78 Carbohydrate content in *S. caseolaris* fruit pulp and prepared jelly

(BDL) in the prepared jelly. These values are compared with the standard level of some fruits to evaluate the nutritional value of the mangrove fruit (Table 3.13).

Discussion

In the recent time people have become very health conscious and take quality food, less fat, sugar, low calories and maintain safety and optimum nutritional balance. Table 3.14 represents the important role of some nutrients in the human body. Protein is one of the nutrients in the human body. It is made up of amino acid chains which are linked by the peptide bonds. Most of the fruits contain low amount of protein and only 0.45% protein was found in the *S. caseolaris* fruit's pulp. The peptide bonds were broken down when the pulp was heated and therefore protein percentage in the jelly was reduced to 0.27%.

Table 3.13 Comparison of the major and trace elements and vitamin C

Fruits	Vitamin C (mg/100 g)	Major elements (ppm)				Trace elements (ppm)			
		Na	K	Mg	Ca	Zn	Cu	Co	Mo
Lemon	44.5	20	1,160	70	220	0.5	0.31	–	–
Lime	19.5	10	680	40	220	0.7	0.44	–	–
Orange	69.7	–	2,370	130	520	0.9	0.59	–	–
Grape fruit	79.1	–	3,200	180	280	1.6	1.08	–	–
S. caseolaris fruit pulp (our study)	437.41	9291.12	17,431.86	1,444.44	2,720.47	23.42	13.27	0.39	0.05
S. caseolaris jelly (our study)	46.83	2,996.93	3,007.87	1,162.40	75.37	16.64	11.15	BDL	BDL

"BDL" means below detectable level; "–" means absence of data in standard literature

Table 3.14 Importance of vitamin C and major and trace elements

Ingredients		Functions
Vitamin C		Vitamin C is a water soluble antioxidant and has a great role to prevent cough and cold. Deficiency of vitamin C can lead to anemia, scurvy, infections, bleeding gums, muscle degeneration, delayed wound healing and neurotic disturbances
Major elements	Na	Na is one of the essential elements required in an appropriate amount in daily diet to regulate the blood pressure and nerve processes for transmitting impulses through the body as well as stabilize the water in human body cells
	K	K is one of the essential elements of human diet, and plays important role in vital cellular mechanism. It acts as a cofactor to catalyze the conversion of the ADP to ATP and also plays a role in excitability of the nerve
	Mg	Mg plays an important role in nervous system stability, muscle contraction as activator of alkaline phosphatase
	Ca	Ca helps in the development and growth of skeletal system e.g. bones and also teeth. It acts as coenzyme in metabolic regulations of biomolecules
Trace elements	Zn	Zn mainly acts as a coenzyme of about 200 enzymes involved in immunity, new cells growth, acid base regulation etc. Lack of sufficient amount of Zn results in reduced activity of related enzymes e.g. carbonic anhydrase
	Cu	Cu is an essential trace element and is required as cofactor in different oxidative and reductive enzymes
	Co	Co is an indispensable element and contributes to vitamin B_{12}
	Mo	Epidemiologic findings have implicated molybdenum deficiency in the incidence of esophageal cancer in Africa, China and Russia. Cancer is often caused by xenobiotic compounds. The molybdoenzymes xanthine oxidase, aldehyde oxidase and sulfite oxidase may be involved in the detoxification of xenobiotic compounds

Carbohydrate is one of the most important macromolecules and plays a significant role in the human nutrition. Metabolic energy comes from carbohydrates. In the present programme, we observed that 69.53% carbohydrate is present in the prepared jelly, which is comparatively higher than the raw fruits due to addition of sugar while mixing the pulp extract.

Vitamin C, an antioxidant helps to fight against cough and cold. It aids in the wound healing operation through the formation of protein in the skin, tendons, ligaments etc. In the present study we observed that vitamin C level in the *S. caseolaris* fruit pulp is comparatively higher than other citrus fruits (Table 3.14). However, this level is reduced in the prepared jelly due to structural deformation of vitamin C at high temperature, although the value is slightly higher than lemon (Table 3.13).

Sodium (Na) is an electrolyte which is required in the human body in small amount. It maintains the balance of water and minerals and also helps in the muscle contraction and nerve transmission. Too much intake of sodium causes high blood pressure, heart diseases etc. The standard range of sodium in the fruits is 200 ppm (in apricot) to 4,800 ppm (in olive pickles) (http://wholefoodcatalog.info/nutrient/sodium/fruits/8). Both pulp and prepared jelly contains higher amount sodium than other fruits.

Potassium (K) mainly maintains the fluid balance by regulating the action of kidneys. The standard range of potassium level in the fruits is 570.00–1,407.20 ppm (http://wholefoodcatalog.info/nutrient/potassium/fruits/8). In the prepared jelly, this level is exceeded may be due to addition of potassium meta-bisulphite as preservative.

Magnesium (Mg) is the second most abundant intracellular cation which plays an important role in the cellular function. More than 300 enzymatic reactions in the human body are regulated by magnesium. It contributes to the structural development of bone and is also involved in the DNA and RNA synthesis. The Mg level in the mangrove fruits is higher than other citrus fruits (Table 3.13).

Most fruits are rich in calcium, which is very much important for the physiological and biological functions in the human body. It has a big role in blood clotting and regulates the normal heart rhythms. The calcium content in the *S. caseolaris* fruit pulp is much higher than many citrus fruits.

Trace elements are present in small amount in the fruits and they act as coenzymes, but their concentrations above the permissible level in food items may pose an adverse impact on human health. In the present study, the prepared jelly contains 16.64 ppm Zn and 11.15 ppm Cu and there is no trace of Co and Mo confirming the fruit to be free of heavy metals. Zinc content in the *S. caseolaris* pulp is 23.42 ppm, which is relatively higher than many other citrus fruits. Zn has a great role in wound healing, bone mineralization etc.

Conclusion

The present study generates few core findings as highlighted here.

1. *Sonneratia caseolaris* has considerable high concentration of vitamin C compared to other citrus fruits, which is also reflected in the jelly prepared from the fruit pulp extract.
2. The fruit as well as the jelly are rich sources of major and trace elements.
3. The low concentrations of Cu and complete absence of Co and Mo in the jelly minimizes the chance of toxic effects of these elements on human beings.
4. *Sonneratia caseolaris* is available only during the monsoon and early post-monsoon season; hence there may be a shortage of raw material if the jelly preparation from pulp is considered from the industrial point of view. However, this small-scale cottage industry can be translated into reality if storage facilities can be developed in the islands of Sundarbans. To provide sustainability to this alternative livelihood several backup nurseries of *S. caseolaris* need to

be created through peoples' participation. During nursery raising of *S. caseolaris*, care should be taken to create a congenial habitat. As the species thrive luxuriantly with *Avicennia* spp., therefore it is recommended that the nursery should be created in low saline soil considering the optimum biotic and abiotic parameters required for the growth of the species.

3.3 Mangrove Associate Based Alternative Livelihoods

We have already stated that mangrove associate flora like saltmarsh grass (*Porteresia coarctata*) and several other halophytes like *Suaeda* sp. and *Salicornia* sp. constitute an integral part of mangrove ecosystem and render a wide array of ecosystem services like erosion control, protection against hypersalinity, wave attenuation, stability of intertidal mudflats, bioremediation, carbon sequestration etc. However, in this section we will discuss few innovative approaches, which have not been carried out so far in the domain of mangrove-centric livelihoods. These approaches are still in pilot scale, but have excellent potential to upgrade the island economy of Indian Sundarbans.

3.3.1 Prawn Feed from Salicornia Brachiata

Introduction

Prawn (*Macrobrachium rosenbergii*) farming is a commonly used practice in all fresh water bodies of Sundarbans. The species is often cultured with carp varieties. Traditional way of culture is widespread, but there is lack of proper scientific backup, which often causes deterioration of water bodies and loss in production. The fresh water prawn culture has attracted majority of the villagers in recent times due to less risk and low cost involved in the farming process. To make the venture more profit oriented, we initiated fresh water prawn farming at Jharkhali and Bony Camp with specially formulated feed of mangrove origin. The feed formulation was developed on the basis of protein and carotenoid level of the floral components to boost up the prawn production volume and make the cultured species disease resistant. Such model can be a pathway of alternative livelihood programme for the fringe population of Indian Sundarbans, who otherwise enter illegally into the forest for procuring wood, timber, honey, wax, fishes, prawn seeds etc.

An overview of the project

The present experiment was initiated at two different areas of Indian Sundarbans namely Jharkhali (22°05′52.82″ N; 88°41′47.25″ E) and Bony camp (21°49′42.90″ N; 88°37′13.70″ E). Jharkhali is situated in the central sector of Indian Sundarbans, which is at the fringe of Sundarban Tiger Reserve. Most of the people of this region are illegal intruders, which often poses threat to the adjoining mangrove forest. Culture of freshwater prawn (*Macrobrachium rosenbergii* Fig. 3.79) with specially formulated

Fig. 3.79 Freshwater prawn *Macrobrachium rosenbergii*

floral (*Salicornia brachiata*) based feed in this region during 2017 was an attempt to defray the island dwellers from entering into the forest and get involved in the alternative livelihood scheme. Upgradation of aquatic (pond) environment through replacement of animal component of the fish feed with endemic floral ingredient was another aim of the programme.

Bony Camp, under Block Ajmalmari 7 is situated on the bank of the river Sundarikathi, a distributary of Thakuran River. This is a tourist spot and the pond in this camp was treated as control (Fig. 3.80).

The unique feature of the programme is the blending of biotechnology with the livelihood component of the island dwellers. The animal ingredients (trash fish dust, shrimp dust, squids, etc.) of the traditional prawn feed (which usually deteriorate the water quality) were replaced by *Salicornia brachiata*. The dust (in dried form) of this mangrove associate species was used as the major source of protein and carotenoid (Table 3.15). The specially formulated prawn feed has the hallmark with respect to the following points:

- Replacement of the animal ingredients of the fish feed with floral components upgrades the water quality in terms of nutrients and ammonia.
- Enhancement of the growth and immune system of the cultured species through incorporation of protein and carotenoid from *Salicornia brachiata*.

Carotenoids are a class of natural fat-soluble pigments found principally in plants, algae and photosynthetic bacteria, where they play a crucial role in the photosynthetic process. They are also found in some non-photosynthetic bacteria, yeasts and

Fig. 3.80 Control pond in Bony camp under Block Ajmalmari 7 is situated on the bank of the river Sundarikathi, a distributary of Thakuran River in central Indian Sundarbans

Table 3.15 Prawn feed composition

Component	Control (g)	Experimental (g)	%
Rice bran	250	250	50
Wheat bran	250	250	
Fish meal	150	*Salicornia brachiata* dust (dried form) = 300	30
Shrimp meal	150		
Sun flower oil cake	50	50	10
Mustard oil cake	50	50	
Vitamin + Min. mixture	50	50	5
Binder	30	30	3
Shark oil	20	20	2

molds where they may carry out a protective function against damage by light and oxygen. Although animals appear to be incapable of synthesizing carotenoids, many animals incorporate carotenoids from their diet. Within animals, carotenoids provide bright colouration, serve as antioxidants and can be a source for vitamin A activity (Ong and Tee 1992; Britton et al. 1995). Carotenoids are defined by their chemical structure. The majority of carotenoids are derived from carbon polyene chain which could be considered as the backbone of the molecule. The hydrocarbon carotenoids are known as carotenes, while oxygenated derivatives of hydrocarbons are known as xanthophylls. β-carotne, the principal carotenoid in carrots is a carotene, while lutein, the major yellow pigment of marigold petals is a common xanthophyll. In our experimental feed, carotenoids have been incorporated from the leaves of *Salicornia brachiata.*

Protein is usually the most expensive nutrient and protein content in feed ranges from around 18% to over 45%. The three important routes through which protein boosts up the growth of the cultured species are:

1. Essential amino acid requirements
2. Overall digestibility of dietary proteins and
3. Feed consumption level.

The most common protein sources used in commercial feed are fish meal and shrimp meal. These are reasonably well digested by prawns. Ideally protein, which is the most expensive component in feed, is used primarily for growth while cheaper components are used to meet the energy demands. If carbohydrate level in the feed is low, prawns use proteins to support their metabolic function other than growth. Hence protein supplement plays a *vital* role in prawn feed. It is interesting to note that *Salicornia brachiata* contains about 18% protein, which was mixed with the feed as a protein supplement instead of fish meal used in traditional/commercial feed.

Methodology

Two ponds were selected for the present programme in which the culture tenure was six months (March 2017–August 2017). Pond P_1 at Bony Camp was treated as control pond and pond P_2 at Jharkhali island was treated as experimental pond. Stocking density of 2 individual/m^2 was maintained in these two ponds (P_1 and P_2) having an approximate area of 1,800 and 1,550 m^2 respectively. Out of the two selected ponds, the control pond (P_1) was provided with traditional prawn feed and the experimental pond was provided with *Salicornia brachiata* dust (Table 3.15) based fish feed. Information on the economics of freshwater prawn farming is limited and only very few studies have been conducted (Senthilathiban et al. 2001). Hence a cost–benefit analysis of the programme was also carried out after harvesting the cultured species to ensure the viability of such approach.

Output of the pilot project

The two ponds considered in the present study exhibited significant variations with respect to production, survival rate, FCR value, Condition Index (CI) value and physico-chemical variables (Table 3.16). The lower nitrate, phosphate and organic carbon level in the experimental pond may be related to the replacement of the animal feed component (trash fish dust) by plant protein (obtained from *Salicornia brachiata*). The weight of the prawn in the experimental pond was also higher (50 g) in comparison to the control pond (42 g).

The cost-benefit analysis (Table 3.17) shows a higher profit per unit area in the experimental pond (Rs. 13.31) in comparison to the control pond (Rs. 8.07).

Table 3.16 Hydro-biological parameters of the culture ponds

Item	Control (P1)	Experimental (P2)
Stocking density	2/m^2	2/m^2
Culture period	6 months	6 months
Weight at harvest	42 g	50 g
Feed conversion ratio (FCR)	1.24	1.19
Condtion index (C. I)*	1.23	1.72
Dissolved oxygen (D.O) (mg/l)*	4.5	6.2
Salinity (‰)*	4.5	4.6
pH*	8.2	8.0
Temperature (°C)*	29.8	30.2
Transparency (cm)*	16.1	25.2
Soil organic carbon (SOC) (%)*	3.85	2.14
Nitrate (μg at/l)*	28.29	18.10
Phosphate (μg at/l)*	6.10	1.25
Silicate (μg at/l)*	69.26	28.62

* Average of 6 months; March, 2017 to August, 2017

Table 3.17 Cost benefit analysis of the project (expected)

	Items	P_1 (Area = 1,500 m^2)	P_2 (Area = 1,800 m^2)
Cost	Number of seeds	3,000	3,600
	Seed cost (in Rs)	1,350.00	1,620.00
	Feed quantity (in kg)	269.01	375.73
	Feed cost (in Rs)	9,415.35	10,520.44
	Experimental cost	6,000.00	6,000.00
	Labor/management cost	2,000.00	2,000.00
	Total cost (in Rs)	18,765.35	20,140.44
	Total unit cost (in Rs/m^2)	12.51	11.19
Benefit	Return (in kg)	88.20	126.00
	Return (@ Rs. 350/kg)	30,870.00	44,100.00
	Expenditure (in Rs)	18,765.35	20,140.44
	Total profit/pond (in Rs)	12,104.65	23,959.56
	Profit/unit area (in Rs/m^2)	8.07	13.31

Conclusion

The Sundarbans mangrove region is a threatened ecosystem due to multitude of factors like prolonged over-exploitation of its natural resources, its use as a sink of anthropogenic wastes, industrial and maritime wastes generated in the upstream of the rivers flowing through the region, high population pressure around the region and the resultant shrinkage of the area brought about by clearing of forest land for agriculture and tiger prawn culture and lack of proper ecological management. The present programme has immense ecological and economic relevance to combat these threats, due to its connection to the following lanes:

- Upgradation of only the freshwater system (ponds, ditches etc.) and therefore clearance of mangrove areas for the culture of *Penaeus monodon* may be totally avoided. *Marcobrachium rosenbergii* can also be cultured in the rain water harvested pond.
- Involvement of the local people in three livelihood tiers: preparation of fish feed, ecofriendly culture practice and nursery development of *Salicornia brachiata* for raw material backup to sustain the floral based fish feed industry. It is expected that such involvement will restrict a sizable fraction of the people from intruding into the forest.
- Scientific utilization of mangrove floral resources for sustainable pisciculture practice in the area.
- Improvement of aquatic health in terms of physico-chemical parameters due to replacement of animal ingredients (like trash fish dust, shrimp dust etc.) in the traditional fish feed with floral components.
- Introduction of a new technology befitted to the area.
- Economic upliftment of the local people.

3.3.2 Mangrove Associate Floral Based Popular Indian Street Food Item (Golgappe/Phuchka)

Introduction

Golgappe is one of street foods commonly known as 'pani puri' in Indian sub-continent. Depending on the region, it has been coined in different names. In North India it is known as Golgappe, in West Bengal as 'phuchka', in Maharashtra, Gujarat, Madhya Pradesh, Karnataka, Tamil Nadu as 'pani puri', in Haryana as 'paanike patashe', in eastern part of Uttar Pradesh as 'phulki', in Odisha, Jharkhand, Chhatisgarh, Hyderabad as 'gup chup', in Aligarh as 'padaka' etc. It is a small sized spherical spicy flour based item called puri (diameter 4–6 cm) that consists of a mixture of potato, onion, chilli, chaat masala (spices), chickpeas, tamarind chutney (tamarind juice mixed with spices). The dry puri is dipped in tamarind water and served in paper plates or folded sal (*Shorea robusta*) leaves. In the state of West Bengal, the phuchka vendors use gandhoraj lebu (a special type of lemon of Bengal) for flavoring. Mangrove associate floral based food items have a significant role in the fulfillment of nutrition of the island dwellers of Indian Sundarbans (Biswas et al. 2019). In the present pilot project, we have used two varieties of mangrove associate species namely *Suaeda maritima* and *Salicornia brachiata* collected from Indian

Sundarbans. They were used in the development of phuchka or golgappe that can be established as an alternative livelihood for island dwellers.

Suaeda maritima (Fig. 3.81) is a salt marsh plant and salty in taste in raw form and hence it is cooked with other types of vegetables to reduce the salty taste (Tanaka 1976). In some region this species is used in salad, curry with crabs and various other types of chilli based preparations. Somewhere again (like in South Indian states) it is pickled in vinegar and used for cooking (Patra et al. 2011). It is locally called as 'giria shak' or 'gire shak'. *S. maritima* is basically non-toxic and contains antioxidant. Polyphenolic compounds are the major bioactive components in the species which are responsible for antioxidation and antiproliferation.

Salicornia brachiata is widely distributed in the supra-littoral zone of the marine and estuarine ecosystems (Fig. 3.82). It is mainly used as an edible food product preferably in salad. Like *S. maritima*, this species is salty in taste. *Salicornia* sp. has been identified to be the right candidate species for reclamation of barren lands, salt flats, and sea shores. In short, they can be deemed for seawater agriculture.

The present pilot-scale mangrove associate based alternative livelihood programme aims to prepare phuchka by using these two mangrove associate species.

Fig. 3.81 *Suaeda maritima* at the collection site

Fig. 3.82 *Salicornia brachiata* at the collection site

Materials and methods

Collection site

In central Indian Sundarbans, the Chotomollakhali island (22°10′40.00″ N; 88°54′26.71″ E) was selected for the collection of two selected mangrove associate species. This part of Indian Sundarbans is hypersaline in nature due to siltation of the Bidyadhari since the fifteenth century (Chaudhuri and Choudhury 1994; Mitra 2000, 2013, 2019; Banerjee et al. 2013; Raha et al. 2013; Sengupta et al. 2013; Mitra and Zaman 2014, 2016, 2020) and hence offers a favourable environment for the growth of these mangrove associate species.

Suaeda maritima and *S. brachiata* were collected during the low tide phase from the study site during December, 2020 and cleaned with luke warm water and processed for the next step.

Preparation of Phuchka

The collected *S. maritima* and *S. brachiata* were dried in hot air oven and then grinded into powder form and kept in separate containers (Fig. 3.83). Dough was made with flour and baking soda and divided into 4 portions and each portion was transformed to a round structure by using small handy kitchen wood roller. These were then fried in oil to prepare spherical puris. Boiled potatoes, onion, ginger, chilli powder, spices, peas, coriander leaves and salt were added in appropriate quantity and mixed well. *S. maritima* and *S. brachiata* powder were mixed separately and the phuchkas were finally prepared (Fig. 3.84). The flowchart of the entire process is presented in Fig. 3.83.

A. Preparation of Puri

Preparation of dough by mixing flour, baking soda, salt and water

⇓

Kneaded the dough and kept for 15 mins by covering a wet cloth

⇓

Dough was divided into 4 portions and each portion was turned to a round structure by using small handy kitchen wood roller

⇓

Round pieces were fried in oil to make small spherical puris and kept in a bowl

B. Preparation of Puri Masala

Potatoes were boiled and the outer skin was peeled off and mashed properly

⇓

Onion, ginger, chilli powder, spices, peas, salt were added to mashed potatoes and mixed well and divided into two parts

⇓

Collected *S. maritima* and *S. brachiata* were dried and grinded separately and made into powder form

⇓

Dried powder of *S. maritima* and *S.brachiata* were added separately

⇓

Tamarind chutney and coriander leaves were added to both parts to make food item tasty

C. Preparation of spicy water

Mint and coriander leaves were rinsed in water and grinded with ginger, lemon juice, chilli powder

⇓

Transferred to a large bowl and little sugar, spice powder, black salt, tamarind and water were added and stirred and mixed properly

⇓

A hole was made on the top portion of the each spherical puri using thumb and stuffed with the mixed spice inside it and dipped into prepared spicy water and served

Fig. 3.83 Flowchart of the preparation of *S. maritima* and *S. brachiata* based phuchka/golgappe

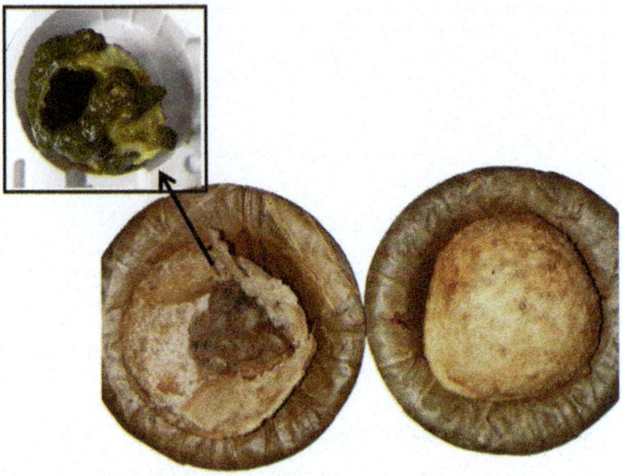

Fig. 3.84 Mangrove associate based phuchka/golgappe

Analysis of biochemical constituents

Both the collected halophyte samples *S. maritima* and *S. brachiata* and two types of phuchka prepared from them were analyzed in the laboratory for the estimation of biochemical constituents namely protein (Lowry et al. 1951), carbohydrate (Dubois et al. 1956), fat (AOAC 2000), fibre, calcium as per the standard protocols. Sodium and potassium were also analyzed by Flame Photometric Method.

Results and Discussion

Our laboratory analysis showed the presence of all the selected biochemical constituents in relatively higher concentrations in mangrove associate based phuchka compared to the normal phuchka (considerd as control) (Table 3.18; Figs. 3.85 and 3.86).

The food products developed from *Suaeda* sp. and *Salicornia* sp. are not only rich in nutrients (preferably Na, K and Ca), but can upgrade the snack industries of Indian Sundarbans at cottage level. This can be an alternative livelihood for the

Table 3.18 Biochemical constituents of halophyte species and popular Indian street food item (golgappe/phuchka)

Halophyte and food type	Protein (g/100 g)	Carbohydrate (g/100 g)	Fat (g/100 g)	Fibre (g/100 g)	Ca (mg/g)	Na (mg/g)	K (mg/g)
Normal phuchka	1.1	17.23	1.23	1.77	0.31	1.17	0.91
S. maritima	3.55	8.33	1.67	3.05	3.37	21.07	5.98
S. maritima based phuchka	4.47	23.89	2.19	4.31	3.4	21.55	6.11
S. brachiata	3.15	5.13	1.42	2.19	3.02	17.92	2.96
S. brachiata based phuchka	4.06	21.35	2.03	3.14	3.17	18.13	3.47

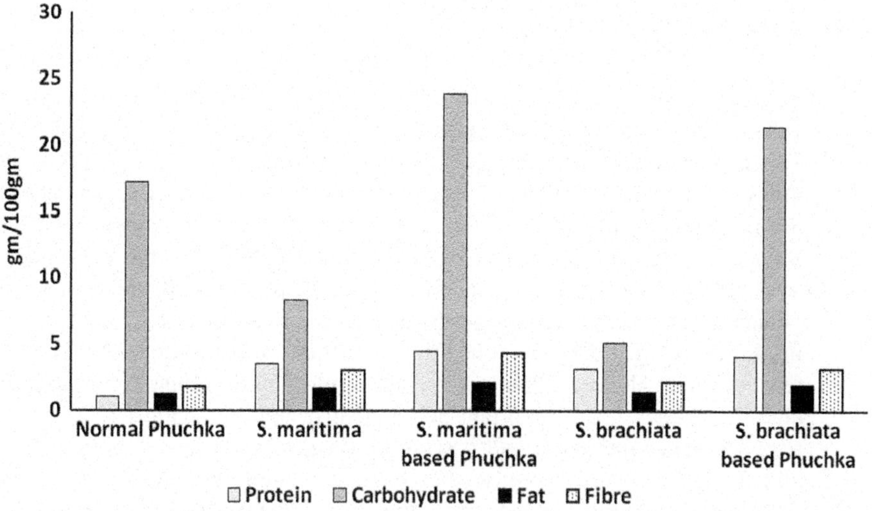

Fig. 3.85 Concentration (in g/100 g) of biochemical parameters in species and prepared products

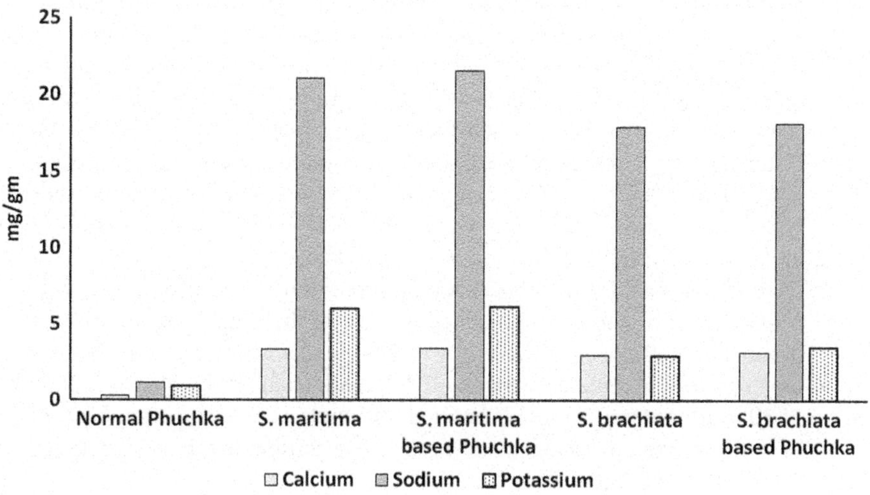

Fig. 3.86 Concentration (in mg/g) of minerals in species and prepared products

island dwellers and local inhabitants of the deltaic complex. Thus the present pilot project may serve as the foundations of both health and economic security in the regional livelihood sector of blue economy.

3.4 Take Home Messages

[A] According to the World Resources Institute, about 39% of the World's population (which is around 7.9 billion in 2022), is approximately 3.0 billion people. This considerable chunk of global population lives on or within 100 km (60 miles) of the seashore. Their lives and livelihoods are totally dependent on the resources available from the marine and estuarine ecosystems. It has been documented that poverty stricken people (preferably the poor island dwellers) of Indian Sundarbans are more susceptible to adverse impact of climate change. They have no proper shelter, no fund for resettlement and no insurances against their lives, health and properties. Institutional help hardly reaches them. Hence, reduction of poverty through development of sustainable livelihoods is an important component in fighting against temperature rise, sea water intrusion, and disease outbreaks, which are the clutches of climate change that frequently, scratch the economics of the region.

[B] In the estuarine region and islands of India Sundarbans, there are large number of poor marine fishermen who primarily depend on the Hooghly-Matla estuarine complex and the adjacent sea for their livelihoods. A wide variety of commercially important fishes are available in and around the region preferably in the seas, bays, estuaries, creeks and inlets, but several risk/cost factors are involved in the sector of capture fisheries like (i) man-animal conflict, (ii) natural calamities like cyclones, (iii) wave actions due to high wind in the deep sea, (iv) excessive rise of fossil fuel cost in recent times etc. The economic benefit is greater in aquaculture practice than traditional fishing and hence there is a general tendency of fisherman population to shift from capture fishery to aquaculture. Fishermen along with other dwellers of Indian Sundarbans have adopted brackishwater farming with remarkable success which has become an important tool for livelihood development in such a vulnerable eco-region. Monoculture of shrimp is practiced in most of the states of India for its high productivity. *Penaeus monodon*, which is highly susceptible to diseases, is now replaced by *Penaeus vannamei*. We think from our long term experience of 30 years that polyculture is befitted for Sundarban deltaic complex because of low risk from diseases, better production, minimum environmental issues and year round production.

[C] Mangroves constitute a small fragment of the total global forest cover, but still provide a wide range of livelihoods. The principal goods and services provided by mangroves are very wide in dimension, but we have considered here only few innovative mangrove floral based livelihoods, not addressed so far by any researchers in context to the frame work of Indian Sundarbans. These include preparation of mangrove (*Sonneratia apetala*) fruit based coconut ladoo (Narkel Naru) and *Sonneratia caseolaris* fruit based jelly, which may sound 'odd man out', but has tremendous potential to develop eco-friendly sustainable mangrove based livelihood for the people of Sundarbans.

[D] Mangrove associate flora like saltmarsh grass and several other halophytes like *Suaeda* and *Salicornia* constitute an integral part of mangrove ecosystem and render a wide array of ecosystem services like erosion control, protection against hyper salinity, wave attenuation, stability of intertidal mudflats, bioremediation, carbon sequestration etc. However, these associate species can also generate few innovative alternative livelihoods like prawn feed from *Salicornia brachiata*, mangrove associate floral based popular Indian street food item (Golgappe/Phuchka) etc., which are befitted to the region. However, a sound supply chain and a nursery backup are needed to make these innovations a success, which are still in pilot scale.

References

AOAC (1990) Association of official analytical chemists, official methods of analysis of AOAC International. Washington DC, U.S.A, p 77

AOAC (1995) Official methods of analysis of the association of official analytical chemists

AOAC (2000) Official methods of analysis, 17 edn. The Association of Official Analytical Chemists, Gaithersburg, MD, USA

Banerjee K, Sengupta K, Raha A, Mitra A (2013) Salinity based allometric equations for biomass estimation of Sundarban mangroves. Biomass Bioenergy (Elsevier) 56:382–391

Barua P, Mitra A, Banerjee K, Shah Nawaz Chowdhury M (2011) Seasonal variation of heavy metal accumulation in water and oyster (*Saccostrea cucullata*) inhabiting central and western sector of Indian Sundarbans. Environ Res J 5(3):121–130

Bhattacharyya SB, Panigrahi A, Mitra A, Mukherjee J (2010) Effect of physico-chemical variables on the growth and condition index of the rock oyster *Saccostrea cucullata* (Born) in the Sundarbans India. Indian J Fish 57(3):13–17

Biswas S, Pal N, Zaman S, Mitra A (2019) Samosa and Kachuri from Mangrove associate species: an innovative utilization of coastal flora. Int J Res Appl Sci Eng Technol 7(1):73–79

Britton G, Liaaen-Jensen S, Pfander H (1995) Carotenoids today and challenges for the future. In: Britton L-J, Pfander H (eds) Carotenoids: isolation and analysis, vol 1A. Birkhauser, Basel

Chakraborty A, Datta U, Zaman S, Pal N, Pramanick P, Mitra A, Mitra A (2016) Spatio-temporal variation of stored carbon in *Sonneratia apetala*: a case study from Indian Sundarbans. Int J Innov Res Sci Eng Technol 5(3):2975–2980

Chaudhuri AB, Choudhury A (1994) Mangroves of the Sundarbans. IUCN, India

Chopin T, Yarish C (1999) Seaweeds must be a significant component of aquaculture for an integrated ecosystem approach. Bull Aquacult Assoc Can 99:35–37

Dubois M, Gilles KA, Hamilton JK, Rebers PA, Smith F (1956) Colorimetric methods for determination of sugars and related substances. J Anal Chem 28:350–356

Dutta P, Pramanick P, Zaman S, Mitra A (2020) Stored carbon in the phytoplankton community of western and central Indian Sundarbans. In: Mitra A, Calma MM, Chakrabarty SP (eds) Proceedings: natural resources and their ecosystem services. HSRA Publication, pp 146–151. ISBN: 978-81-947216-7-3

Enin U (1994) Length–weight parameters and condition factor of two West African prawns. Rev Biol Trop 27:121–127

Food and Agriculture Organization (FAO) (1992) Manual of quality control microbiological analysis. 4(1):1–338

Granada L, Sousa N, Lopes S, Lemos MFL (2016) Is integrated multitrophic aquaculture the solution to the sectors' major challenges?—a review. Rev Aquacult 8:283–300. https://doi.org/10.1111/raq.12093

http://wholefoodcatalog.info/nutrient/potassium/fruits/8

http://wholefoodcatalog.info/nutrient/sodium/fruits/8

Islam S, Biswas S, Zaman S, Mitra A (2018) Vulnerability assessment in the Western Indian Sundarbans ecosystem. Technol Int J Health Eng Man Sci 2:1–4

Kazi NA, Yadav JP, Agale MG (2015) Nutritional value of fruits. S Res J Interdiscip Stud III(XVI):2937–2943

Lowry OH, Rosebrough NJ, Farr AL, Randall RJ (1951) Protein measurement with the Folin-Phenol reagents. J Biol Chem 193:265–275

Madhav A, Pushpalatha PB (2002) Quality degradation of jellies prepared using pectin extracted from fruit wastes. J Trop Agric 40:31–34

Martínez-Porchas M, Martínez-Córdova LR, Porchas-Cornejo MA, López-Elías JA (2010) Shrimp polyculture: a potentially profitable, sustainable, but uncommon aquacultural practice. Rev Aquac 2:73–85

Mehta GS (2009) Inland fisheries development the way forward in West Bengal. Fish Chimes 29(7):49–59

Mitra A (2000) The north west coast of the Bay of Bengal and Deltaic Sundarbans. Seas at the millennium: an environmental evaluation, U.K., vol II, pp 145–160

Mitra A (2013) Sensitivity of mangrove ecosystem to changing climate. Springer, New Delhi, Heidelberg, New York, Dordrecht, London. ISBN-10: 8132215087; ISBN-13: 978-8132215080; ISBN 978-81-322-1509-7 (eBook)

Mitra A (2019) Estuarine pollution in the lower Gangetic delta, vol XVI. Springer International Publishing, p 371. ISBN 978-3-319-93305-4

Mitra A, Bhattacharyya DP (2003) Environmental issues of shrimp farming in mangrove ecosystem. J Indian Ocean Stud 11(1):120–129

Mitra A, Pramanick P, Zaman S, Fazli P, Pal N, Mitra A (2015) Response of *Sonneratia Apetala* to salinity in the frame work of Indian Sundarbans. Int J Innov Stud Aquat Biol Fish 2(1):1–5

Mitra A, Zaman S (2014) Carbon sequestration by coastal floral community. The Energy and Resources Institute (TERI) TERI Press, India. ISBN: 978-81-7993-551-4

Mitra A, Zaman S (2015) Blue carbon reservoir of the blue planet, vol XII. Springer, pp 299. https://doi.org/10.1007/978-81-322-2107-4. ISBN: 978-81-322-2106-7

Mitra A, Zaman S (2016). Basics of marine and estuarine ecology. Publisher Springer India, pp 1–481. ISBN: 978-81-322-2707-6

Mitra A, Zaman A (2020) Environmental science—a ground zero observation on the Indian subcontinent, vol XIV. Springer, Chem, pp 478. https://doi.org/10.1007/978-3-030-49131-4. eBook ISBN: 978-3-030-49131-4

MPEDA (2017) http://mpeda.gov.in/MPEDA/cms.php?id=eWVhci13aXNlLXNwZWNpZXMtd2lzZS1zdGF0ZS13aXNl. Accessed 30 Jun 2017

Mukherjee P, Nayak B, Mukherjee S, Aman AA, Paul M, Roy M, Mitra A (2017) Bioaccumulation of Fe, Zn, Cu and Pb in the mangroves of western Indian Sundarbans. Int J Comput Sci Eng Technol 2(5):74–77

Naidu BG, Shirke KJ, Shekhar A (2012) A study of a galactogouge and protein rich–multigrain ladoo

Ong ASH, Tee ES (1992) Natural sources of carotenoids from plants and oils. Methods Enzym 213:142–152

Pal N, Amin G, Zaman S, Biswas P, Mitra A (2016) Spatial variation of stored carbon in *Avicennia alba* seedlings of Indian Sundarbans. Int J Trend Res Dev (4):100–103. ISSN: 2394-93333

Pal N, Zaman S, Pramanick P, Mitra A (2017) Impact of aquatic salinity on mangrove seedlings: a case study on *Heritiera fomes* (Common Name: Sundari). Biom J Sci Tech Res 1(4) BJSTR.MS.ID.000348. https://doi.org/10.26717/BJSTR.2017.01.000348

Patra KJ, Dhal KN, Thatoi NH (2011) In Vitro bioactivity and phytochemical screening of *Suaeda maritima* (Dumort), a mangrove associate from Bhitarkanika India. Asian Pac J Trop Med 4:727–734

Petrell RJ, Mazhari Tabrisi K, Harrison PJ, Druehl LD (1993) Mathematical model of Laminaria production near a British Columbian salmon sea cage farm. J Appl Phycol 5:1–14

Pramanick P, Zaman S, Bera D, Raha AK, Mitra A (2014a) Mangrove fruit products: a search for alternative livelihood for Island dwellers of Gangetic delta. Int J Pharm Res Sch 3(1):131–137

Pramanick P, Zaman S, Mitra A (2014b) Processing of fruits with special reference to *S. apetala* fruit jelly preparation. Int J Univ Pharm Bio Sci 3(5):36–49

Raha AK, Bhattacharyya SB, Zaman S, Banerjee K, Sengupta K, Sinha S, Sett S, Chakraborty S, Datta S, Dasgupta S, Roy Chowdhury M, Ghosh R, Mondal K, Pramanick P, Mitra A (2013) Carbon census in dominant mangroves of Indian Sundarbans. J Energy Environ Sci 127:345–354

Saha A, Mitra A, Pal N, Zaman S, Mitra A (2019) Spatial variation of soil texture in the world heritage site of Indian Sundarbans. Acta Sci Microbiol 2(8):120–122

Saha SB, Bhattacharyya SB, Mitra A, Choudhury A (2001) Quality of shrimp culture farm effluents and its impact on the receiving environment. Bangladesh J Zool 29:139–149

Sahu S, Jana AK, Sarkar S, Dora KC, Chowdhury S (2012) Econometric modelling of shrimp (*Penaeus monodon*, fabricius) farming at Nandigram-II block, Purba Medinipur district (W.B.). Int J Innov Res Sci Eng Technol 1(1):121–124

Sengupta K, Roy Chowdhury M, Bhattacharyya SB, Raha A, Zaman S, Mitra A (2013) Spatial variation of stored carbon in Avicennia alba of Indian Sundarbans. Discov Nat 3(8):19–24

Senthilathiban R, Rajakumar M, Selvaraj P (2001) An Economic appraisal of freshwater prawn farming in Thanjavur district. Fish Technol 38(1):56–58

Tanaka T (1976) Tanaka's cyclopedia of edible plants of the world. Keigaku Publishing Co., Tokyo

Trivedi S, Datta U, Das M, Zaman S, Mitra A (2018) Spatio-temporal variations of heavy metals in edible crabs collected from the lower Gangetic delta region. Pet Petrochem Eng J 2(1):1–8

Zvaigzne G, Karklina D, Seglina D, Krasnova I (2009) Antioxidants in various citrus fruit juices. Chem Technol 3(52):56–61

Chapter 4
Threats to Blue Economy in Indian Sundarbans

Abstract Blue Economy has emerged as a new paradigm in recent times, which is mostly based on living and non-living resources of the marine and estuarine ecosystem. The vertical has great potential in terms of upgradation of the economic profile of the region if it is backed up with sustainability and improved environmental condition. Unfortunately, the domain of blue economy is surrounded with threats of various categories, which need to be overcome with proper scientific technologies and policies. The entire array of threats connected to blue economy has been highlighted in this chapter under two broad headings. The manmade threats include construction of shrimp farms and tourism units at the cost of mangroves, release of untreated wastes from fish landing stations, industries and shrimp farms in the estuaries, overfishing, release of low quality oil and diesel from fishing vessels and trawlers, destructive fishing practice like bottom trawling, plastic pollution from the tourism units and tourist boats, shrimp seed collection, illegal timber, fuel wood, honey and wax collection from the mangrove forest, pouching of wildlife etc. The most important natural threat on blue economy is related to climate change that causes sea level rise, erosion, elevation of near surface atmospheric carbon dioxide level, salt water intrusion in the islands, acidification, increased frequency of cyclones etc. All these threats to blue economy have been highlighted in this chapter with case studies backed up with three-decade real time data bank.

4.1 Natural Threats to Blue Economy

The present *era* is witnessing an escalating interest in the sector of blue economy, which encompasses the dependency of lives and livelihoods of human beings on the living and non-living resources of the marine and estuarine ecosystems. Whilst the goods and services provided by the blue economy have conservatively been valued at over USD 2.5 trillion per annum (Hoegh-Guldberg et al. 2015), the IPCC's Special Report on the Oceans and Cryosphere in a Changing Climate, 2018 Living Planet Report and 2019 Intergovernmental Science-Policy Platform on Biodiversity and Ecosystem Services (IPBES) report provide strong proofs that the climate change coupled with human activities are severely deteriorating the ocean health and

A. Mitra et al., *Blue Economy in Indian Sundarbans*,
https://doi.org/10.1007/978-3-031-07908-5_4

eroding the resource base on which human society and business depend. Maintaining sustainability has become a matter of concern in the vertical of blue economy.

The major natural threats that are negatively impacting the vertical of blue economy may be categorized in to three major types namely Sea Level Rise (SLR), acidification of the marine and estuarine ecosystems and weather extremes.

Each of these natural threats is discussed here in points.

4.1.1 Sea Level Rise

Climate change induced global warming is triggering the rise of global mean sea level through two possible routes. First, glaciers and ice sheets worldwide are melting and adding water to the existing volume of ocean water. Second, as per the law of physics on thermal expansion of liquid, the volume of the oceans, seas, bays and estuaries are expanding as their aquatic phase is getting warmed due to rise of atmospheric temperature.

The Intergovernmental Panel on Climate Change (IPCC) states that "most of the observed increase in globally averaged temperatures since the mid-twentieth century is very likely due to the observed increase in anthropogenic greenhouse gas concentrations (http://www.ipcc.ch/SPM2feb07.pdf)", which leads to warming of the surface and lower atmosphere by accelerating the greenhouse effect. Natural phenomena such as solar variation combined with volcanoes may probably have a small warming effect from pre-industrial times to 1950, but a small cooling effect since 1950.

There are still lots of uncertainty in forecasting over the degree by which temperatures will actually increase and what the effects of these increases will be on the planet as a whole. What is not in doubt is that the Earth's temperature has increased by 0.5 °C over the past century, and that recent years have been among the hottest on record. The bitter truth of global warming has serious implications on the environment and ecosystems of various categories. A 20 cm rise in sea level by 2030 is expected to result from glacial melting and from the thermal expansion of the oceans as water temperatures rise. This may inundate some coastal regions, and increase the risk of flooding in many areas. In case of Indian sub-continent, population density often exceeds 1000 persons per sq. km within Low Elevation Coastal Zone (LECZ), due to which India has been identified as one of the most vulnerable nations to climate change induced sea level rise (SLR) (Mitra 2013).

The potential of a rise in sea levels due to global warming would wipe out existing shoreline; contaminate freshwater/low saline aquatic systems with salt water, diminish the minimum water flow required for fishes in river and dry up floodplain wetlands, a rich habitat for fishes. The services that aquatic ecosystem provide to humans such as fisheries, flood protection, recreation and biodiversity conservation, which if converted in monetary terms would be in millions of rupees. A recent global assessment of the status of aquatic ecosystems showed that their capacity to provide goods and services would be drastically degraded due to SLR.

The warning bell of SLR has also been heard in Indian Sundarbans, at the apex of Bay of Bengal. *The Independent* (a news agency) reported in December 2006 that the first island claimed by the global warming induced SLR in Indian Sundarbans was Lohachara Island. However several scientists have contradicted the correlation between erosion of Lohachara Island and global warming and forwarded tectonic cause as one of the important reasons behind the loss of landmass of Lohachara Island (Mitra 2013). In the central part of Indian Sundarbans, many of the estuaries are becoming hypersaline due to SLR. This has considerable adverse impact on the flora and fauna of the region. The stunted growth of the mangrove floral species or the dominancy of trash fishes in the catch basket of central Indian Sundarbans are prominent indicators of the adverse impact of SLR on the biotic communities of the mangrove dominated Indian Sundarbans. The main causes behind global warming in Indian Sundarbans are excessive rise of near surface atmospheric carbon dioxide (Fig. 4.1) due to deforestation of mangroves for constructing shrimp farms and tourism units, urban developments, fish landing stations, fishing boats and trawlers repairing sites etc. In addition to these, emissions from industries, vessels and trawlers and brick kilns add substantial amount of carbon dioxide to Sundarban atmosphere.

One of the major adverse impacts of SLR in Indian Sundarbans (which is around 3.15 mm/year) is the rise of salinity, which is very much pronounced in the central sector of Indian Sundarbans due to complete blockage of fresh water as a result of the massive siltation of the Bidyadhari River (Chaudhuri and Choudhury 1994). Because of this, the Matla estuary in the central sector receives only the saline water during high tide from the Bay of Bengal in the south. The event of salinification of the Matla estuary was documented by us since 1984–2021 through three seasons as presented in Table 4.1.

With this salinity data of the central part of Indian Sundarbans, we carried out a time series modelling to visualize the trend of salinity level using a nonlinear

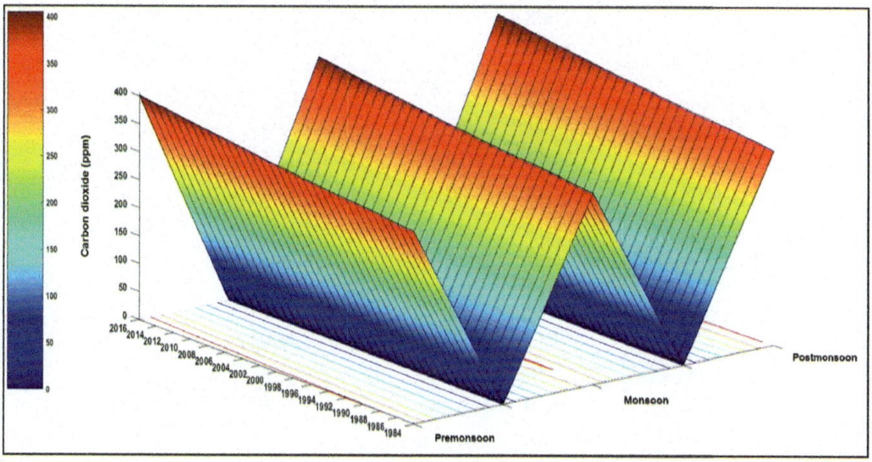

Fig. 4.1 Near Surface atmospheric carbon dioxide level (in ppm) in Indian Sundarbans

Table 4.1 Rising trend of salinity (in psu) in the Matla estuary of central Indian Sundarbans

Year	Premonsoon	Monsoon	Postmonsoon
1984	22.07	20.67	21.81
1985	23.59	20.74	22.08
1986	24.34	20.80	22.02
1987	24.22	20.14	22.67
1988	25.04	20.03	22.40
1989	25.69	20.35	23.08
1990	25.84	21.02	23.54
1991	25.01	21.67	23.57
1992	24.72	21.73	23.96
1993	25.60	21.86	23.77
1994	26.00	21.74	24.21
1995	26.04	21.99	24.99
1996	26.08	21.61	25.05
1997	26.83	22.01	24.92
1998	26.81	22.06	25.43
1999	26.75	20.64	25.41
2000	26.98	21.37	25.51
2001	27.04	22.60	25.96
2002	27.13	22.92	26.42
2003	27.45	22.98	26.48
2004	27.20	23.04	26.45
2005	27.18	23.16	27.56
2006	27.86	23.78	27.38
2007	27.69	23.83	27.69
2008	28.34	23.92	27.72
2009	34.68*	22.87	28.30
2010	29.87	22.96	28.54
2011	30.13	23.37	28.89
2012	30.08	23.52	28.72
2013	30.46	23.69	28.92
2014	30.73	24.04	29.28
2015	30.81	23.89	29.80
2016	31.09	24.03	29.88
2017	30.17	25.03	29.30
2018	30.85	25.41	30.15
2019	31.03	26.55	30.44
2020	31.38	27.10	30.87

(continued)

Table 4.1 (continued)

Year	Premonsoon	Monsoon	Postmonsoon
2021	31.50	27.74	31.24

Source Mitra and Zaman, 2021; In: Estuarine Acidification: Exploring the Situation of Mangrove Dominated Indian Sundarban Estuaries, published by Springer, Chem. https://doi.org/10.1007/978-3-030-84792-0

Note The abrupt rise of surface water salinity value during premonsoon 2009 (represented by *) is the effect of the Super Cyclone AILA that hit Indian Sundarbans during 25th May, 2009 causing huge intrusion of sea water from the Bay of Bengal

autoregressive model (NAR) treating seasonal salinity values (Fig. 4.2) as inputs and observed significant rising trend in salinity with a base of 70% training, 15% validation and 15% testing (Fig. 4.3). This increased salinity due to reduced fresh water inflow in the central part of Indian Sundarbans coupled with sea level rise has significantly impacted the spawning and larval development of Indian shad, *Tenualosa ilisha.*

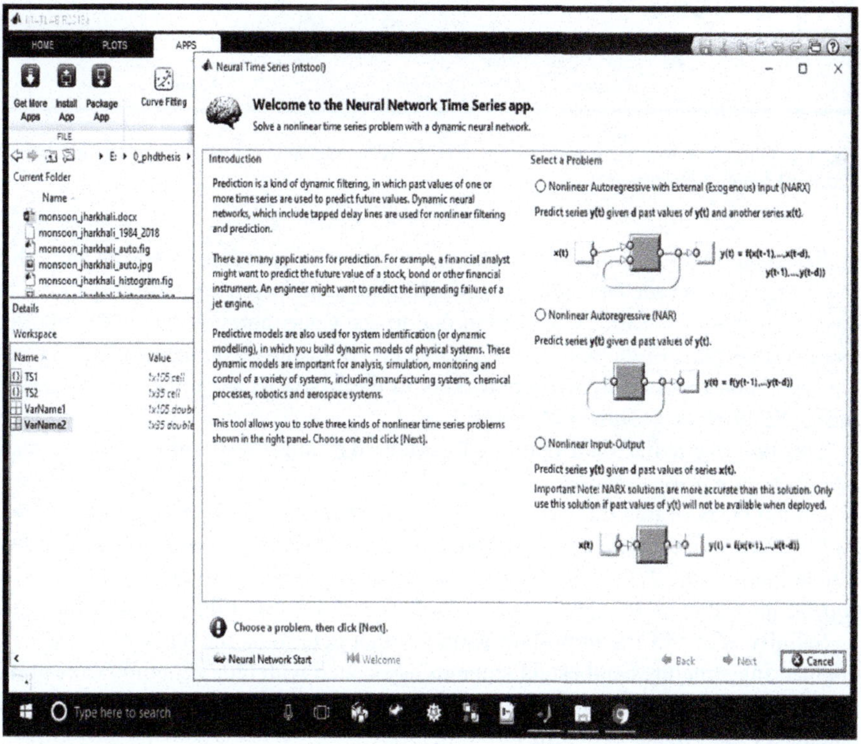

Fig. 4.2 Nonlinear Auto Regression (NAR) model for evaluating the salinity trend in the Matla estuary of central Indian Sundarbans

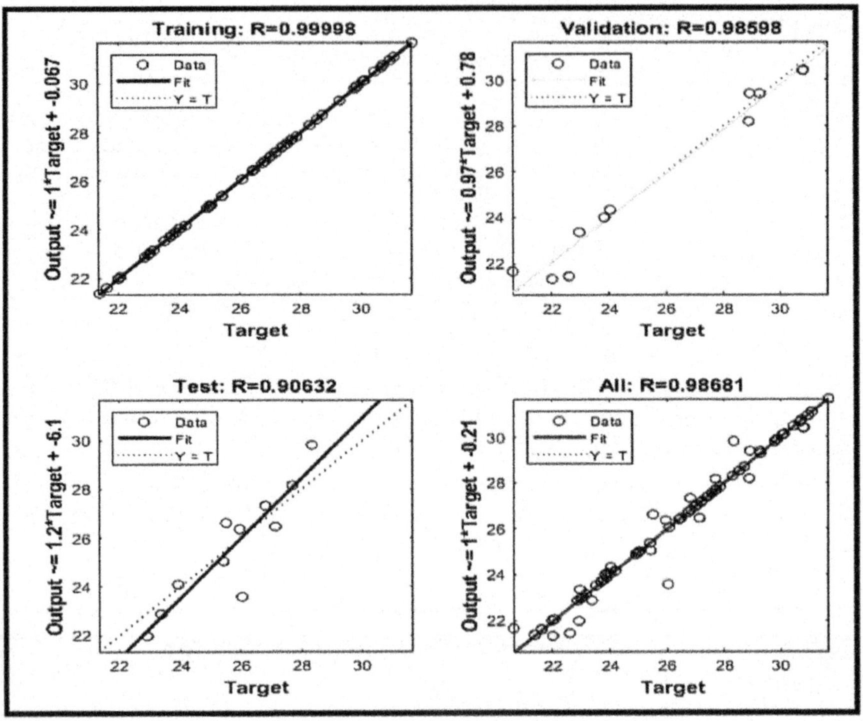

Fig. 4.3 Outputs of Nonlinear Auto Regression (NAR) model showing the trend of aquatic salinity in central Indian Sundarbans

SLR is a direct arm of climate change, which may inundate wetlands and other low-lying lands, erode beaches, intensify flooding, and increase the salinity of rivers, bays and estuaries. Salinity alteration due to seawater intrusion has been reported from various parts of the world. The rising trend of salinity over a period of 27 years has also been reported by earlier workers in the Matla River in the central part of Indian Sundarbans (Mitra et al. 2009).

The biogeographic distribution of certain fish species is influenced by such alteration of aquatic salinity.

On the basis of this three decade data bank (real time data), we used Nonlinear Autoregressive Neural Network Model to forecast the salinity levels in central Indian Sundarbans (with 29 years ahead time steps) for three seasons and observed a significant rise in all the three seasons (Tables 4.2, 4.3, 4.4; Figs. 4.4, 4.5, 4.6).

Salinity is one of the important variables that determine the shelf life of several engineering structures and establishments associated with blue economy (like ports, harbours, hotels, recreational units, ship building yards, aquaculture farms, oil rigging machinery etc.) (Dhar et al. 2021).

Table 4.2 Forecasting of salinity (in psu) in Central Indian Sundarbans during premonsoon using Autoregressive Neural Network Model

Year	Predicted salinity (in psu) in premonsoon
2022	30.87
2023	30.86
2024	30.86
2025	30.86
2026	30.86
2027	30.86
2028	30.86
2029	30.86
2030	30.86
2031	30.86
2032	30.86
2033	30.86
2034	30.86
2035	30.86
2036	30.86
2037	30.86
2038	30.86
2039	30.86
2040	30.86
2041	30.86
2042	30.86
2043	30.86
2044	30.86
2045	30.86
2046	30.86
2047	30.86
2048	30.86
2049	30.86
2050	30.86

The nonlinear autoregressive model (NAR) run with the seasonal training of salinity data for more than three decades clearly shows that the hard engineering structures at Jharkhali are extremely susceptible to corrosion on account of the increasing trend in salinity which might touch 32.6 psu during premonsoon (Fig. 4.4), 29.5 psu during monsoon (Fig. 4.5) and 32.0 psu during postmonsoon (Fig. 4.6) in 2050.

Jharkhali is situated in the central part of the Indian Sundarban delta. The tourists from different corners of the world enter to enjoy the beauty of the mangrove forest of Indian Sundarban through the gateway of Jharkhali. The area has several concrete buildings for accommodations and tourism. A high saline environment can accelerate

Table 4.3 Forecasting of salinity (in psu) in Central Indian Sundarbans during monsoon using Autoregressive Neural Network Model

Year	Predicted salinity (in psu) in monsoon
2022	27.05
2023	24.41
2024	30.28
2025	17.89
2026	23.60
2027	19.99
2028	21.15
2029	22.29
2030	22.62
2031	22.30
2032	21.81
2033	21.58
2034	21.73
2035	21.85
2036	21.79
2037	21.71
2038	21.74
2039	21.78
2040	21.77
2041	21.75
2042	21.75
2043	21.76
2044	21.76
2045	21.75
2046	21.75
2047	21.76
2048	21.76
2049	21.76
2050	21.76

the corrosion of these structures. Corrosivity of natural water is directly proportional to salinity and thus the structures in Jharkhali are more prone to corrosion, which may retard the pace of success of blue economy in this region.

In addition to adverse impact on hard engineering structures, there are considerable evidences of negative impact of salinity rise on the fisheries and aquaculture sector.

The impacts of SLR induced erosion and salinity alteration on the most relevant sectors of blue economy pertinent to the framework of Indian Sundarbans are presented in Tables 4.5 and 4.6.

Table 4.4 Forecasting of salinity (in psu) in Central Indian Sundarbans during postmonsoon using Autoregressive Neural Network Model

Year	Predicted salinity (in psu) in postmonsoon
2022	31.45
2023	29.12
2024	31.90
2025	28.83
2026	32.29
2027	28.58
2028	32.61
2029	28.43
2030	32.81
2031	28.38
2032	32.91
2033	28.36
2034	32.95
2035	28.35
2036	32.97
2037	28.35
2038	32.97
2039	28.35
2040	32.97
2041	28.35
2042	32.97
2043	28.35
2044	32.97
2045	28.35
2046	32.97
2047	28.35
2048	32.97
2049	28.35
2050	32.97

4.1.2 Acidification of the Marine and Estuarine Ecosystems

Acidification has crept in the major estuaries of Indian Sundarbans as confirmed through our data bank of Hooghly and Matla estuaries in the western and central sectors respectively (Tables 4.7 and 4.8).

The root cause behind acidification of Sundarban estuaries is mostly anthropogenic (Mitra and Zaman 2021). The rapid pace of industrialization, urbanization, deforestation and change in land-use pattern has increased the carbon dioxide level

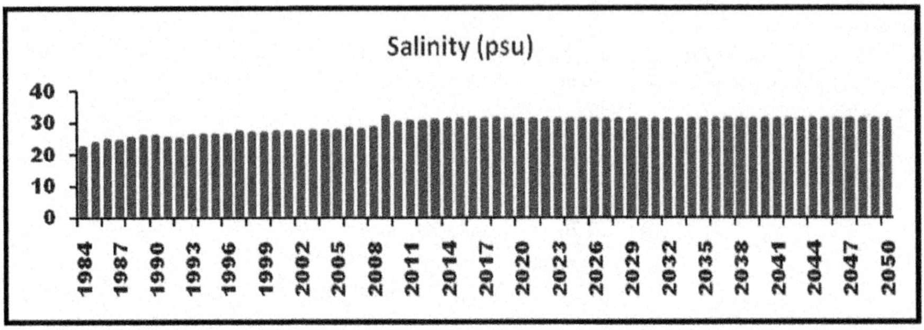

Fig. 4.4 Predicted salinity (in psu) in central Indian Sundarbans during premonsoon using Nonlinear Autoregressive Neural Network Model; real time data from 1984 to 2021 has been used as training in the model

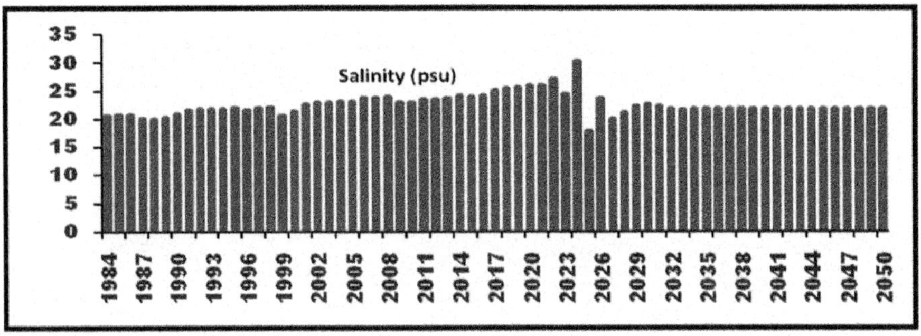

Fig. 4.5 Predicted salinity (in psu) in Central Indian Sundarbans during monsoon using Nonlinear Autoregressive Neural Network Model; real time data from 1984 to 2021 has been used as training in the model

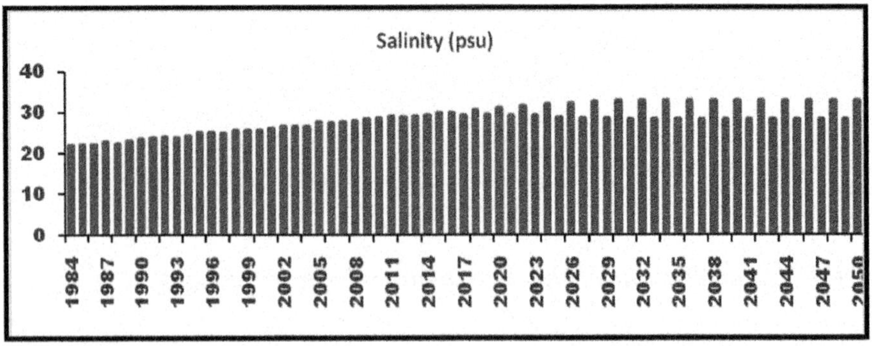

Fig. 4.6 Predicted salinity (in psu) in Central Indian Sundarbans during postmonsoon using Nonlinear Autoregressive Neural Network Model; real time data from 1984 to 2021 has been used as training in the model

Table 4.5 Impact of SLR induced erosion in three sectors of Indian Sundarbans

Sector	Western Sector	Central Sector	Eastern Sector	Remark
Fisheries and Aquaculture	Moderate Stress	High Stress	Lack of information	Maximum erosion and subsequent siltation are observed in the central sector in comparison to the western part, due to which the natural feed of fishes like the plankton cannot survive in turbid condition; No fisheries and aquaculture activities are allowed in the eastern sector as it is under Reserve Forest (RF) area
Shipping and Transportation of goods	High Stress	High Stress	Moderate Stress	Erosion and subsequent siltation decreases the optimum draft required for navigation as seen in the western and central sectors; the presence of luxuriant mangroves in the eastern sector arrest the soil particles and thus the draft is moderately maintained
Estuary/Sea-based recreation and tourism	Moderate Stress	High Stress	Less Stress	Erosion and subsequent siltation pose adverse impact on the movement of tourist vessels as seen in the western and central sectors; erosion of embankments in the central sector also put the tourists in great trouble during low tide condition while getting on the embankment to reach the destination/lodges; the eastern sector being under Reserve Forest zone has luxuriant mangrove cover due to which the erosion is controlled and tourists can enjoy the wilderness and beauty of wildlife in the dense mangrove forest
Biodiversity and conservation	High Stress	High Stress	Less Stress	Biodiversity and conservation of natural resources are negatively affected due to erosion and siltation in the western and central sectors although other factors like pollution, industrial activities also act in synergy; the eastern sector being under Reserve Forest zone has luxuriant mangrove cover due to strict vigilance by the staffs of West Bengal Forest Department
Non-conventional renewable energy sector	Moderate Stress	High Stress	Lack of information	Non-conventional renewable energy like tidal energy can get moderate benefit in the western Indian Sundarbans due to high tidal amplitude unlike the central sector where the water is extremely turbid due to erosion; the eastern sector being under Reserve Forest zone perhaps does not permit any implementation of hard structures and hence there is no information in this sector with respect to initiation of non-conventional renewable energy units

Index:

High Stress	Moderate Stress	Less Stress	Lack of information
(red)	(orange)	(green)	(white)

Table 4.6 Impact of SLR induced salinity alteration in three sectors of Indian Sundarbans

Sector	Western Sector	Central Sector	Eastern Sector	Remark
Fisheries and Aquaculture	Less Stress	High Stress	Lack of information	Gradual lowering of salinity in the western sector promotes the growth, survival and migration of several commercially important fish species; Hypersaline condition as observed in the central sector of Indian Sundarbans due to blockage of freshwater causes reproductive failure and diversion of migratory route of fishes like *Tenualosa ilisha*; No fisheries and aquaculture activities are allowed in the eastern sector as it is under Reserve Forest (RF) area
Shipping and transportation of goods	Less Stress	Moderate Stress	Less Stress	Hypersalinity in the central sector accelerates corrosion of the hard metallic structures of ships, trawlers, vessels and jetties; settlement of biofoulers (like oysters, barnacles *etc*., which are very common in this sector) also decreases their shelf life; western and eastern sectors are relatively hyposaline and does not pose any major problem on shipping and transportation of goods in terms of salinity
Estuary/Sea-based recreation and tourism	Moderate Stress	High Stress	Less Stress	The central Indian Sundarbans is highly vulnerable due to salinity rise and the ground water has even become saline (~5.0 psu), which is the source of drinking water for tourists in majority of the lodges; stunted growth of mangroves has also been reported by many researchers; the eastern sector being under Reserve Forest zone has luxuriant mangrove cover due to which the erosion is controlled and tourists can enjoy the wilderness and beauty of wildlife in the dense mangrove forest
Biodiversity and conservation	Moderate Stress	High Stress	Less Stress	Biodiversity and conservation are negatively affected due to salinity rise in the central sector; high salinity not only pose an adverse impact on the growth and survival of mangroves particularly for the species like *Sonneratia* spp., *Heritiera fomes*, *Nypa fruticans etc.*, but the diversity of fish community is also impacted with more dominancy of low cost trash fishes; the eastern sector being under Reserve Forest zone with congenial salinity has luxuriant mangrove cover due to strict vigilance by the staffs of West Bengal Forest Department
Non-conventional renewable energy sector	Less Stress	High Stress	Lack of information	Structures and machineries related to non-conventional renewable energy generation like tidal energy has high probability to face adverse impact in the central sector as hypersalinity accelerates the process of corrosion unlike the western Indian Sundarbans where the salinity is relative low due to fresh water input from the Hooghly River; the eastern sector being under Reserve Forest zone perhaps does not permit any implementation of hard structures and hence there is no information in this sector with respect to initiation of non-conventional renewable energy units

Index:

High Stress	Moderate Stress	Less Stress	Lack of information

Table 4.7 Mean value of surface water pH at Kakdwip (21°52′26.50″ N; 88°08′04.48″ E) along the Hooghly estuary during 1984–2021

Year	Kakdwip	
	High tide	Low tide
1984	8.34	8.32
1985	8.32	8.31
1986	8.33	8.32
1987	8.32	8.31
1988	8.32	8.31
1989	8.32	8.31
1990	8.32	8.31
1991	8.30	8.29
1992	8.31	8.30
1993	8.31	8.30
1994	8.30	8.29
1995	8.31	8.29
1996	8.29	8.28
1997	8.29	8.28
1998	8.29	8.28
1999	8.27	8.26
2000	8.28	8.26
2001	8.28	8.27
2002	8.26	8.25
2003	8.28	8.25
2004	8.27	8.26
2005	8.25	8.24
2006	8.25	8.24
2007	8.24	8.23
2008	8.25	8.23
2009*	8.30	8.28
2010	8.22	8.20
2011	8.21	8.20
2012	8.18	8.16
2013	8.14	8.13
2014	8.12	8.11
2015	8.13	8.10
2016	8.12	8.10
2017	8.11	8.09
2018	8.07	8.05
2019	8.09	8.05
2020**	8.16	8.14

(continued)

Table 4.7 (continued)

Year	Kakdwip	
	High tide	Low tide
2021	8.12	8.10

Note The abrupt rise of surface water pH value during 2009 (represented by *) is the effect of the Super Cyclone AILA that hit Indian Sundarbans during 25th May, 2009 causing huge intrusion of sea water from the Bay of Bengal; another peak observed during 2020 (represented by **) is the effect of COVID-19 lockdown phase during which the near surface atmospheric carbon dioxide reduced considerably in the region resulting in the hike of aquatic pH

of the atmosphere. The increase of atmospheric carbon dioxide with the passage of time has resulted in the enhancement of Dissolved Inorganic Carbon (DIC) in the aquatic phase, which has dropped the pH value of the aquatic phase by few units as per the sequence stated here.

Step 1: In the first step, the atmospheric CO_2 dissolves in the aquatic phase to form aqueous CO_2 along with H_2CO_3

$$CO_2 + H_2O \leftrightarrow H_2CO_3$$

Step 2: In this step H_2CO_3 dissociates to produce HCO_3^- and proton (H^+)

$$H_2CO_3 \leftrightarrow HCO_3^- + H^+$$

Step 3: In this step HCO_3^- has the tendency to dissociate into $CO_3^=$ and protons (H^+), which results in the lowering of pH

$$HCO_3^- \leftrightarrow CO_3^= + H^+$$

It is to be noted in this context that the total Dissolved Inorganic Carbon (DIC) in aquatic phase is the sum total of aqueous CO_2, HCO_3^- and $CO_3^=$. Thus,

$$DIC = CO_2(aq) + HCO_3^- + CO_3^=$$

The more the concentration of protons (H^+) in the water bodies, the lower will be the pH value. In this unit we have taken a closer look on the impact of acidification on the major sectors of blue economy in Indian Sundarbans (Table 4.9).

4.1.3 *Weather Extremes*

Weather extremes refer to significant deviation of environmental variables like wind speed, rainfall, air temperature etc. from their normal values that usually persist

Table 4.8 Mean value of surface water pH at Canning (22°19′03.20″ N; 88°41′04.43″ E) along the Matla estuary during 1984–2021

Year	Canning	
	High tide	Low tide
1984	8.34	8.31
1985	8.33	8.30
1986	8.33	8.31
1987	8.33	8.30
1988	8.33	8.30
1989	8.32	8.30
1990	8.31	8.30
1991	8.31	8.28
1992	8.32	8.29
1993	8.32	8.29
1994	8.31	8.28
1995	8.31	8.28
1996	8.31	8.27
1997	8.30	8.27
1998	8.31	8.27
1999	8.30	8.25
2000	8.31	8.26
2001	8.31	8.26
2002	8.30	8.24
2003	8.32	8.26
2004	8.30	8.25
2005	8.30	8.23
2006	8.29	8.23
2007	8.29	8.22
2008	8.30	8.22
2009*	8.33	8.31
2010	8.29	8.20
2011	8.29	8.19
2012	8.29	8.20
2013	8.29	8.21
2014	8.29	8.20
2015	8.28	8.20
2016	8.27	8.19
2017	8.28	8.20
2018	8.25	8.18
2019	8.30	8.19
2020**	8.29	8.25

(continued)

Table 4.8 (continued)

Year	Canning	
	High tide	Low tide
2021	8.26	8.21

Note The abrupt rise of surface water pH value during 2009 (represented by *) is the effect of the Super Cyclone AILA that hit Indian Sundarbans during 25th May, 2009 causing huge intrusion of sea water from the Bay of Bengal; another peak observed during 2020 (represented by **) is the effect of COVID-19 lockdown phase during which the near surface atmospheric carbon dioxide reduced considerably in the region resulting in the hike of aquatic pH

in a particular region. In the framework of Indian Sundarbans, weather extreme is observed in terms of wind speed during the period of cyclones; extreme deviation in rainfall and air temperature is observed in the summer and rainy seasons. Here we discuss the impact of cyclonic depressions on the major sectors of regional blue economy (Table 4.10).

4.2 Anthropogenic Threats to Blue Economy

It is really difficult to predict the success of blue economy as the inter-relationship between environmental drivers and blue economy is highly dynamic. We give a very simple example here. If the coastal industries need to be flourished for the well being of coastal population, one cannot circumvent the 2nd law of thermodynamics and pollution will surely occur. Now if these pollutants are bioaccumulated in the fish tissue, the costs of these contaminated fishes go down. Thus at the end, one sector may suffer for giving boost to another sector.

We have also seen that sectors like tourism particularly eco- tourism is purely nature based, but in many regions mangroves are sacrificed to promote aquaculture, which has tremendous prospect to earn foreign exchange. Thus at times, it is difficult to judge that which sectors should be given priority. To sum up we can state that balancing between various sectors is the key to unlock a sustainable blue economy.

The anthropogenic factors play a crucial role in sustaining the blue economy as the threats arising from anthropogenic activities are so intricately related that one has to very cautiously and wisely use the technology to harness the resources without posing much threat to the environment.

In the deltaic complex of Sundarban estuaries the major anthropogenic threats encompass emission of Green House gases (GHGs), aquatic pollution and over exploitation of resources.

Table 4.9 Impact of climate change induced acidification in three sectors of Indian Sundarbans

Sector	Western Sector	Central Sector	Eastern Sector	Remark
Fisheries and Aquaculture				The pace of acidification is high in the western sector of Indian Sundarbans due to presence of different point sources of pollution. The phenomenon of acidification causes detrimental effects on the fish resources with significant effect observed on the oyster *Cassostrea cucullata e.g.,* thinning of shell, mortality of the species etc. In the central sector, the effect is moderate because the pace of acidification is not very high; again due to presence of luxuriant mangrove vegetation and associate species in the eastern sector, the pH value does not exhibit a significant decrease and hence the sector is relatively stress free in terms of acidification. The inter-relationship between mangrove biomass and DIC is significantly negative (-0.9038; $p < 0.01$), while between mangrove biomass and the ambient aquatic pH is significantly positive (0.9635; $p < 0.01$). The datasets on which these inter-relationships have been computed are presented in Annexure 4A. These datasets were collected during our field trip in Indian Sundarbans in February 2022 covering 24 sampling stations. These relations clearly confirm the unique buffering capacity of halophytes due to which the pace of acidification is relatively low in the mangrove dominated eastern sector
Shipping and transportation of goods				The lowering of pH in the western sector is high due to which there is maximum probability of corrosive action; in the central and eastern sectors the rate of lowering of pH is not significantly high enough to corrode the hard structures of ships, vessels and trawlers. Again, the mangrove vegetations in these two sectors are acting as natural buffers
Estuary/Sea-based recreation and tourism				Because of the rapid pace of acidification in the western sector, the hard engineering structures related to tourism like jetties, vessels, trawlers, buildings, resorts *etc.* easily get corroded, which is not evident in the central Indian Sundarbans; in the eastern sector there is no information on the adverse impact of acidification on buildings and hard engineering structures related to tourism. However, corrosion is observed in several watch towers and forest rest houses which may be the effect of continuous exposure to saline environment

(continued)

Table 4.9 (continued)

Biodiversity and conservation				In the western sector, the spectrum of biodiversity is not very broad due to intense industrial, urban, fish landing and tourism activities. The pace of acidification is moderately affecting the region because several conservation related activities in terms of afforestation are under progress. The central sector is extremely vulnerable as the mangrove vegetation is poor and stunted due to impact of high salinity. However, information on the impact of acidification is scanty for this region; the eastern sector being under the Reserve Forest sustains luxuriant mangrove vegetation, which plays a vital role in the buffering action and thus checks the lowering of pH
Non-conventional renewable energy sector				Corrosion may adversely affect the hard structures needed to install tidal power generation projects in the estuaries of western sector due to high rate of acidification; the rate of lowering of pH is not very high in the central sector due to which the vulnerability of corrosion is moderate; no hard structures are supposed to be installed in the eastern sector due its location under the Reserve Forest area

Index:

High Stress	Moderate Stress	Less Stress	Lack of information

4.2.1 *Emission of GHGs*

GHGs are emitted from various point and non-point sources. Considering the matrix of Indian Sundarbans the main sources are the industries around the region (like Haldia Industrial belt), fishing vessels and trawlers, shrimp farms, tourism units, brick kilns, domestic emission by burning mangrove sourced fuel woods and mangrove forests preferably the marsh vegetation (Table 4.11).

4.2.2 *Aquatic Pollution*

The foundation of blue economy needs to stand on the strong pillars of environmental quality particularly the water quality of the oceans, seas, bays and estuaries, which supports the resource base of the blue economy. In context to Indian Sundarbans, the picture of water quality is to some extent gloomy as recorded by us during our ground zero observations on *Liza parsia* muscle tissues sampled and analyzed since more than two decades in the month of September for every year from four different sampling stations in and around Indian Sundarbans (Table 4.12, 4.13, 4.14, 4.15,

Table 4.10 Impact of climate change induced cyclonic depression in three sectors of Indian Sundarbans

Sector	Western Sector	Central Sector	Eastern Sector	Remark
Fisheries and Aquaculture	High Stress	High Stress	Moderate Stress	During cyclonic depressions, the western and central sectors suffer most due to less abundance of mangroves; in the eastern sector the luxuriant mangrove vegetation acts as the natural shield
Shipping and transportation of goods	High Stress	Moderate Stress	Moderate Stress	The tidal amplitude and wave actions are maximum in the western sector which get intensified due to input of high wind energy that may deviate the stability of the ships, vessels and trawlers; the wave actions are not very high in the central and eastern sectors due to which there is moderate effect of storms on the movement of ships, vessels and trawlers
Estuary/Sea-based recreation and tourism	High Stress	High Stress	Lack of information	During super cyclones, tidal surges and wave actions become extremely high due to which the resorts, lodges and vessels are severely damaged in the western sector. Many of the resorts and lodges in Indian Sundarbans have been constructed within 500 m from the high tide line, which violates the Coastal Regulation Zone (CRZ) norms; there is no information for eastern sector mostly due to absence of resorts and lodges in the Reserve Forest area
Biodiversity and conservation	High Stress	High Stress	Less Stress	Due to high tidal amplitude and wave actions in the western sector, biodiversity is adversely affected during stormy weather resulting in the uprooting of mangrove vegetation; the central Indian Sundarbans being an erosion prone zone is greatly affected during super cyclones; due to presence of luxuriant mangrove vegetation in the eastern sector the impact of cyclones is greatly minimized. The dense mangrove cover along with their root systems help attenuate the waves
Non-conventional renewable energy sector	High Stress	Moderate Stress	Lack of information	The turbines and other hard engineering structures required to install tidal and wind based power units may be damaged due to high wave actions during cyclonic depressions in the western sector; increase of erosion and subsequent siltation during cyclones may pose adverse impact on non-conventional power generation units in the central sector; power generation units from non-conventional units are difficult to install in the eastern sector due to its location within the protected Reserve Forest area

Index:

High Stress	Moderate Stress	Less Stress	Lack of information

Table 4.11 Point and non-point sources and types of GHGs emitted from three sectors of Indian Sundarbans

Source and types of GHGs	Western sector	Central sector	Eastern sector	Remarks
 *Source* Industrial zone adjacent to Indian Sundarbans Major GHGs: CO_2, SO_2, N_2O and volatile organic compounds	+++	–	–	1.Haldia Industrial belt situated adjacent to western Indian Sundarbans is the major source of GHGs. In central and eastern sectors, there are no industries that emit GHGs
 *Source* Fishing vessels and trawlers Major GHGs: CO_2, SO_2 and volatile organic compounds	+++	++	+	2.Fishing vessels and trawlers are mostly seen in the western sector for passenger movements, fishing, tourism activities etc. The magnitude of tourism intensifies during mid January to take the holy bath in the Hooghly River mouth adjacent to Kapil Muni Temple. In central sector, fishing and tourism activities occur in moderate scale, while in the eastern sector only tourist vessels are mostly seen

(continued)

Table 4.11 (continued)

Source and types of GHGs	Western sector	Central sector	Eastern sector	Remarks
 *Source* Shrimp farms Major pollutants: CO_2, CH_4, and N_2O from mangrove soil receiving shrimp farm effluents	+++	+++	–	Shrimp farms are mostly concentrated in the western and central sectors; aquaculture is strictly prohibited in the eastern sector owing to its location within the Reserve Forest (RF) area
 *Source* Tourism units Major GHGs: CO_2, SO_2 and volatile organic compounds	+++	++	–	Tourism units are mostly concentrated in the western and central Indian Sundarbans

(continued)

Table 4.11 (continued)

Source and types of GHGs	Western sector	Central sector	Eastern sector	Remarks
 *Source* Brick Kilns Major GHGs: CO_2 and SO_2	+++	–	–	Brick kilns are all concentrated in and around the Kakdwip and Namkhana regions of western Indian Sundarbans
 *Source* Domestic emission Major GHGs: CO_2 and SO_2	+++	+++	–	3.Domestic emission generates CO_2 and SO_2 gases by burning fuel woods illegally procured from the mangrove forests. The activities are common in both western and central sectors

(continued)

Table 4.11 (continued)

Source and types of GHGs	Western sector	Central sector	Eastern sector	Remarks
 *Source* Mangrove vegetation and saltmarsh grass habitat Major GHGs: oxides of nitrogen, CO_2 and CH_4	+	++	+++	Marsh grasses are the primary sources of methane emission, which grow luxuriantly in the Reserve Forest area of eastern Indian Sundarbans; decomposition of mangrove vegetations generate oxides of nitrogen and CO_2 through microbial degradation

4.16, 4.17; Fig. 4.7). Similar reports have also been published by many researchers in the present geographical locale (Das et al. 2003; Mitra et al. 2010, 2011a, b; Bhattacharyya et al. 2013; Mitra 2013, 2019, 2020; Banerjee et al. 2014; Chakraborty et al. 2014, 2017; Mitra and Ghosh 2014; Mitra and Zaman 2014, 2015, 2020, 2021; Pramanick et al. 2015a, b, 2016; Ghosh et al. 2016; Nayak et al. 2016, 2017; Mukherjee et al. 2017a, b, c; Agarwal and Mitra 2018; Dutta et al. 2018).

The main sources of aquatic pollution in and around the mangrove ecosystem are the industries (adjacent to western sectors of Indian Sundarbans), fishing vessels and trawlers, tourism units, shrimp farms, fish landing stations etc. (Table 4.18). In addition, nutrient run–off from the adjacent agricultural fields also contribute appreciable amount of nitrate and phosphate in the estuaries of Sundarbans. Ballast discharge from the ships and tankers is also a threat to the estuarine species of Indian Sundarbans. The species released from ballast discharge are mostly invasive in nature and may displace the native species, which may pose an adverse impact on the existing food chain of the Sundarban estuaries.

Microbial pollution is also a major issue in the domain of blue economy as the fin fishes, shrimps, crabs and other edible sea food items get contaminated from the

Table 4.12 Dissolved zinc concentrations (in ppb)

Year	Stn. 1	Stn. 2	Stn. 3	Stn. 4
2001	292.57	211.74	114.41	97.56
2002	293.32	246.67	122.85	117.61
2003	317.46	256.10	128.18	121.67
2004	319.24	263.21	128.77	122.90
2005	322.98	269.89	129.65	111.11
2006	331.63	271.89	131.41	102.16
2007	352.13	282.44	132.32	130.66
2008	346.93	269.12	100.78	89.20
2009	363.74	253.38	108.34	103.42
2010	358.32	290.89	127.26	120.67
2011	362.12	291.88	138.54	133.00
2012	357.13	296.42	149.36	144.76
2013	365.00	303.05	153.85	150.17
2014	372.16	315.47	160.69	153.42
2015	384.57	325.94	166.96	161.15
2016	392.34	343.44	173.62	166.04
2017	403.75	354.65	177.05	162.48
2018	414.67	351.07	172.31	164.89
2019	426.10	358.16	181.10	171.782
2020	299.47	206.85	97.04	86.11
2021	333.61	236.64	121.61	115.39

Table 4.13 Zinc concentrations (in ppm dry wt.) in *Liza parsia* muscle

Year	Stn. 1	Stn. 2	Stn. 3	Stn. 4
2001	82.13	64.37	34.24	9.89
2002	90.75	68.60	35.69	11.55
2003	101.35	72.04	37.02	11.57
2004	107.80	80.11	38.99	11.84
2005	114.48	83.29	41.13	13.77
2006	113.57	87.04	42.88	13.13
2007	126.18	94.07	44.67	13.46
2008	147.31	100.62	47.23	12.90
2009	180.62	102.51	48.66	12.95
2010	197.05	110.29	48.86	11.84
2011	212.24	118.01	49.95	12.90
2012	228.48	125.92	51.66	13.83
2013	238.56	134.76	56.00	18.95
2014	243.94	141.45	53.55	21.12
2015	256.15	145.14	61.51	23.54
2016	241.74	138.90	64.18	25.11
2017	250.33	152.22	68.99	27.79
2018	252.67	155.32	70.79	26.71
2019	258.44	160.11	73.94	30.77
2020	104.10	100.31	42.32	12.06
2021	158.14	122.08	50.45	15.87

ambient habitat. In Indian Sundarbans, most of microbial pollution is sourced from shrimp farms, followed by the effluents released from the fish landing and tourism units. We studied the microbial load in a shrimp farm situated in the Frazergaunge region of western Indian Sundarbans during 2000 to 2021 (Table 4.19). The sampling of both the water and sediment was carried out during the month of June in every year at the time of harvesting the shrimps after a culture period of three months (March to May of every year).

The water of the shrimp farms gets contaminated with microbes by the run-off from the adjacent land masses, uneaten feed and shrimp excreta. The source water (Hooghly estuary in this case) used to fill the culture ponds also transports microbes to the aquatic phase (Figs. 4.8, 4.9, 4.10).

The uneaten food particles and excreta of the cultured species deposit on the pond bottom, which are the sources of microbial pollution in the pond sediment (Figs. 4.11, 4.12, 4.13).

Table 4.14 Dissolved copper concentrations (in ppb)

Year	Stn. 1	Stn. 2	Stn. 3	Stn. 4
2001	69.24	43.78	18.85	11.48
2002	69.12	46.98	18.84	12.66
2003	75.34	49.67	22.43	11.04
2004	74.13	48.11	22.76	11.66
2005	72.80	48.12	21.76	12.54
2006	73.23	42.82	26.42	12.12
2007	76.12	46.01	25.28	18.95
2008	79.11	52.89	28.76	18.11
2009	80.21	51.77	31.07	19.86
2010	77.62	59.88	27.26	17.96
2011	82.13	62.43	29.93	14.88
2012	87.21	61.12	29.09	17.67
2013	89.27	65.98	33.41	23.21
2014	94.77	69.02	38.99	22.34
2015	97	62.87	40.51	28.80
2016	89.26	70.01	44.04	30.21
2017	95.18	75.23	46.93	32.12
2018	100.27	77.01	52.40	35.23
2019	103.68	79.38	57.89	36.77
2020	75.27	48.45	25.46	17.52
2021	82.34	60.06	36.71	21.04

4.2.3 *Overexploitation of Resources*

People of Indian Sundarbans mostly dwell below the poverty line. They have limited access to proper education facility, health care system and various resource bases that are needed to lead a sound life with sustainable livelihoods. The region, being vulnerable to natural disasters like cyclones, wave actions and tidal surges exaggerate the sufferings of the people to a greater degree. These miserable conditions mostly force them to exploit the existing natural resources without any limitations. Apart from illegal procurement of timber, fuel wood, honey and wax from the deep mangrove forests in the eastern sector of Indian Sundarbans, many local people also over exploit the molluscan shells as sources of lime. Extraction of blue blood of horseshoe crabs (*Carcinoscorpious rotundicauda* and *Tachypleus gigas*) is also reported from Sagar Island (Mitra 2000). The pattern of overexploitation exhibits significant variation between sectors depending on the availability of resources (Table 4.20).

In the frame work of Indian Sundarbans, the over exploitation of resources is the outcome of four basic causes, namely (i) poverty, (ii) vulnerability at regional scale

Table 4.15 Copper concentrations (in ppm dry wt.) in *Liza parsia* muscle

Year	Stn. 1	Stn. 2	Stn. 3	Stn. 4
2001	54.30	37.98	19.88	10.10
2002	57.82	51.74	20.52	10.56
2003	60.92	55.78	20.10	10.58
2004	68.78	56.77	22.66	10.63
2005	78.10	66.22	28.75	10.67
2006	79.99	70.29	28.77	10.72
2007	82.44	74.79	29.26	11.17
2008	85.11	77.19	29.78	11.55
2009	86.34	82.45	30.66	11.67
2010	76.32	84.01	31.06	11.69
2011	85.42	86.11	32.26	12.07
2012	94.22	88.56	32.87	16.37
2013	99.25	90.23	40.99	19.54
2014	102.57	89.35	43.85	16.87
2015	106.69	94.54	48.90	23.21
2016	113.49	97.06	50.62	27.39
2017	111.88	91.18	54.3	29.34
2018	117.21	98.16	56.89	30.90
2019	120.57	105.32	61.04	32.16
2020	71.95	51.19	22.63	12.88
2021	89.37	60.04	28.14	19.65

to natural disaster, (iii) lack of orientation towards biodiversity conservation and (iv) anomalous institutional support.

These four causes exhibit significant spatial variation, e.g., poverty is observed maximum in the central Indian Sundarbans compared to the western and eastern sectors. This is mainly because people of western Indian Sundarbans have wide options for livelihoods like shrimp farming, brick manufacturing, fish landing related activities (like net making, vessel repairing, icing the landed fishes, loading/unloading, transporting the landed fishes, deep sea fishing etc.) and tourism. There is also the provision of income for the local people during the religious gathering to take the holy bath at the time of Makar Sankranti in the mid January of every year in the Hooghly estuary adjacent to Kapil Muni temple (Fig. 4.14).

The western sector, being nearer to the megacity of Kolkata and connected through road ways provide a great advantage to the people of the area as they often migrate to the city to take up the jobs of house keeper, cleaner, gardener and vendor in the local trains. These opportunities of livelihoods are not available to the people of central Indian Sundarbans mainly due their connectivity problem with Kolkata and lack of job opportunities in fish landing activities, religious tourism, brick manufacturing

Table 4.16 Dissolved lead concentrations (in ppb)

Year	Stn. 1	Stn. 2	Stn. 3	Stn. 4
2001	13.14	12.70	7.01	5.82
2002	12.68	13.93	6.77	6.12
2003	13.21	13.55	6.08	5.30
2004	13.69	13.74	7.52	8.00
2005	14.39	12.43	9.53	9.77
2006	13.55	16.86	8.31	8.56
2007	14.48	13.51	6.76	5.39
2008	16.05	12.87	6.75	6.71
2009	14.90	17.94	6.98	5.33
2010	16.57	15.12	9.31	8.67
2011	17.03	13.89	10.26	9.69
2012	17.77	12.24	5.85	5.42
2013	21.36	16.56	7.08	6.23
2014	25.57	19.79	7.81	6.80
2015	32.17	21.77	9.17	6.87
2016	34.25	29.94	11.10	8.74
2017	36.79	34.79	14.61	9.61
2018	40.01	37.06	16.40	9.82
2019	43.66	42.10	18.77	10.27
2020	20.05	13.94	9.14	4.84
2021	29.78	26.18	13.27	7.05

units etc. Most of the islands of central Indian Sundarbans have no electricity, health care units, educational institutes or any industries although there is no scarcity of natural resources like mangrove based honey, wax, seaweeds, oyster that can generate several innovative livelihood to the local people.

The hypersaline condition in the central sector does not provide a congenial environment for commercially valuable fishes like *Tenualosa ilisha* to breed and stay in the estuaries in this region, thus forcing the island dwellers towards more poverty and subsequent over exploitation of natural resources from the Reserve Forest / Core area in the eastern sector of Indian Sundarbans adjacent to Bangladesh Sundarbans.

Natural calamities in the form of cyclones /super-cyclones (Table 4.21) and excessive rainfall (Table 4.22) cause major impact on the lives and livelihoods of Sundarban people not only by damaging the embankments, but also by disrupting the agricultural production and fresh water fish farming that cannot withstand high salinity due to salt water intrusion during cyclones that occur very frequently (Mitra et al. 2011a, b).

The table here reflects the frequency of occurrence of cyclonic storm in the Indian Sundarbans during 1990 to 2020 (Table 4.21).

Table 4.17 Lead concentrations (in ppm dry wt.) in *Liza parsia* muscle

Year	Stn. 1	Stn. 2	Stn. 3	Stn. 4
2001	11.30	7.45	4.99	2.74
2002	13.88	9.56	5.03	2.76
2003	16.00	10.55	5.05	2.77
2004	16.93	11.91	5.07	3.79
2005	19.79	12.95	5.13	3.81
2006	20.13	13.88	5.19	3.82
2007	20.62	13.99	5.34	3.85
2008	21.17	15.09	5.36	3.89
2009	22.13	16.19	5.48	3.91
2010	23.72	16.99	5.56	3.94
2011	24.32	12.19	5.64	3.86
2012	25.13	12.99	5.86	3.87
2013	29.24	16.67	7.94	4.01
2014	31.04	17.65	8.89	4.72
2015	36.01	19.11	8.12	5.65
2016	36.28	23.34	11.19	5.81
2017	39.24	27.82	12.20	7.79
2018	44.20	26.4	16.15	8.25
2019	45.68	29.51	18.21	8.96
2020	22.01	12.24	9.78	4.77
2021	25.82	16.45	12.27	5.82

Table 4.22 highlights the average rainfall data for three decades (1981–2018). The average monthly data of rainfall has been computed on the basis of secondary data available from Canning station (*Source* IMD -Indian Meteorological Department) and it is observed that monsoon in the lower Gangetic delta region starts from mid-June and continues till October.

We present here the data of alteration of physico-chemical variables at 12 different sites in and around Indian Sundarbans (Fig. 4.15) that occurred during Amphan and Yash, the major cyclones that hit Sundarban during 2020 and 2021 respectively. During these cyclonic depressions high wave actions coupled with tidal surges devastated the mangrove habitat of Indian Sundarbans. The salinity increased considerably during both the cyclonic events (Tables 4.23 and 4.24; Figs. 4.16 and 4.17).

We observed that the recovery from the damage is much faster in the western and eastern sectors compared to the central part of Indian Sundarbans. Apart from road connectivity that transports the relief materials very quickly in the western part, the presence of large number of school buildings that act as the shelter sites for local population, has made the western sector less vulnerable to natural calamities. The eastern sector with rich mangrove vegetation acts as the natural shield against

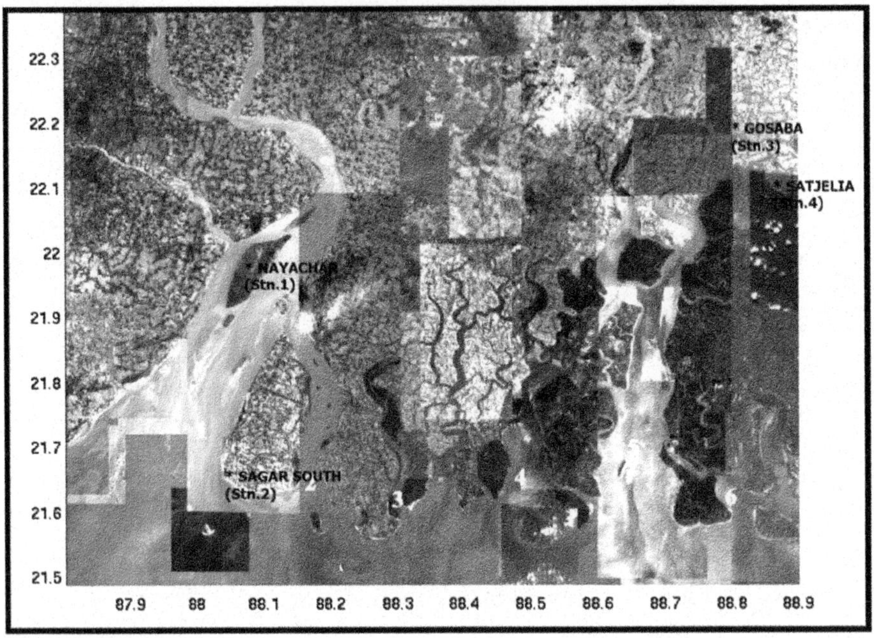

Fig. 4.7 Location of sampling stations in and around Indian Sundarbans

cyclones and provides adequate protection to the embankments. The institutional support is not very high in the central sector as the main administrative offices and Government lodges are mostly concentrated in places like Kakdwip, Namkhana, Sagar Island etc., which are in the western sector.

In blue economy, few major actors like job opportunity, institutional support, biodiversity conservation, alternative livelihood, poverty scenario are crucial factors that determine the sustainability of the vertical. We present here a sector-wise comparative picture of these actors in the framework of Indian Sundarbans (Table 4.25).

4.3 Take Home Messages

A. The major natural threats that are negatively impacting the vertical of blue economy may be categorized in to three major types namely Sea Level Rise (SLR), acidification of the marine and estuarine ecosystems and weather extremes. The warning bell of SLR has also been heard in Indian Sundarbans, at the apex of Bay of Bengal. *The Independent* (a news agency) reported in December 2006 that the first island claimed by rising sea levels caused by global warming in Indian Sundarbans was Lohachara Island. However several scientists

Table 4.18 Major sources of pollution in Indian Sundarban estuaries

Source of aquatic pollution and major pollutants	Western sector	Central sector	Eastern sector	Remarks
 *Source* Industrial zone adjacent to Indian Sundarbans Major pollutants: Heavy metals, oil and grease	+++	–	–	The western Indian Sundarbans is exposed to effluents released from the Haldia port-cum-industrial complex adjacent to the Hooghly estuary
 *Source* Fishing vessels and trawlers Major pollutants: Heavy metals from antifouling paints, oil and grease	+++	+++	–	Fishing vessels, passenger vessels and trawlers are frequently coated with antifouling paints (source of heavy metals) to get rid of corrosion and biofoulers; except movement of tourist vessels such activities are absent in the eastern sector
 *Source* Shrimp farms Major pollutants: Organic wastes, NH_3, NO_3, PO_4, H_2S, microbes, antibiotics	+++	+++	–	Shrimp farms are abundant in western and central sectors; aquacultural activities are absent in the eastern sector due to its presence in Reserve Forest (RF) area

(continued)

have contradicted the correlation between erosion of Lohachara Island and global warming and forwarded tectonic cause as one of the important reasons behind the loss of landmass of Lohachara Island. The main causes behind global warming in Indian Sundarbans are excessive rise of near surface atmospheric carbon dioxide due to deforestation of mangroves for shrimp farms, tourism units, urban developments, fish landing stations, fishing boats and trawlers repairing sites coupled with emissions from industries, vessels and trawlers and brick kilns.

Table 4.18 (continued)

Source of aquatic pollution and major pollutants	Western sector	Central sector	Eastern sector	Remarks
 *Source* Tourism units Major pollutants: Organic wastes, NH_3, NO_3, PO_4, H_2S, microbes, diesel	+++	+++	–	Maximum tourism activities are seen in the western sector due to presence of beaches, landing stations, temples etc.; central sector has moderate flow of tourists and in absence of any staying facilities in the eastern sector, pollution from tourism units is almost absent in this Reserve Forest area
 *Source* Religious gathering Major pollutants: sewage, microbes, diesel	+++	–	–	Pollution from religious gatherings is mostly seen in the western sector due presence of several temples and religious staying facilities. The huge gatherings to take the holy bath in the Hooghly estuary adjacent to Kapil Muni temple every year during Makar Sankranti (in mid-January) deteriorate the water quality
 *Source* Fish landing station Major pollutants: Heavy metals, oil, grease, sewage, microbes, diesel	+++	+	–	Fish landing stations are concentrated in the western sector in places like Kakdwip, Namkhana, Frazergunge etc.; few unorganized fish landing occurs in the central sector and eastern sector mostly under the Reserve Forest does not permit such activities

Table 4.19 Microbial load in the water and sediment of a shrimp farm in western Indian Sundarban

Year	Water TBC (CFU/ml)	Water TC (MPN/100 ml)	Water TFC (MPN/100 ml)	Sediment TBC (MPN/100 g)	Sediment TC (MPN/100 g)	Sediment TFC (MPN/100 g)
2000	12,000,000	230	150	89,000,000	25	30
2001	14,000,000	310	210	94,000,000	35	18
2002	90,00,000	400	270	98,000,000	42	22
2003	16,000,000	530	280	85,000,000	50	19
2004	20,000,000	560	260	90,000,000	53	23
2005	22,000,000	590	230	96,000,000	56	26
2006	26,000,000	670	250	99,000,000	59	28
2007	28,000,000	600	320	12,000,000	63	32
2008	32,000,000	720	370	14,000,000	70	36
2009	29,000,000	970	330	16,000,000	85	40
2010	25,000,000	1150	380	19,000,000	92	43
2011	34,000,000	930	400	15,000,000	78	46
2012	40,000,000	1020	430	18,000,000	84	48
2013	43,000,000	1250	450	20,000,000	98	51
2014	44,000,000	1550	480	17,000,000	100	55
2015	42,000,000	1290	530	15,000,000	120	60
2016	45,000,000	1200	510	22,000,000	130	54
2017	47,000,000	1350	550	23,000,000	145	57
2018	52,000,000	1390	580	24,000,000	150	53
2019	57,000,000	1430	610	63,000,000	175	34
2020	6,000,000	530	200	43,000,000	65	16
2021	9,000,000	710	240	59,000,000	88	21

The root causes behind acidification of Sundarban estuaries is mostly anthropogenic. The rapid phase of industrialization, urbanization, deforestation and change in land-use pattern has increased the carbon dioxide level of the atmosphere. The increase of atmospheric carbon dioxide with the passage of time has resulted in the enhancement of Dissolved Inorganic Carbon (DIC) in the aquatic phase, which has dropped the pH value of the aquatic phase by few units.

Weather extremes refer to significant deviation of environmental variables like wind speed, rainfall, air temperature etc. from their normal values that usually persist in a particular region. In the framework of Indian Sundarbans, weather extreme is observed in terms of wind speed during the period of cyclones; extreme deviation in rainfall and air temperature is observed in the summer and rainy seasons.

B. The foundation of blue economy needs to stand on the strong pillars of environmental quality particularly the water quality of the oceans, seas, bays and estuaries, which supports the resource base of the blue economy. In context

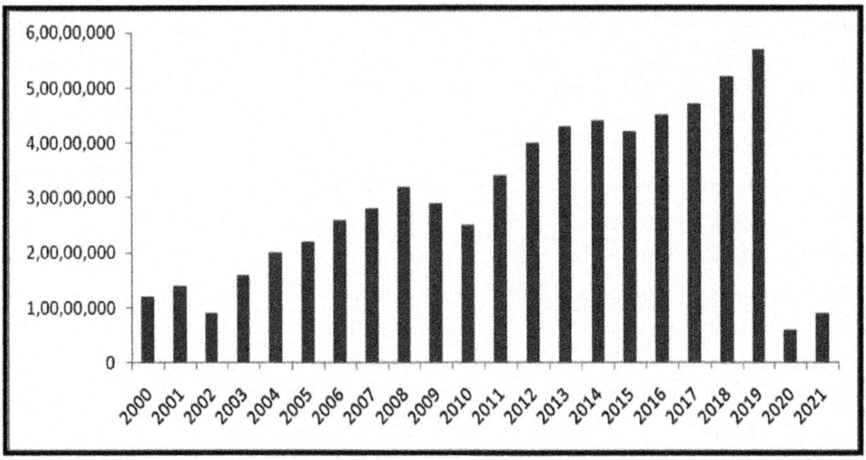

Fig. 4.8 Total Bacterial Count (TBC) in the water of a shrimp farm in western Indian Sundarbans

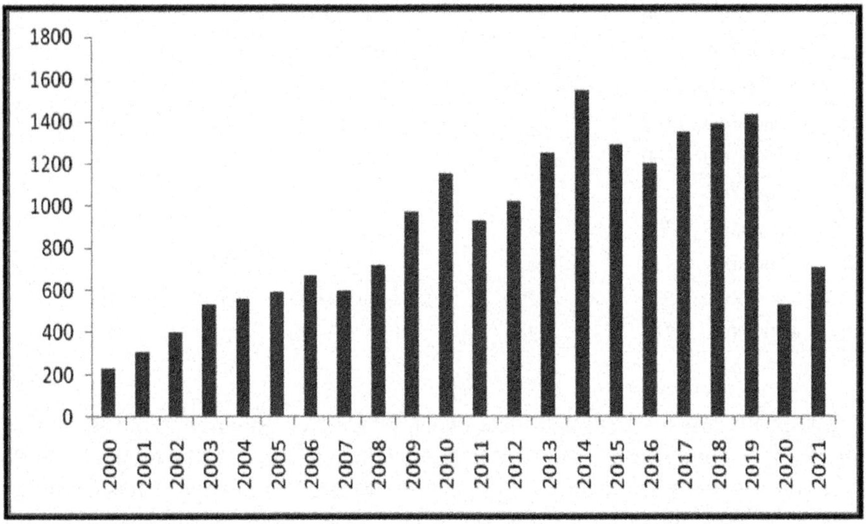

Fig. 4.9 Total Coliform (TC) in the water of a shrimp farm in western Indian Sundarbans

to Indian Sundarbans, the picture of water quality is to some extent gloomy. The main sources of aquatic pollution in and around the mangrove ecosystem are the industries (adjacent to western sectors of Indian Sundarbans), fishing vessels and trawlers, tourism units, shrimp farms, fish landing stations etc. In addition, nutrient run–off from the adjacent agricultural fields also contribute appreciable amount of nitrate and phosphate in the estuaries of Sundarbans. Ballast discharge from the ships and tankers is also a threat to the estuarine

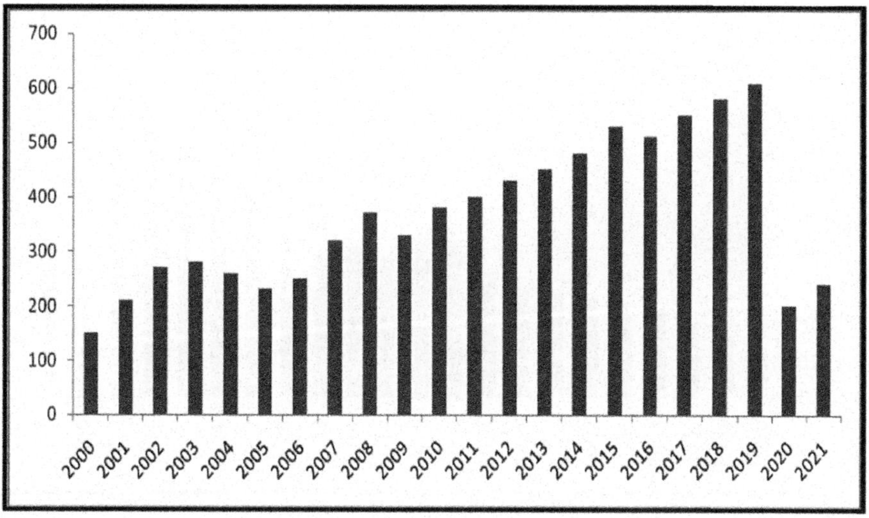

Fig. 4.10 Total Faecal Coliform (TFC) in the water of a shrimp farm in western Indian Sundarbans

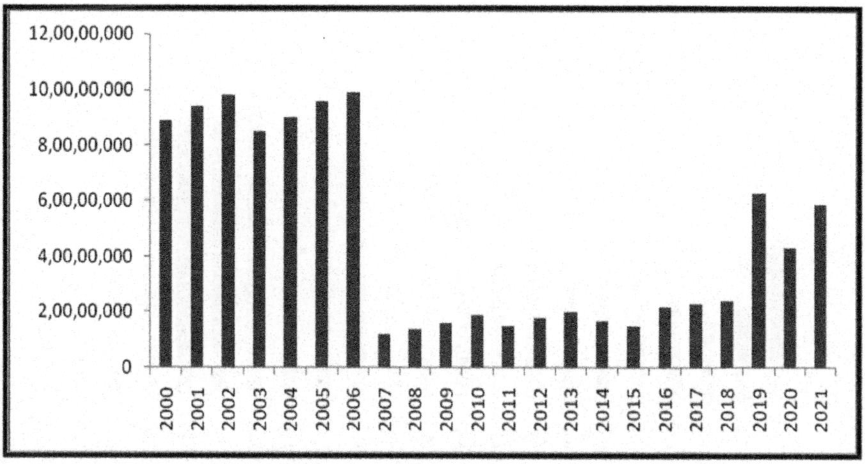

Fig. 4.11 Total Bacterial Count (TBC) in the pond bottom sediment of a shrimp farm in western Indian Sundarbans

species of Indian Sundarbans. The species released from ballast discharge are mostly invasive in nature and may displace the native species, which may pose an adverse impact on the existing food chain of the Sundarban estuaries. Microbial pollution is also a major issue in the domain of blue economy as the fin fishes, shrimps, crabs and other edible sea food items get contaminated from the ambient habitat. In Indian Sundarbans, most of microbial pollution is sourced

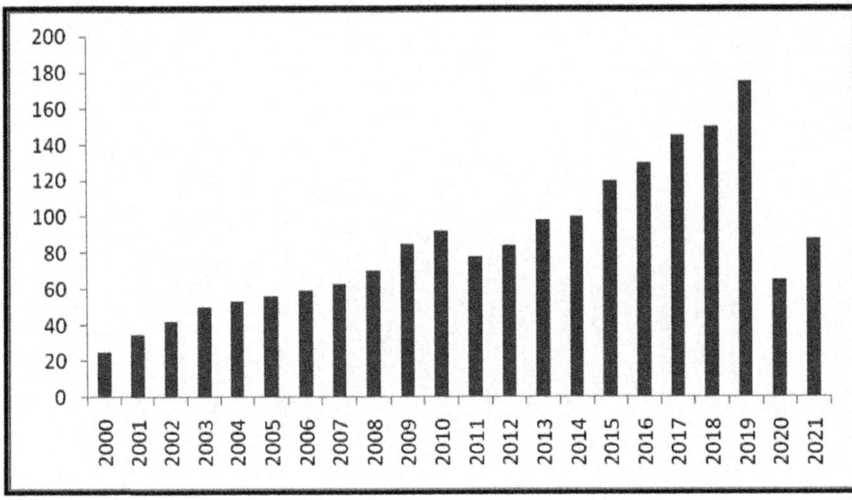

Fig. 4.12 Total Coliform (TC) in the pond bottom sediment of a shrimp farm in western Indian Sundarbans

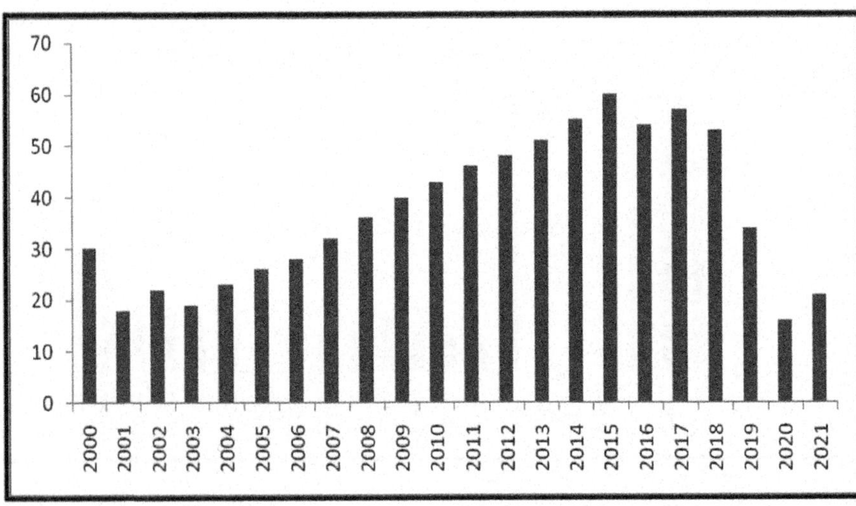

Fig. 4.13 Total Faecal Coliform (TFC) in the pond bottom sediment of a shrimp farm in western Indian Sundarbans

from shrimp farms, followed by the effluents released from the fish landing and tourism units.

C. People of Indian Sundarbans mostly dwell below the poverty line. They have limited access to proper education facility, health care system and various resource bases that are needed to lead a sound life with sustainable livelihoods. The region, being vulnerable to natural disasters like cyclones, wave actions and

Table 4.20 Sector-wise overexploitation of resources in Indian Sundarbans

Resource exploitation	Western sector	Central sector	Eastern sector	Remarks
 Timber from mangroves	+	+++	+	Cutting mangroves for timber is illegally carried out in central Indian Sundarbans and rarely in the Reserve Forest area of eastern sector; in the western sector, the activity is least because of different alternative options and poor mangrove vegetation
 Fuel wood from mangroves	+++	+++	+	In the absence of LPG in most of the households, cooking is done by fuel woods procured from the mangrove forests

(continued)

Table 4.20 (continued)

Resource exploitation	Western sector	Central sector	Eastern sector	Remarks
 Honey and wax from mangroves	–	++	+++	Honey and wax are illegally derived from the core region of eastern Indian Sundarbans and to some extent from the central sector; such activity is not common in the western sector due to absence of dense mangrove forest and several type of livelihood options (like aquaculture, tourism, fish landing related jobs etc.) available to the local population
 Saltmarsh grass as fodder	+++	+++	–	*Porteresia coarctata* is the common fodder provided to the cattle, which is frequently cut in the central sector, but not very common in western and eastern sectors

(continued)

Table 4.20 (continued)

Resource exploitation	Western sector	Central sector	Eastern sector	Remarks
 Fishes from the estuaries	+++	–	–	Over exploitation of fishes is done in the western and central sectors using various type of gears and nets; bottom trawling is done off western sector, which destroys the benthic habitats; in eastern sector, it is completely absent due to strict vigilance by the staffs of forest department
 Ichthyoplankton from the estuaries	+++	+	–	Overexploitation of ichthyoplankton is very common in the western sector because of tiger prawn seed collection, which is very low in the central sector and fully restricted in the eastern sector

(continued)

Table 4.20 (continued)

Resource exploitation	Western sector	Central sector	Eastern sector	Remarks
 Molluscan shell from the intertidal zone	–	+++	–	Overexploitation of molluscs is done at random in the central sector, but completely absent in the western and eastern sectors
 Horseshoe crabs from the estuaries and intertidal mudflats	+ + +	-	-	Horseshoe crabs have become endangered as the arthropods are caught for extracting their blue blood, which is the source of Carcinoscorpius Amoebocyte Lysate (CAL) and Tachypleus Amebocyte Lysate (TAL) and have high demand in the pharmaceutical industries for testing the quality of intravenous injections; the activity is very common in the western sector

Fig. 4.14 Kapil Muni Ashram at Sagar Island adjacent to Hooghly estuary is a hot spot of religious tourism in Indian Sundarbans

tidal surges exaggerate the sufferings of the people to a greater degree. These miserable conditions mostly force them to exploit the existing natural resources without any limitations. Apart from illegal procurement of timber, fuel wood, honey and wax from the deep mangrove forests in the eastern sector of Indian Sundarbans, many local people also over exploit the molluscan shells as sources of lime. Extraction of blue blood of horseshoe crabs (*Carcinoscorpious rotundicauda* and *Tachypleus gigas*) is also reported from Sagar Island. The pattern of overexploitation exhibits significant variation between sectors depending on the availability of resources.

D. Resource base, job opportunities and economic profile in Indian Sundarbans exhibit significant seasonal variations. The western sector, being nearer to the megacity of Kolkata and connected through road ways provide a great advantage to the people of the area as they often migrate to the city to take up the jobs of house keeper, cleaner, gardener and vendor in the local trains. These opportunities of livelihoods are not available to the people of central Indian Sundarbans mainly due their connectivity problem with Kolkata and lack of job opportunities in fish landing activities, religious tourism, brick manufacturing units etc. Most of the islands of central Indian Sundarbans have no electricity, health care units, educational institutes or any industries although there is no scarcity of natural

Table 4.21 A brief summary of extremely severe cyclonic storms (ESCSs) over the Bay of Bengal region during 1990–2020

S. no	Year	Name year-wise	Lifetime	MSW (knots)	Duration during ESCS
1	1990[a]	1990	6 days 6 h	127	60 h
2	1991[a]	1991	5 days 12 h	127	36 h
3	1992[b]	1992	5 days 12 h	102	48 h
4	1993[b]	1993	3 days 6 h	90	3 h
5	1994[a]	1994	3 days 3 h	102	18 h
6	1995[b]	1995	3 days 21 h	102	15 h
7	1997[a]	1997	4 days 6 h	90	12 h
8	1999[b]	1999	4 days	90	3 h
9	1999[b]	Orissa SuCS	5 days 21 h	140	24 h
10	2000[b]	2000	4 days 3 h	102	12 h
11	2000[b]	2000	5 days 9 h	90	3 h
12	2004[a]	2004	3 days	90	6 h
13	2006[a]	Mala	4 days 15 h	100	24 h
14	2007[b]	Sidr	4 days 18 h	115	63 h
15	2008[a]	Nargis	6 days 6 h	90	9 h
16	2009[a]	Aila	2 days	67	24 h
17	2010[b]	Giri	2 days 18 h	105	6 h
18	2013[b]	Phailin	6 days 3 h	115	54 h
19	2014[b]	Hudhud	7 days 6 h	100	27 h
20	2019	Bulbul	3 days	72	2 h
21	2019[a]	Fani	8 days 9 h	116	27 h
22	2020[a]	Amphan SuCS	5 days 12 h	102	60 h
23	2021[a]	Yash	5 days 4 h	112	12 h

VSCS very severe cyclonic storm (wind speed between 64 and 89 knots); ESCS extremely severe cyclonic storm (wind speed between 90 and 119 knots); SuCS super cyclonic storm (wind speed more than 120 knots)
[a]During premonsoon period (March–May); [b]During monsoon and postmonsoon period (October–December); One knot is equal to 1.8 kmph

resources like mangrove based honey, wax, seaweeds, oyster that can generate several innovative livelihood to the local people.

The hypersaline condition in the central sector does not provide a congenial environment for commercially valuable fishes like *Tenualosa ilisha* to breed and stay in the estuaries in this region, thus forcing the island dwellers towards more poverty and subsequent over exploitation of natural resources from the Reserve Forest/Core area in the eastern sector of Indian Sundarbans adjacent to Bangladesh Sundarbans.

Natural calamities in the form of cyclones/super-cyclones and excessive rainfall cause major impact on the lives and livelihoods of Sundarban people not only by

Table 4.22 Average monthly rainfall data (in mm) collected for Canning region, 24 Parganas, South (West Bengal)

Year	Jan	Feb	Mar	Apr	May	Jun	Jul	Aug	Sep	Oct	Nov	Dec
1981	–	–	–	–	–	–	–	–	–	–	0	0.38
1982	0	1.69	1.65	2.77	0.85	5.69	8.49	10.4	4.23	0.36	0.87	0
1983	0.14	1.67	1.2	1.08	2.51	11.22	6.42	17.79	9.96	3.61	0.25	0.18
1984	0.57	0.03	0	2.29	3.19	23.81	7.26	15.55	6.69	3.28	0	0
1985	0.88	0.21	0.2	1.12	5.11	9.25	9.25	12.2	7.69	3.86	0	0
1986	0.33	0	0.05	1.04	6.2	6.34	10.66	3.35	27.91	8.55	7.54	0.05
1987	0	0.7	0.6	5.8	4.21	4.73	11.55	13.41	9.58	0.81	2	0.32
1988	0	0.92	0.12	0.71	5.49	20.29	17.4	9.28	6.38	4.89	4.66	0
1989	0.07	0.25	0.32	0	5.15	2.83	9.1	10.35	11.34	6.63	0.39	0.08
1990	0	2.48	6.35	3.27	4.79	10.18	16.6	10.77	10.4	5.72	2.41	0.06
1991	2	0.64	0.55	1.13	0.49	16.21	8.64	8.07	8.02	4.56	0.02	0.57
1992	0.92	1.73	0	0.59	6.1	13.31	12.13	13.51	16.1	3.8	0	0
1993	0.03	0	2.75	0.77	4.61	10.34	11.34	11.55	17.73	3.97	0.43	0
1994	0.11	0.84	0.25	3.82	2.35	8.21	14.93	9.12	6.46	2.52	1.92	0
1995	0.23	1.36	0.08	0.61	7.39	11.37	17.6	10	17.84	3.08	6.68	0
1996	0.29	0.3	0.07	0.89	2.56	18.97	11.88	16.72	5.09	9.94	0	0
1997	0.57	0.49	1.36	6.64	3.41	7.12	12.98	15.78	9.36	0.16	0.49	0.48
1998	0.98	0.14	5.97	2.13	5.86	4.94	12.07	12.09	10.41	5.31	5.84	0
1999	0.08	0	0.01	0.05	6.72	8.94	16.47	13.38	14.91	5.96	0.07	0
2000	0.05	1.96	0	2.29	10.79	7.79	15.21	6.34	11.69	4.22	0.06	0.06
2001	0.09	0.01	0.8	3.43	6.8	14.54	10.63	6.87	5.84	5.14	0.62	0
2002	0.85	0	0.53	2.66	4.01	19.81	8.78	10.05	8.31	3.97	4.62	0
2003	0	0.51	0.9	2.26	4.36	10.65	11.9	11.49	9.37	17.09	0.67	1.29
2004	0.15	0.25	0.13	2.26	2.42	8.96	13.35	14.93	7.75	8.77	0	0
2005	1.47	0	4.25	1.24	2.27	8.3	17.89	9.17	11.73	16.74	0	0.03
2006	0	0	0.11	2.06	3.9	4.08	21.16	14.55	17.35	1.045	0.03	0
2007	–	2.8	0.07	1.74	5.63	5.32	20.25	8.97	21.39	3.9	2.13	0
2008	3.46	0.34	0.15	1.4	2.26	12.22	7.16	8.96	9.62	6	–	–
2009	0	0	0.53	0.03	6.52	1.79	13.19	11.47	10.05	4.93	0.34	0
2010	0	0.13	0	0.72	4.25	8.44	8.53	7.56	7.24	4.51	0.12	0.36
2011	0	0.24	1.2	7.31	3.34	14.09	8.98	22.56	7.86	3.1	–	0
2012	1.57	1.01	0.17	3.15	1.52	6.08	10.2	14.78	8.35	4.28	1.86	1.15
2013	0.12	0.18	0.15	0.63	6.36	8.25	8.65	22.08	14.84	9.3	0	0
2014	0	2.33	1.77	0	–	5.78	7.35	–	11.37	1.7	0	0
2015	0.14	0.19	1.83	0.74	4.24	8.21	6.80	13.14	20.14	2.85	0	0

(continued)

Table 4.22 (continued)

Year	Jan	Feb	Mar	Apr	May	Jun	Jul	Aug	Sep	Oct	Nov	Dec
2016	0	0.07	1.09	0	5.10	4.98	5.93	9.68	12.65	3.10	0.12	0
2017	0	0.14	0	0	1.26	11.02	7.80	12.65	9.89	4.02	0	0.14
2018	0.10	1.02	1.21	0.35	3.58	13.65	11.66	14.20	8.42	5.17	0	1.02
2019	1.23	1.45	1.66	0.41	2.97	17.04	13.45	12.06	9.17	8.03	1.05	1.33
2020	1.42	1.27	1.54	0.66	1.56	13.65	12.04	9.28	10.44	11.55	2.40	1.91
2021	1.18	1.34	0.22	0.38	1.11	15.39	19.56	13.76	12.15	14.59	3.26	2.54

Source IMD (Indian Meteorological Department; '–' means data not available; Jan—January; Feb—February; Mar—March; Apr—April; May—May; Jun—June; Jul—July; Aug—August; Sep—September; Oct—October; Nov—November; Dec—December

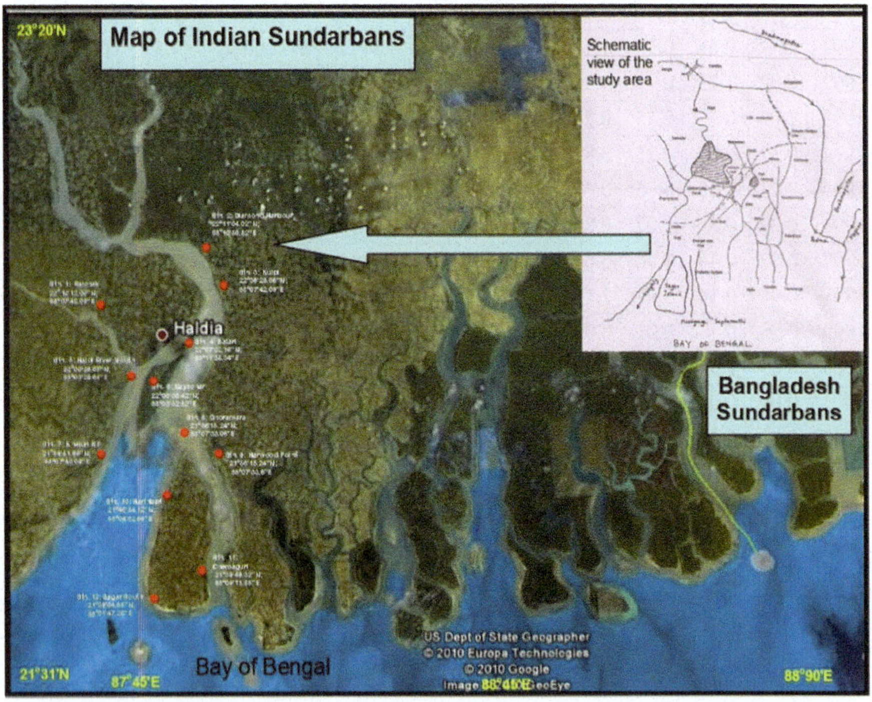

Fig. 4.15 Map showing the locations of 12 sampling stations in and around Indian Sundarbans

damaging the embankments, but also by disrupting the agricultural production and fresh water fish farming that cannot withstand high salinity due to salt water intrusion during cyclones that occur very frequently. All these threats have put sticks in the smooth running of blue economy in Indian Sundarbans.

Table 4.23 Alteration of salinity between three phases of Amphan that hit lower Gangetic delta in 2020

Station	Pre-Amphan phase	Amphan phase	Post-Amphan phase
Stn. 1	3.80	5.07	4.31
Stn. 2	5.27	6.94	5.82
Stn. 3	6.49	8.54	7.04
Stn. 4	13.56	17.33	14.84
Stn. 5	12.44	17.96	15.64
Stn. 6	14.23	18.24	14.87
Stn. 7	15.94	20.41	16.84
Stn. 8	14.82	19.05	15.73
Stn. 9	17.61	22.59	18.82
Stn. 10	18.72	25.31	20.91
Stn. 11	20.97	26.94	22.86
Stn. 12	23.47	30.78	25.53

Table 4.24 Alteration of salinity between three phases of Yash that hit lower Gangetic delta in 2021

Stations	Pre-Yash phase	Yash phase	Post-Yash phase
Stn. 1	4.82	8.98	5.49
Stn. 2	7.29	9.92	5.99
Stn. 3	7.51	13.52	12.16
Stn. 4	12.58	20.31	14.94
Stn. 5	11.46	18.94	13.74
Stn. 6	13.25	21.22	14.97
Stn. 7	13.96	23.39	16.94
Stn. 8	14.02	22.03	15.83
Stn. 9	15.63	25.57	18.92
Stn. 10	19.74	28.29	21.01
Stn. 11	19.02	29.92	21.96
Stn. 12	22.49	35.76	29.63

4.4 Annexure 4A

Above Ground Biomass (AGB) of dominant five mangroves floral species* along with DIC and pH values of the ambient aquatic phase of 24 sampling stations in Indian Sundarbans.

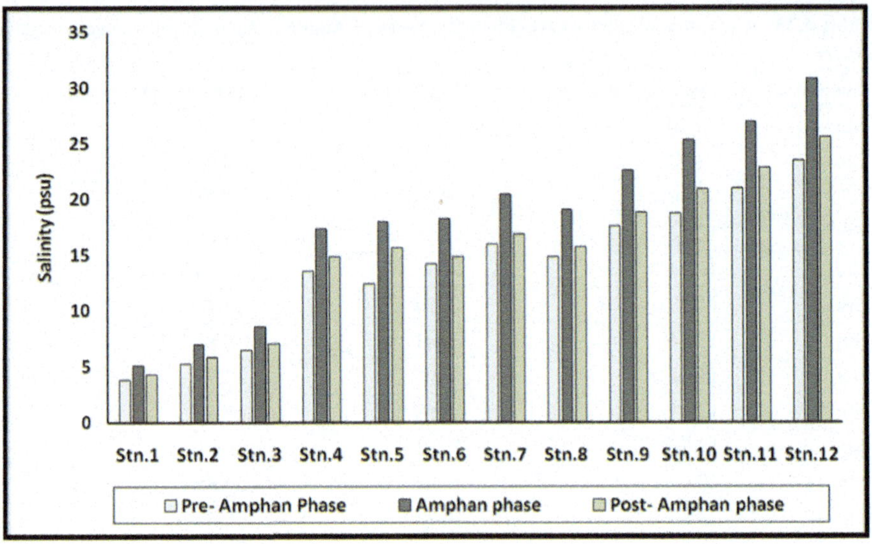

Fig. 4.16 Alteration of salinity during three phases of Amphan that hit lower Gangetic delta during 2020

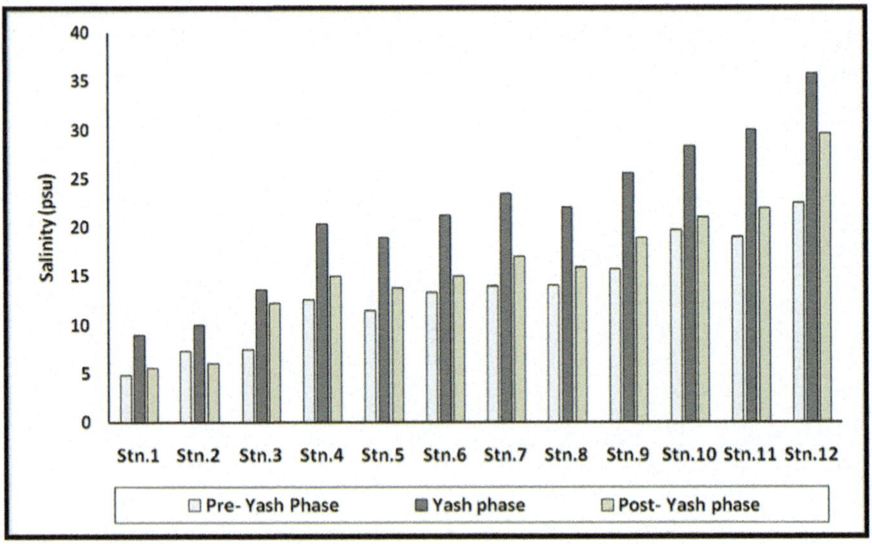

Fig. 4.17 Alteration of salinity during three phases of Yash that hit lower Gangetic delta during 2021

Table 4.25 Sector-wise actors related to sustainability of blue economy in Indian Sundarbans

Major actor	Western sector	Central sector	Eastern sector	Remark
Job opportunity				Job opportunities are maximum in the western sector in shrimp farms, fish landing stations, brick kilns, deep sea fishing, ice factory, tourism units and vessels/trawlers repairing units. In addition, the religious tourism centered around Kapil Muni temple also boost up the local economy during the postmonsoon season (preferably in January of every year); Opportunity for job is moderate in the central sector and limited to shrimp farms and occasional tourism that take a hike during the postmonsoon period (December – February); the eastern sector is under the Reserve Forest zone and limited access is given for honey collection
Institutional Support				Pilgrims, tourism units, administrative offices, fish landing stations are mostly concentrated in the western sector due to which the region receives maximum institutional support; the eastern sector being under the Reserve Forest zone receives extra attention from the state Forest department in terms of protection of natural resources; institutional support is minimum in the central sector due to lack of connectivity and administrative set ups except in Gosaba
Biodiversity conservation				Maximum effort on species conservation is given in the eastern part owing to its location within the core zone of Indian Sundarban Biosphere Reserve; the people of central Indian Sundarbans try to conserve the biodiversity as their livelihood is very much dependent on tourism; biodiversity faces an adverse impact in the western part because of extreme anthropogenic activities in this sector
Alternative livelihood				Alternative livelihood is given great stress in the central sector as there is minimum provision of getting entry in the main stream livelihood options like deep sea fishing, fish landing related jobs, religious tourism, beach tourism, large scale shrimp farming etc. that
				are opted by the people of western sector; in the eastern part there is no set up for alternative livelihood or any mangrove resource – based cottage industry due to restrictions imposed by the West Bengal State forest department
Poverty				Poverty is minimum in the western sector due to several categories of job options ranging from shrimp farming to fish landing. Religious tourism concentrated in the Sagar Island has added an extra wing to the local economic profile; due to restricted job opportunities (except in few shrimp farms and occasional tourism), most of the people of central Indian Sundarbans live below the poverty line; eastern Indian Sundarbans is a restricted zone where there is no inhabitation

Index:

Maximum in magnitude	Moderate in magnitude	Minimum in magnitude	No information

Stn. no	Sampling station	Coordinates	AGB (tonnes/ha) Date type: Primary	DIC (μmol/L)	pH value
1	Muriganga	21° 38′ 25.86″ N; 88° 08′ 53.55″ E	184.055	1927	8.04
2	Saptamukhi	21° 36′ 02.49″ N; 88° 23′ 47.18″ E	184.289	1869	8.09
3	Thakuran	21° 49′ 43.17″ N; 88° 33′ 21.57″ E	184.366	1808	8.21
4	Herobhanga	21° 59′ 34.32″ N; 88° 41′ 46.52″ E	184.482	1805	8.27
5	Ajmalmari	21° 51′ 34.72″ N; 88° 39′ 00.68″ E	184.345	1821	8.19
6	Dhulibasani	21° 47′ 06.62″ N; 88° 33′ 48.20″ E	184.466	1812	8.25
7	Chulkathi	21° 41′ 53.62″ N; 88° 34′ 10.31″ E	184.478	1744	8.26
8	Arbesi	22° 11′ 43.14″ N; 89° 01′ 09.04″ E	184.555	1781	8.32
9	Jhilla	22° 09′ 51.53″ N; 88° 57′ 57.07″ E	184.543	1775	8.30
10	Pirkhali	22° 06′ 00.97″ N; 88° 51′ 06.04″ E	184.539	1779	8.30
11	Panchmukhani	21° 59′ 41.58″ N; 88° 54′ 14.71″ E	184.541	1784	8.31
12	Harinbhanga	21° 57′ 17.85″ N; 88° 59′ 33.24″ E	184.525	1786	8.29
13	Khatuajhuri	22° 03′ 06.55″ N; 89° 01′ 05.33″ E	184.528	1806	8.29
14	Chamta	21° 53′ 18.56″ N; 88° 57′ 11.40″ E	184.495	1873	8.28
15	Matla	21° 53′ 15.30″ N; 88° 44′ 08.74″ E	184.408	1871	8.14
16	Chandkhali	21° 51′ 13.59″ N; 89° 00′ 44.68″ E	184.417	1875	8.18
17	Goashaba	21° 43′ 50.64″ N; 88° 46′ 41.44″ E	184.41	1801	8.17
18	Gona	21° 41′ 15.44″ N; 88° 54′ 31.09″ E	184.48	1805	8.27
19	Chhotahardi	21° 44′ 42.24″ N; 88° 44′ 17.79″ E	184.477	1797	8.26
20	Baghmara	21°39′ 04.45″ N; 89° 04′ 40.59″ E	184.486	1788	8.28
21	Mayadwip	21° 35′ 50.23″ N; 88° 47′ 09.95″ E	184.49	1782	8.29

(continued)

(continued)

Stn. no	Sampling station	Coordinates	AGB (tonnes/ha) Date type: Primary	DIC (μmol/L)	pH value
22	Jambu Island	21° 35′ 42.03″ N; 88° 10′ 22.76″ E	184.037	2030	7.98
23	Lothian	21° 38′ 21.20″ N; 88° 20′ 29.32″ E	184.562	1796	8.31
24	Sagar Island	21° 38′ 51.55″ N; 88° 02′ 20.97″ E	183.929	2138	7.97

**Avicennia alba, Avicennia marina, Avicennia officinalis, Excoecaria agallocha and Sonneratia apetala* are the five dominant true mangrove floral species in Indian Sundarbans that contribute about 63% of the total true mangrove floral species

References

Agarwal S, Mitra A (2018) Bioaccumulation pattern of heavy metals in water, sediment and barnacle (*Balanus balanoides*) in coastal West Bengal. Int J Sci Basic Appl Res 8(11):683–693

Banerjee K, Amin G, Chakraborty S, Pramanick P, Ghosh R, Fazli P, Zaman S, Biswas P, Pal N, Mitra A (2014) Bioaccumulation pattern of Heavy metals in saltmarsh grass (*Porteresia coarctata*) of Indian Sundarbans. J Energy Environ Carbon Cred (STM) 4(3):1–10

Bhattacharyya SB, Roychowdhury G, Zaman S, Raha AK, Chakraborty S, Bhattacharjee AK, Mitra A (2013) Bioaccumulation of heavy metals in Indian white shrimp (*Fenneropenaeus indicus*: a time series analysis). Int J Life Sci Biotech Pharm Res 2(2):97–113

Chakraborty S, Zaman S, Fazli P, Mitra A (2014) Bioaccumulation pattern of heavy metals in three mangrove species of *Avicennia* inhabiting lower Gangetic delta. J Chem Biol Phys Sci 4(4):3884–3896

Chakraborty S, Zaman S, Mitra A (2017) Bioaccumulation of heavy metals by dominant oyster in Indian Sundarbans. J Sci Eng Health Manage, Published by Techno India University, West Bengal 1(2):34–39

Chaudhuri AB, Choudhury A (1994) Mangroves of the Sundarbans. IUCN, India

Das J, Halder KC, Banerjee K, Bhattacharyya DP, Mitra A (2003) Bioaccumulation of Zn, Cu and Pb in Catenella repens inhabiting the northern tip of Sagar Island, Sundarbans. Sea Explorers 6:63–68

Dhar I, Sengupta G, Biswas S, Sinha M, Mitra A (2021) Salinity: a major environmental factor in sustainability of the Blue Carbon. J Mech Conti Mathe Sci 16(11):34–42. https://doi.org/10.26782/jmcms.2021.11.00004

Dutta P, Zaman S, Mitra A (2018) Bioaccumulation of selected heavy metals in tissue and organs of *Oreochromis niloticus*. Int J Basic Appl Res 8:54–59

Ghosh I, Mitra A, Rudra T, Pramanick P, Biswas PP (2016) Bioaccumulation pattern of heavy metals by gastropods: a case study from lower Gangetic delta. J Environ Sci Comp Sci Eng Technol 5(1):99–102

Hoegh-Guldberg O et al (2015) Reviving the ocean economy: the case for action. WWF International, Gland, Switzerland, pp 60

http://www.ipcc.ch/SPM2feb07.pdf

Mitra A (2000) The Northwestern Bay of Bengal and deltaic Sundarbans. In: Sheppard C (ed) Seas at the millennium: an environmental evolution. The Netherlands, Elsevier Science ltd.

Mitra A, Ghosh R (2014) Bioaccumulation pattern of Heavy metals in commercially important fishes on and around Indian Sundarbans. Glob J Anim Sci Res 2(1):33–45

Mitra A, Banerjee K, Sengupta K, Gangopadhyay A (2009) Pulse of climate change in Indian Sundarbans: a myth or reality? Nat Aca Sci Let 32(1 & 2):19–25

Mitra A, Banerjee K, Ghosh R, Ray SK (2010) Bioaccumulation pattern of heavy metals in the shrimps of the lower stretch of the River Ganga. Mesopo J Mar Sci 25(2):1–14

Mitra A, Banerjee K, Sinha S (2011a) Shrimp tissue quality in the lower Gangetic delta at the apex of Bay of Bengal. Toxicol Environ Chem 93(3):565–574

Mitra A, Halder P, Banerjee K (2011b) Changes of selected hydrological parameters in Hooghly estuary in response to a severe tropical cyclone (Aila). Indian J Geo Mar Sc 40(1):32–36

Mitra A (2013) Sensitivity of mangrove ecosystem to changing climate. Published by Springer New Delhi Heidelberg, New York Dordrecht London, ISBN-10: 8132215087; ISBN-13: 978–8132215080. ISBN 978–81–322–1509–7 (eBook)

Mitra A (2019) Estuarine pollution in the lower Gangetic delta. Published by Springer International Publishing, ISBN 978–3–319–93305–4, XVI:371

Mitra A (2020) Mangrove forests in India exploring ecosystem services. Springer, Chem, eBook ISBN 978–3–030–20595–9, XV, pp. 361. https://doi.org/10.1007/978-3-030-20595-9

Mitra A, Zaman S (2014) Carbon sequestration by coastal floral community. Published by The Energy and Resources Institute (TERI) TERI Press, India. ISBN 978–81–7993–551–4

Mitra A, Zaman S (2015) Blue carbon reservoir of the blue planet. Published by Springer, ISBN 978–81–322–2106–7, XII, pp. 299. https://doi.org/10.1007/978-81-322-2107-4

Mitra A, Zaman S (2016) Basics of marine and estuarine ecology. Publisher Springer India, ISBN 978–81–322–2707–6, pp 1–481

Mitra A, Zaman A (2020) Environmental science: a ground zero observation on the Indian subcontinent. Springer, Chem, eBook ISBN 978–3–030–49131–4, XIV, pp. 478. https://doi.org/10.1007/978-3-030-49131-4

Mitra A, Zaman S (2021) Estuarine acidification: exploring the situation of mangrove dominated Indian Sundarban Estuaries. Springer, Chem, e-Book ISBN 978–3–030–84792–0, XII, pp. 402. https://doi.org/10.1007/978-3-030-84792-0

Mukherjee P, Nayak B, Mitra A, Roy M (2017a) A comparative analysis of bioaccumulation of Fe, Zn, Cu and Pb between the vegetative parts of *Suaeda nudiflora*, a dominant mangrove associate species of Indian Sundarbans. J Sci Eng Health Manage Published by Techno India University, West Bengal 1(1):16–19

Mukherjee P, Nayak B, Mukherjee S, Aman Ansari A, Paul M, Roy M, Mitra A (2017b) Bioaccumulation of Fe, Zn, Cu and Pb in the mangroves of western Indian Sundarbans. Int J Comp Sci Eng Technol 2(5):74–77

Mukherjee P, Zaman S, Mitra A, Roy M (2017c) Bioaccumulation of Zn, Cu, Pb and Cd in the root and rhizosphere soil of *Avicennia marina*: a case study from Indian Sundarbans. J Sci Eng Health Manage Published by Techno India University, West Bengal 1(4):40–42

Nayak B, Roy M, Chaudhuri TR, Zaman S, Mitra A (2016) Bioaccumulation of Heavy metals in sand binder *Ipomoea pes-caprae*: a case study from lower Gangetic delta region. Int J Trend Res Develop 3(2):358–361

Nayak B, Roy M, Roy S, Mondal Rana S, Zaman S, Mitra A (2017) Bioaccumulation factor (BAF) of heavy metals in *Ipomoea pes-caprae*: a case study from Indian Sundarbans. Int J Adv Res 5(3):10–15

Pramanick P, Zaman S, Rudra T, Guha A, Mitra A (2015a) Heavy metals in a dominant seaweed species from the islands of Indian Sundarbans. Int J Life Sci Pharm Res 5(2):64–71

Pramanick P, Zaman S, Trivedi S, Bera D, Mitra A (2015b) Seasonal variation of heavy metals in *Catenella repens* in Indian Sundarbans. Ame J Biol Pharma Res 2(4):149–152

Pramanick P, Zaman S, Rudra T, Guha A, Mitra A (2016) Comparative account of Heavy metal level in two species of Chlorophyceae inhabiting Indian Sundarbans. Int J Res Chem Environ 6(2):65–68

Chapter 5
Sustainable Management of Resource Base in Indian Sundarbans

Abstract The deltaic complex of Indian Sundarbans is an aggregation of some 105 islands girdled and inter-connected by estuaries, tidal rivers, creeks and inlets. The resource base of the region consists of true mangrove flora, mangrove associate species, and several types of fauna, many of which are endangered and vulnerable. The ecosystem services provided by the living and non-living resources of the deltaic region offer economic security, food security and even health security to the local population. This chapter explores several scientific, technological and policy level approaches to manage the resource base that is connected to blue economic sectors of the region after a comprehensive SWOT analysis. Sustainability of blue economy also depends on the magnitude and mode of utilization of endemic resources and carbon footprint associated with different anthropogenic activities including livelihoods. Considering this background, the chapter also highlights the carbon sequestration potential of endemic mangrove floral species that are relevant to blue conservation policies.

5.1 Technology Based Management

Blue economy stands on the pillars of wide variety of biotic and abiotic resources of the oceans, seas, bays and estuaries, which could determine the course of development and sustainability of this vertical. For centuries, resources of the marine and estuarine ecosystems are serving as the lifelines that have the potential to absorb the rising trend of atmospheric carbon dioxide. The property of absorbing carbon dioxide from the atmosphere not only regulates the balance of gaseous substances in the Earth's atmosphere, but also controls the temperature, which is an important switch to trigger various categories of natural hazards. All the coastal nations in the world are dependent on oceans for fishing, minerals, oil and gas, rare earth metals, renewable energy and other living and non-living resources for earning livelihood, achieving holistic growth, empowering native coastal communities and attaining greater social and economic inclusion. The spirit of blue economy which aims at optimum and sustainable use of aquatic resources for growth and development is extremely crucial for the coastal nations especially for the small island developing states.

A. Mitra et al., *Blue Economy in Indian Sundarbans*,
https://doi.org/10.1007/978-3-031-07908-5_5

Indian Sundarbans in the lower Gangetic delta region is one such cluster of some 105 islands, which is enriched with genetic diversities scattered in the mangrove forest, underlying saline soil, estuarine water and intertidal mudflats. Apart from the biotic resources, the tidal range, wind velocity, solar radiation, extensive beaches are important abiotic resources of the region that can provide momentum to the regional blue economy. However, several natural and anthropogenic factors often add sticks to the spokes of blue economy causing retardation of the progress in the sectors of fisheries and aquaculture, shipping and transportation of goods, estuary/sea based recreation and tourism, marine biodiversity conservation and non-conventional energy sector etc.

We carried out a SWOT analysis during 2021 (Tables 5.1, 5.2, 5.3 and 5.4) in three sectors of Indian Sundarbans (viz. western, central and eastern sectors), which are markedly different in terms of salinity and geomorphological features that impact the biotic resources of the region (Banerjee et al. 2010, 2013, 2017; Mitra et al. 2010a, 2011a, b, c; Bhattacharjee et al. 2013a; Mitra 2013, 2018, 2020; Zaman et al. 2013; Mitra and Zaman 2014, 2015, 2016, 2020, 2021; Ghosh and Mitra 2015; Trivedi et al. 2016; Pal et al. 2017).

To obtain data for SWOT analysis, stratified random sampling technique was employed and data was collected from 988 respondents comprising of 375 fisherman, 264 vessel owners, 170 hotel/tourism unit owners and 179 researchers using questionnaires.

Table 5.1 reflects that the western sector of Indian Sundarbans has high strength in the fisheries and aquaculture sector owing to the presence of several fish landing stations at Kakdwip, Namkhana, Frazerganj, Bakkhali etc. Shrimp farms have also proliferated in this sector since early nineties due to congenial aquatic salinity and

Table 5.1 Strength of the relevant sectors of blue economy in the framework of Indian Sundarbans

Sector	Western Sector	Central Sector	Eastern sector
Fisheries and Aquaculture	Maximum Strength	Less Strength	No Information
Shipping and transportation of goods	Maximum Strength	Less Strength	Moderate Strength
Estuary/Sea-based recreation and tourism	Maximum Strength	Moderate Strength	Maximum Strength
Marine biodiversity and conservation	Less Strength	Moderate Strength	Maximum Strength
Non-conventional renewable energy sector	Moderate Strength	Less Strength	No Information

Strength Index

Maximum Strength	Moderate Strength	Less Strength	No Information

Table 5.2 Weakness of the relevant sectors of blue economy in the framework of Indian Sundarbans

Sector	Western Sector	Central Sector	Eastern sector
Fisheries and Aquaculture	Moderate Weakness	Maximum Weakness	No Information
Shipping and transportation of goods	Less Weakness	Maximum Weakness	Less Weakness
Estuary/Sea-based recreation and tourism	Less Weakness	Moderate Weakness	Less Weakness
Marine biodiversity and conservation	Maximum Weakness	Maximum Weakness	Less Weakness
Non-conventional renewable energy sector	Less Weakness	Maximum Weakness	No Information

Weakness Index

Maximum Weakness	Moderate Weakness	Less Weakness	No Information

Table 5.3 Opportunity of the relevant sectors of blue economy in the framework of Indian Sundarbans

Sector	Western Sector	Central Sector	Eastern sector
Fisheries and Aquaculture	Maximum Opportunity	Less Opportunity	No Information
Shipping and transportation of goods	Maximum Opportunity	Less Opportunity	Maximum Opportunity
Estuary/Sea-based recreation and tourism	Maximum Opportunity	Less Opportunity	Maximum Opportunity
Marine biodiversity and conservation	Less Opportunity	Less Opportunity	Maximum Opportunity
Non-conventional renewable energy sector	Moderate Opportunity	Less Opportunity	No Information

Opportunity Index

Maximum Opportunity	Moderate Opportunity	Less Opportunity	No Information

high nutrient load. It is to be mentioned in this context that the Hooghly estuary in the western sector of Indian Sundarbans is the migratory route of *Tenualosa ilisha* for breeding in the upstream region at almost zero salinity. This Indian shad has high value in the export basket of the country and thus has important role in strengthening

Table 5.4 Threat to relevant sectors of blue economy in the framework of Indian Sundarbans

Sector	Western Sector	Central Sector	Eastern sector
Fisheries and Aquaculture	[Moderate Threat]	[Maximum Threat]	[No Information]
Shipping and transportation of goods	[Less Threat]	[Maximum Threat]	[Less Threat]
Estuary/Sea-based recreation and tourism	[Less Threat]	[Moderate Threat]	[Less Threat]
Marine biodiversity and conservation	[Maximum Threat]	[Moderate Threat]	[Less Threat]
Non-conventional renewable energy sector	[Moderate Threat]	[Maximum Threat]	[No Information]

Threat Index

Maximum Threat	Moderate Threat	Less Threat	No Information

the regional blue economy. The central Indian Sundarbans with a hypersaline environment lacks commercially important fish diversity and sustains mostly low valued trash fishes (Fig. 5.1). Fish landing activities are also carried out sporadically in an unorganized manner due to lack of proper landing facilities, icing, storage facility and transportation. Thus this sector is not up to the mark at present state to link it with the regional blue economy.

The eastern sector is under the Reserve Forest area and no fishing and aquaculture activities are permitted in this zone. Although there is no information on commercial fishing and aquaculture in this zone due to heavy restrictions, but from the sustainability point of view, the zone has high importance as the natural supply of fishes

Fig. 5.1 Trash fishes with low commercial values are dominant in central Indian Sundarbans

originate from this core area where there are no anthropogenic disturbances/activities and the aquatic phase has favorable salinity, nutrients and pH for breeding, growth and survival of the fishes.

The success of shipping and transportation of goods through the estuaries of Indian Sundarbans depends primarily on the draft of the navigational channel. In the western sector of Indian Sundarbans, there is a sufficient draft due to which the ships and vessels can have a free passage to Haldia and Kolkata ports. However, in the central sector, due to heavy siltation and embankment erosion, the draft has decreased to a great extent. We even had the experience that during low tide the fishing vessels, trawlers, tourist boats and passenger vessels get stuck with the underlying sediment bed. The estuaries in the eastern sector have considerable draft for the movements of ships, tankers and large vessels.

The estuary/sea-based recreation and tourism has great scope in the western sector because of the presence of several tourism units, sandy beaches, fish landing stations, mangrove patches etc. The Kapil Muni Ashram at Sagar Island attracts about 10–15 lakhs people every year during Makar Sankranti, in mid January, when people from different corners of the country assemble at the confluence of the Hooghly estuary and the Bay of Bengal to take the holy bath. This has placed the site in the map of religious tourism. The central sector does not attract the tourists due to absence of standard lodging facilities and dense mangrove forests. The embankment erosion is a great issue in this sector, which negatively impacts the tourist flow in this region. The eastern sector has luxuriant mangrove vegetations with a variety of wild animals like Royal Bengal Tigers (*Panthera tigris*), deer, crocodiles, monkeys, snakes, wild boars and several species of migratory birds. All these attract the tourist from different corners of the world to enjoy the wilderness of the region. This sector also attracts the researchers, academicians, students and nature lovers and people from different strata of the society to enjoy the aesthetic beauty and learn the art of nature.

The western sector of Indian Sundarbans is exposed to pollutions from fish landing stations, tourism units, shrimp farms, brick kilns and industries mostly concentrated in the Haldia region, adjacent to this sector. This has caused a negative impact on the biodiversity of the region, unlike the central Indian Sundarbans, where the anthropogenic activities are relatively less except tourism. The eastern sector is a unique gene pool supporting a wide range of flora and fauna and is under strict protection as per the rule of Reserve Forest. Thus biodiversity and conservation activities have the highest magnitude in the eastern sector of Indian Sundarbans, adjacent to Bangladesh.

In the vertical of non-conventional renewable energy, the generation of tidal and wind energy can emerge with moderate strength in the western sector due to optimum tidal amplitude and wind velocity. However, in the central sector this particular vertical of blue economy cannot be implemented because of siltation problem and frequent erosion of the unstable embankments. The silt particles have high chance to clog the turbines of the tidal energy generator and need a huge recurring expenditure for maintenance. The eastern sector, being under the Reserve Forest area, does not permit any hard engineering construction and hence there is little information on the implementation of non-conventional renewable energy related activities in this region.

When the weakness point (Table 5.2) is considered in the fisheries and aquaculture sector, we see that the western Indian Sundarbans has moderate weakness, although there are large numbers of fish landing centers and aquaculture farms. The main reason behind this is the pollution level of the estuarine water in terms of heavy metals, oil and grease, pesticides etc. (Mitra 2013, 2019). This sector of Indian Sundarbans directly faces the Haldia port—cum—industrial complex; furthermore all the wastes from the megacity of Kolkata are discharged in the Hooghly estuary in the western Indian Sundarbans. Hence the chance of bioaccumulation of conservative pollutants (like heavy metals, pesticides etc.) in the fish tissues cannot be ruled out. Fisheries and aquaculture sector cannot upgrade/boost the regional blue economy if the scenario of central Indian Sundarbans is considered. This is primarily because the hypersaline aquatic phase in central Indian Sundarbans causes fish mortality, their reproductive failure and absence of many commercially important fishes. It has also been documented that Hilsa fish (*Tenualosa ilisha*) which is a delicacy of fish lovers also avoids the Matla estuary in the central sector because of high salinity. The eastern Indian Sundarbans, being a restricted zone does not permit any type of fishing and aquaculture activities by the State Forest Department, and hence there is lack of information on fisheries and aquaculture status in this sector adjacent to Bangladesh Sundarbans.

The primary weakness of the shipping and transportation sector in the western Indian Sundarbans is the siltation problem although the navigational channel has sufficient draft. However, in the central sector, the shipping and transportation of goods are not expected to flourish at highest level because of the draft problem, which is a major weakness in this region. The eastern sector has sufficient draft and the movements of ships and vessels are easy without much hindrance. This sector, being the border with the neighboring country, Bangladesh is always under strict vigilance by the Border Security Force deployed by the Government of India (Fig. 5.2).

The western sector has great potential to promote estuaries/sea-based recreation and tourism due to presence of several lucrative sites, beaches, religious places (at Sagar Island) and hence the weakness is less. In the central sector, the weakness component of tourism is medium due to lack of lodging facilities and erosion problems. The eastern sector is a heaven for the tourists as the region supports dense mangrove forests with wildlife.

Biodiversity spectrum and its conservation are extremely weak in the western sector because of tremendous anthropogenic activities like fish landing, tourism, religious gatherings etc. Large numbers of brick kilns have come up in this sector, which are the major sources of CO_2 in the atmosphere. People in this sector are so business/commercial minded, preferably who are involved in shrimp culture and fishery activities that they care a little about the environment. Most of the shrimp farms in this region have been constructed at the cost of mangroves. Awareness on conservation of nature is also very limited. In the central sector, the hypersaline environment does not promote the biodiversity to that extent and mangrove vegetation

Fig. 5.2 Vessel of Border Security Force (BSF) deployed by the Government of India in the eastern sector of Indian Sundarbans adjacent to Bangladesh Sundarbans

suffers stunted growth (Mitra et al. 2011b, 2012). Most of the people are below the poverty line and hence they illegally intrude into the adjacent Reserve Forest. The eastern sector is under the Reserve Forest area and conservation is given the highest priority. The minimum weakness witnessed in the vertical marine biodiversity and conservation in this sector is mostly related to the illegal intrusion of the island dwellers into the core area for procuring mangrove timber, fuel wood, fodder, honey, wax etc.

Non-conventional energy sector has minimum weakness if they are installed in the western Indian Sundarbans due to substantial merits of this region like optimum tidal amplitude, stable embankments, continuous power supply, transportation and easy linkage with the megacity of Kolkata through both roadways and railways. In the central sector of Indian Sundarbans, there is little scope of installing the projects on non-conventional renewable energy supply, mainly due to erosion problem that saturate the ambient water with silt particles, which may obstruct the smooth movements of the turbines. The beaches are not so stable to install the windmills. The eastern sector, being under the strict vigilance of State Forest Department is a restricted zone where the permission of setting any hard engineering structure is doubtful. As per the norms of conservation, no constructions are allowed in the core region of Sundarbans.

In SWOT analysis, opportunity is a great factor on which the sustainability of a vertical depends. It may be a fact that in the present situation, a particular vertical (say fisheries and aquaculture) is under threat due to various negative impacts of environmental variables, but if the threat can be gradually minimized and managed, the doors of opportunity open up. Table 5.3 highlights the opportunity of the relevant sectors of blue economy in the framework of Indian Sundarbans.

The fisheries and aquaculture sector in the western Indian Sundarbans has moderate level of weakness on account of pollution from different point and non-point sources, but if the emissions from the brick kilns, vessels and trawlers can be mitigated through implementation of clean and green technologies, and treatment plants are properly installed in shrimp farms (biological treatments using filter feeding bivalves and endemic seaweeds), tourism units and industries, then the opportunities in the fisheries and aquaculture sector would definitely open up, as this region has great strength in terms of landing stations, icing and transportation facilities, congenial salinity favoring the migration of Indian shad, *Tenualosa ilisha* and survival ground of several categories of commercially important fishes. The supply chain is also quite strong due to proximity of the western Indian Sundarbans to one of the biggest marine and estuarine fish market of the state, Diamond Harbour (22° 11′ 4.2″ N; 88° 11′ 22.2″ E). We do not find any considerable opportunity in central Indian Sundarbans mainly due to lack of infrastructural facilities and power supply. The eastern sector has great opportunity for fishery and aquaculture not in terms of exploitation/catch/farming, but in terms of sustainability as this Reserve Forest area serves as the natural breeding ground in the absence of any anthropogenic pressure. Thus, a sustainable natural supply chain of fish juveniles (ichthyoplankton) is maintained by this sector of Indian Sundarbans where the juveniles get nutrition from the mangroves that release nutrients like nitrate and phosphate (which trigger the growth of phytoplankton) and free amino acids in the aquatic phase. These are the natural food for ichthyoplankton that ultimately disperse in various estuaries of Sundarbans and gradually attain their respective marketable sizes. Being a restricted zone, there is no scope of fishing and aquaculture in the eastern sector and as such there is no information available on this vertical.

Opportunity of shipping and transportation of goods is high in the western and eastern sectors due to optimum draft in the navigational route, but erosion of banks, siltation and high salinity in the central Indian Sundarbans are not very favorable to sync the sector of shipping and transportation of goods with the regional blue economy.

There are great possibilities of earnings from the recreation and sea-based tourism in the western and eastern sectors of Indian Sundarbans as there is great opportunity of eco-tourism in these sectors. The religious gathering at Sagar Island in the western Indian Sundarbans during Makar Sankranti in mid-January of every year is a source of income for the local population. However, the central sector of Indian Sundarbans lacks power supply, proper lodging facilities and luxuriant mangrove vegetation, due to which opportunity has been narrowed down in the sector.

Biodiversity is an important pillar of blue economy as it forms the base of sustainability by producing several ecosystem services to the island dwellers and local population of Sundarbans. In the eastern sector of Indian Sundarbans, major thrust is given on conservation and with the active involvement of the stakeholders (like Forest Department, local population, several NGOs etc.), the gene pool is preserved in this sector. There is a great opportunity for biodiversity conservation in this zone unlike in the western and central sectors, where the anthropogenic activities are going on in full swing starting from the expansion of shrimp farms to tourism units at the cost of mangrove habitats.

Non-conventional energy sector has scope to flourish in the western Indian Sundarbans because of comparatively better infrastructural facilities and tidal amplitude compared to the other two sectors. The opportunity to install non-conventional energy projects in central Indian Sundarbans may not be viable due to excessive siltation, erosion and poor infrastructural facilities. In the eastern sector, the opportunity is rather less due to restrictions imposed on heavy engineering constructions as per the norms of Reserve Forest.

The mangrove ecosystem of Sundarbans is exposed to different types of threats (Table 5.4). The threats are of medium scale on fisheries and aquaculture in western Indian Sundarbans due to pollution, intense anthropogenic activities (like fish landing, tourism, religious gatherings etc.) and hence proper mitigative actions are the options to turn these threats into opportunities for flourishing fishery and aquaculture activities. These mitigative measures include installation of bio-treatment plant to reduce the deterioration of water quality in and around the shrimp farms, introduction of biofloc system to accelerate the pace of fish production etc. The establishment of Combined Effluent Treatment Plant (CETP) for treating the wastes generated from several tourism units is a roadmap to upgrade the water quality, where fishes can breed and survive without any chance of bioaccumulation of toxic substances in their body tissues. In central Indian Sundarbans, the hyposaline environment coupled with massive erosion of the embankments are the major threats, while in the eastern sector, no fisheries and aquaculture activities are permitted within the core zone/Reserve Forest.

In the vertical of shipping and transportation of goods, the threat is practically less in the western sector due to presence of optimum draft throughout the navigational channel. Similar condition also exists in the eastern sector adjacent to Bangladesh, where there is continuous movement of vessels carrying fly ash and other goods (Fig. 5.3). However, the central sector is not at all congenial for shipping and transportation of goods due to absence of optimum draft and infrastructural facilities. The zone is also prone to high level of erosion. The waves generated during cyclones in recent times (on 20th May, 2020) left the mark of severe damages in most of the islands (Fig. 5.4) in the central sector.

In western Indian Sundarbans, the magnitude of threat on estuary/sea based recreation and tourism is less because of the presence of large numbers of hotels, tourism units, boats, vessels and other infrastructural facilities. In central Indian Sundarbans, the tourism sector is not very promising because of the threats due to erosion, lack of infrastructural facilities etc. The embankments are highly fragile and hence there is a threat of life risk or any physical injury of the tourist while they climb through the embankments (preferably during the low tide phase) to reach the lodge. However, many entrepreneurs are trying introducing eco-tourism in this sector considering the proximity of the Reserve Forest to this region. The eastern sector is the main site of attraction for the tourists due to the presence of dense mangrove forests along with

Fig. 5.3 Movement of vessels along the estuaries of eastern sector of Indian Sundarbans

Fig. 5.4 Damages of the embankments due to wave actions and tidal surges

the wildlife. Tourist guides are also provided by the State Forest Department to let the people know about the ecology of mangroves and other wildlife.

The western Indian Sundarbans is exposed to maximum threats on biodiversity and conservation due to intense human activities. Conservation of biodiversity is given less importance in this sector unlike the eastern sector because of the dependency of maximum population on livelihoods like shrimp farming, brick kilns etc.

which are carried out at the cost of mangroves. Moderate threats exist in central Indian Sundarbans, which may be related to hypersaline environment and excessive erosion in this sector. Conservation related approaches are high in eastern part due to restrictions imposed by the State Forest Department.

Non-conventional energy sectors like tidal energy and wind energy relevant to the framework of Indian Sundarbans are exposed to threats like siltation and disasters due to frequent cyclones in this deltaic complex. In the western Indian Sundarbans, optimum tidal amplitude, continuous power supply and beach stability have minimized the threat to some extent. The major threats in central Indian Sundarbans are excessive siltation, erosion and poor infrastructural facilities. High salinity in the central Indian Sundarbans also accelerate the process of corrosion where there would be a high recurring expenditure to maintain the accessories. We carried out a survey in three stations of central Indian Sundarbans namely Minakhan (22° 30′ 27.01″ N; 88° 43′ 51.61″ E), Canning (22° 19′ 03.20″ N; 88° 41′ 04.43″ E) and Bali Island (22° 04′ 35.17″ N; 88° 44′ 55.70″ E) over a period of more than three decades (1984–2021) and observed an increasing trend with time primarily related to Bidyadhari Siltation (Tables 5.5, 5.6 and 5.7 and Figs. 5.5, 5.6 and 5.7).

In the eastern sector, viability to introduce non-conventional energy is rather less due to restrictions imposed on heavy engineering constructions as per the norms of Reserve Forest.

Technology based management is the need of the hour to nullify/minimize the threats as highlighted here in brief (Table 5.8).

5.1.1 *Ecofriendly Technology to Prevent Erosion*

The embankments are comparable with the life support system to the people dwelling in the lower Gangetic region preferably in Sundarbans. They protect the people and properties from the tidal inundation that get breached frequently caused by physical, climatic, biotic and several other factors. Such breaching of the embankments poses an adverse impact on the lives, livelihoods and properties of the island dwellers. We observed from a series of field study that the nature of embankment failure in the Indian part of Sundarban is very complicated and is a function of physical, chemical, geological and biological factors coupled with human intervention in the entire failure loop. Four major categories of reasons behind embankment failure have been identified (Table 5.9).

The breaching of the embankments in Sundarbans can be controlled by propagating and expanding the coverage of the intertidal mudflats by *Porteresia coarctata* (Fig. 5.8), which will provide three direct benefits in context to the domain of climate change induced vulnerability, namely (i) carbon sequestration, (ii) erosion control and (iii) aquatic pH regulation.

Also mangroves play a very important role in stabilizing embankments by strongly holding the soil particles with their root system (Fig. 5.9).

Table 5.5 Salinity (in psu) in Minakhan

Year	Premonsoon	Monsoon	Postmonsoon
1984	3.93	2.53	3.67
1985	5.45	2.60	3.94
1986	6.20	2.66	3.88
1987	6.08	2.00	4.53
1988	6.90	1.89	4.26
1989	7.55	2.21	4.94
1990	7.70	2.88	5.40
1991	6.87	3.53	5.43
1992	6.58	3.59	5.82
1993	7.46	3.72	5.63
1994	7.86	3.60	6.07
1995	7.90	3.85	6.85
1996	7.94	3.47	6.91
1997	8.69	3.87	6.78
1998	8.67	3.92	7.29
1999	8.61	2.50	7.27
2000	8.84	3.23	7.37
2001	8.90	4.46	7.82
2002	8.99	4.78	8.28
2003	9.31	4.84	8.34
2004	9.06	4.90	8.31
2005	9.04	5.02	9.42
2006	9.72	5.64	9.24
2007	9.55	5.69	9.55
2008	10.20	5.78	9.58
2009	16.54	4.73	10.16
2010	11.73	4.82	10.40
2011	11.99	5.23	10.75
2012	11.94	5.38	10.58
2013	12.32	5.55	10.78
2014	12.59	5.90	11.14
2015	12.67	5.75	11.66
2016	12.95	5.89	11.74
2017	12.03	6.89	11.16
2018	12.71	7.27	12.01
2019	12.89	8.41	12.30
2020	13.24	8.96	12.73
2021	13.36	9.60	13.10

Table 5.6 Salinity (in psu) in Canning

Year	Premonsoon	Monsoon	Postmonsoon
1984	6.97	5.57	6.71
1985	8.49	5.64	6.98
1986	9.24	5.7	6.92
1987	9.12	5.04	7.57
1988	9.94	4.93	7.30
1989	10.59	5.25	7.98
1990	10.74	5.92	8.44
1991	9.91	6.57	8.47
1992	9.62	6.63	8.86
1993	10.50	6.76	8.67
1994	10.90	6.64	9.11
1995	10.94	6.89	9.89
1996	10.98	6.51	9.95
1997	11.73	6.91	9.82
1998	11.71	6.96	10.33
1999	11.65	5.54	10.31
2000	11.88	6.27	10.41
2001	11.94	7.50	10.86
2002	12.03	7.82	11.32
2003	12.35	7.88	11.38
2004	12.10	7.94	11.35
2005	12.08	8.06	12.46
2006	12.76	8.68	12.28
2007	12.59	8.73	12.59
2008	13.24	8.82	12.62
2009	19.58	7.77	13.20
2010	14.77	7.86	13.44
2011	15.03	8.27	13.79
2012	14.98	8.42	13.62
2013	15.36	8.59	13.82
2014	15.63	8.94	14.18
2015	15.71	8.79	14.70
2016	15.99	8.93	14.78
2017	15.07	9.93	14.2
2018	15.75	10.31	15.05
2019	15.93	11.45	15.34
2020	16.28	12.00	15.77
2021	16.40	12.64	16.14

Table 5.7 Salinity (in psu) in Bali Island

Year	Premonsoon	Monsoon	Postmonsoon
1984	21.77	20.37	21.51
1985	23.29	20.44	21.78
1986	24.04	20.5	21.72
1987	23.92	19.84	22.37
1988	24.74	19.73	22.10
1989	25.39	20.05	22.78
1990	25.54	20.72	23.24
1991	24.71	21.37	23.27
1992	24.42	21.43	23.66
1993	25.30	21.56	23.47
1994	25.70	21.44	23.91
1995	25.74	21.69	24.69
1996	25.78	21.31	24.75
1997	26.53	21.71	24.62
1998	26.51	21.76	25.13
1999	26.45	20.34	25.11
2000	26.68	21.07	25.21
2001	26.74	22.30	25.66
2002	26.83	22.62	26.12
2003	27.15	22.68	26.18
2004	26.9	22.74	26.15
2005	26.88	22.86	27.26
2006	27.56	23.48	27.08
2007	27.39	23.53	27.39
2008	28.04	23.62	27.42
2009	34.38	22.57	28.00
2010	29.57	22.66	28.24
2011	29.83	23.07	28.59
2012	29.78	23.22	28.42
2013	30.16	23.39	28.62
2014	30.43	23.74	28.98
2015	30.51	23.59	29.5
2016	30.79	23.73	29.58
2017	29.87	24.73	29.00
2018	30.55	25.11	29.85
2019	30.73	26.25	30.14
2020	31.08	26.80	30.57
2021	31.20	27.44	30.94

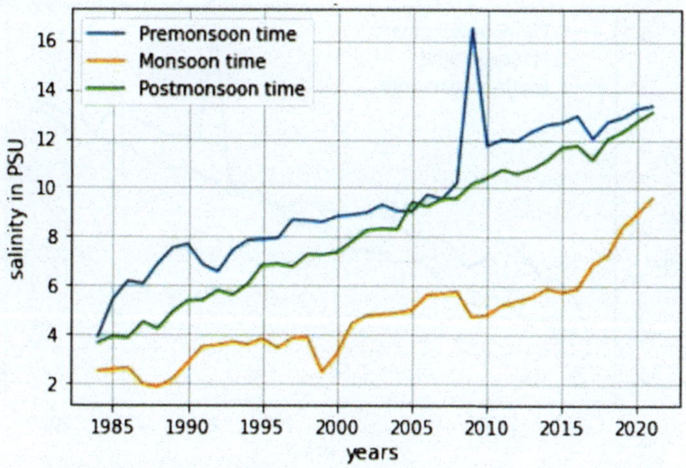

Fig. 5.5 Salinity (in psu) level at Minakhan during 1984–2021

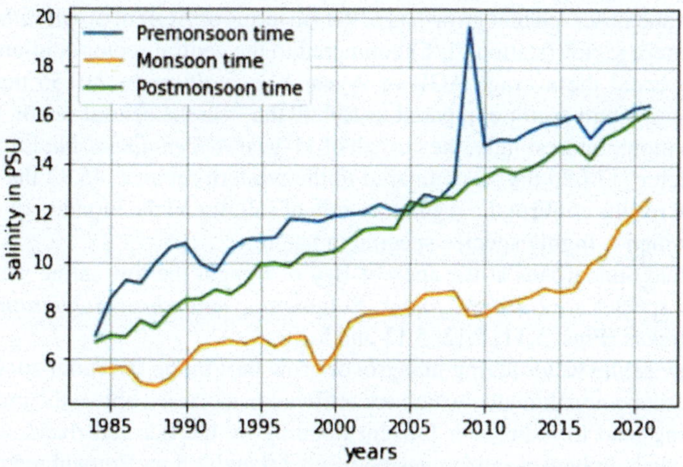

Fig. 5.6 Salinity (in psu) level at Canning during 1984–2021

It is to be noted in this context that all the species of mangroves do not prefer a common salinity regime. Each and every species has a salinity preference of their own and therefore they prefer a distinct niche/microhabitat depending on their distance from the sea. Even congeneric species occupy different coordinates on the same intertidal zones due to differential ability of salt tolerance. These features of mangroves make them ideal ecological models to study the adaptive potential of these halophytes to variable salinity.

We made a survey on the growth of true mangrove species in ten different stations distributed in the western and central sectors of Indian Sundarbans (Table 5.10 and

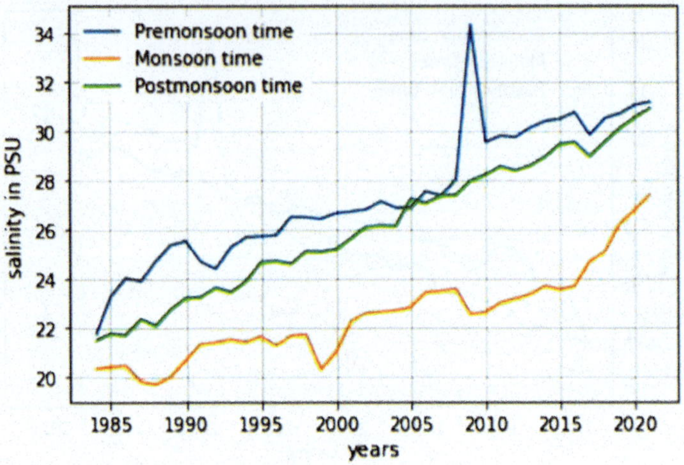

Fig. 5.7 Salinity (in psu) level at Bali Island during 1984–2021

Fig. 5.10) and observed that growth is better (in terms of AGB of *Sonneratia apetala*) in the western sector (stations 1–5) compared to the central sector (stations 6–10).

We observed the average AGB of 9-year *S. apetala* to be 210.55 tha^{-1} in the hyposaline western sector compared to 154.29 tha^{-1} in the central sector, where the salinity is more. Interestingly, the growth of 11-year *Avicennia marina* is more in the central sector (174.26 tha^{-1}) compared to the western sector (145.87 tha^{-1}) (Table 5.11). The results confirm the regulatory role of salinity on the biomass of mangrove species, which is highly species-specific in nature.

In deltaic Sundarbans at the apex of Bay of Bengal the low saline water loving mangrove species are *Heritiera fomes*, *Bruguiera gymnorrhiza*, *Sonneratia* spp. and *Nypa fruticans* (Figs. 5.11, 5.12, 5.13 and 5.14).

The low saline water loving mangroves grow best in the intertidal zone adjacent to the riverine embankments, which are artificial bank raised above the immediately surrounding land to redirect or prevent flooding by the tide fed river/estuary. The height of such embankments ranges between 2.0 and 2.5 m. Embankments, mostly earthen are the lifelines of the residents of Sundarbans. These are erected to protect the lives and properties of some 4.5 million people of Indian Sundarbans from the saline water of the tidal rivers and to hold back the high tides that occur twice daily.

Many species of true mangroves like *Avicennia marina*, *Avicennia alba*, *Avicennia officinalis* and *Excoecaria agallocha* prefer salinity of higher degree (Figs. 5.15, 5.16, 5.17 and 5.18).

The moderate to high saline loving mangroves are the most suitable species to impart stability to the sea facing embankments, where the water salinity ranges between 15.0 and 32.0 psu depending on the season.

It has been observed that structural deformation of embankments occur in zones where the population density of *Excoecaria agallocha* is high. Structural failure of an earthen embankment may take the form of a slide or displacement of material in either

Table 5.8 Major road maps for sector-wise upgradation of blue economy in Indian Sundarbans

Sector	Upgradation road map
Shipping and transportation	1. Dredging at regular intervals to increase the draft in the navigational channel 2. Improvement in waste management sector for incineration of plastic, metals, glass bottles, medical wastes, oil rags, sludge, waste oil etc. considering the circular economy as primary objective 3. Improvement of the efficiency of Oil Water Separator (OWS) by installing emulsion breaking filters on the separators 4. Monitoring the biotic and abiotic parameters of the ballast discharge to prevent the introduction of invasive species 5. Erosion control of the embankment, which may otherwise cause siltation leading to reduction of the depth 6. Salinity triggers the process of corrosion of metallic structures related to shipping and ports. Hence management of aquatic salinity is very effective in the dry docking sites through channelization of harvested rain water
Non-conventional renewable energy sector	1. Erosion of embankment needs to be controlled, which may otherwise clog the turbines with silt, sand and clay particles and pause their mobility in the sector of tidal energy generation 2. Salinity management is extremely essential to check the corrosion of the accessories and machineries required in the wind and tidal energy sectors; inter-linking of rivers may be a 'food for thought' in this context 3. Integration of tidal energy generation sector with other services like mariculture, shoreline development, flood protection etc. 4. Disaster management should be given priority as the region is prone to frequent cyclone of varied magnitude, which may damage the machineries, accessories and the engineering structures related to tidal and wind energy generation sectors. In this context mass scale mangrove afforestation programme is suggested as they serve as unique natural shield
Fisheries and aquaculture	1. Accessibility to the GPS based records by the fisherman to locate the Potential Fishing Zone (PFZ) more precisely. This will not only save the fuel cost, but will also reduce emission and increase profitability 2. Installation of aqua clinic near the shrimp farms for controlling diseases and monitoring the water quality on regular basis 3. Installation of bio-treatment plant using oyster (which are filter feeding organisms under phylum mollusca), seaweeds and salt marsh grass that can absorb nitrogen and phosphorous from the waste water generated from the shrimp farms 4. 'Iron fertilization' may be considered as the 'food for thought' to promote the natural feed supply (phytoplankton) in the polyculture farms of Sundarbans, which is the recent trend in the aquaculture sector of the region

(continued)

Table 5.8 (continued)

Sector	Upgradation road map
Estuary/sea based recreation and tourism	1. Shoreline protection and stability of embankment should be given priority as these are the passages for the tourists to reach the islands; Stability of embankment may be improved by salinity—specific plantation of halophytes 2. Improvement in the tourism waste management sector is important particularly for plastics, glass batteries, medical wastes etc. In this context awareness should be generated to replace plastic bottles and carriers with earthen utensils and jute bags 3. Concept of ecotourism should be imbibed amongst the tourists through proper local guides who would be able to showcase the importance of flora and fauna to the tourists 4. Improvement of the interpretation centers (which are under the control of State Forest Department) by appointing local guides who are well versed with the regional flora and fauna. The folk tales of Bonobibi and other deities of Sundarbans may add new dimension to the tourists if they are highlighted through light and sound technology in the interpretation centers
Marine biodiversity conservation	1. Control of erosion in an ecofriendly way should be given priority to enhance the floral and faunal biodiversity on the intertidal mudflats 2. Pollution from the point- and non-point sources should be controlled by adopting ecofriendly technology (like installation of bio-treatment pond) to provide a congenial environment for reproduction, breeding, survival and growth of the species preferably the finfish and shellfish species 3. Mass scale halophytes should be planted and protected to retard the process of acidification, which may otherwise pose a negative effect on the shelled organisms (like molluscs, gastropods etc.), fishes and various types of fauna present in the estuaries of Sundarbans

the downstream or upstream face. Sloughs, bulges, cracks or other irregularities in the embankment generally are signs of serious instability and may indicate structural failure (Fig. 5.19).

The tidal creeks in the high saline zone of central Indian Sundarbans are studded with *Phoenix paludosa*, which are ideal for providing stability to the earthen embankments (Fig. 5.20) along the creeks in the high saline region of central Indian Sundarbans.

The team of experts from Techno India University, West Bengal visited Bali Island on 5th August, 2021 with officers and staffs of Forest Department and Block Development office to carry out a respondent analysis. The stakeholders such as residents of the island, fisherman, agriculturists, and owners of small scale tourism units and others who are directly or indirectly impacted by erosion of embankments were interacted with to carry out a respondent analysis (Fig. 5.21).

73% of the respondents expressed their views towards the construction of ring embankments and emphasized on hard structures. According to them, the current pattern in Bali Island has posed an adverse impact of high degree even on the concrete path on the embankment and caused severe damage (Fig. 5.22).

Table 5.9 Reasons behind embankment failure in the Indian part of Sundarbans

S. No.	Reason	Description
1	Solution and mud flow	The soils of the dikes are heterogeneous and form solutions of variable nature and composition. Stratification of the soil layers in terms of composition thus plays a major role behind embankment failure. The permeability of the dike body determines the velocity of percolation of water through the dike and therefore its stability too. This can be seen as solution breaching that is most frequent on the sides and top of the dikes. The pH of the ambient water triggers the process of breaching
2	Wash out	In this type, the forces of water currents during forward and retreating tides play a major role, and therefore related to the velocity of the tidal actions. The river side wall gradually becomes narrow in breadth resulting in the loss of stability. In this case also the aquatic pH is a major player as the rate of corrosion increases with the lowering of pH
3	Over topping	This type of embankment breaching occurs with strong upsurge of tide water mostly during the time of cyclonic storms from March to June as witnessed in the lower Gangetic delta region. In this case, the water flows with high force over the top of the embankment and results in the removal of the top soil. The height of the embankment quickly recedes and the overtopped gap expands rapidly on both ends
4	Slump breaching	This is most frequent on walls of the embankments that are exposed to strong tides. Initially the open wall facing the river or the bay is attacked by the high and strong waves of tide water accompanied with strong gush of storms. Such attack erodes the wall almost vertically and as a result, the base of the wall becomes weak that cannot bear the load and falls down straight to the adjacent water mass. The gap in the wall created by such type of breaching gets widened and deepened within a short span of time through which intrusion of saline water occurs with great velocity

Fig. 5.8 Saltmarsh grass *Porteresia coarctata* in Indian Sundarbans

Fig. 5.9 Mangroves with supporting root system strongly hold the soil particles

Many respondents (16%) also forwarded their views on dredging a newly emerged 'char' (piece of land that emerged from the estuary due to siltation) near Bali Island, which according to them is diverting the current towards Bali Island resulting in heavy scouring. The opinions of the stakeholders were duly noted. We designed a biogeoengineering treatment to protect the western part of Yuba Abash in Bali Island (specifically Bali II) located in the central Indian Sundarbans (22° 05′ 44.1″ N and 88° 46′ 38.1″ E) through 3 treatments as stated here:

Treatment 1: Structure or hard measures

Hard engineering structures are mostly designed to resist the energy of the waves and currents and protect the islands. The common seawalls constructed to resist wave energy, groynes designed to increase sediment storage on the shore and concrete hexapod/tetrapod structures fixed in the lower littoral zone to attenuate the energy of waves and currents are examples of hard structures. Considering the geo-morphological features of the shore and current pattern adjacent to our treatment site (western part of Yuba Abash in Bali II), concrete hexapod/tetrapod structures need to be fixed about 50 m down the HTL at a distance of approximately 5 m (Fig. 5.23). This will help in diverting the current, as well as attenuation of waves, which otherwise may wash away the mangrove seedlings.

Treatment 2: Soft measures on the inter-tidal zone

These are measures that are usually taken up synchronizing with nature by altering/restoring the community structure of the endemic floral species present in the region. This approach has economic benefits while minimizing the environmental impact of traditional engineering structures and includes plantations on the intertidal mudflat/shore based on salinity preference of the floral species. In this

Table 5.10 Sampling stations with salient features

Study site	Latitude and longitude	Site description
Harinbari (Stn. 1)	21° 46′ 53.07″ N 88° 04′ 22.88″ E	Situated in the western region of Indian Sundarbans almost in the middle of the Sagar Island; receives the water of the Hooghly River
Chemaguri (Stn. 2)	21° 39′ 42.88″ N 88° 08′ 49.01″ E	Situated on the south-eastern side of Sagar Island and receives the water of the Mooriganga River
Sagar South (Stn. 3)	21° 37′ 49.90″ N 88° 04′ 0.51″ E	Situated on the south-western part of the Sagar Island at the confluence of the River Hugli and the Bay of Bengal. Anthropogenically stressed zone due to presence of passenger jetties, fishing activities and pilgrimage
Lothian island (Stn. 4)	21° 39′ 08.04″ N 88° 19′ 8.47″ E	A Wildlife sanctuary that faces the River Saptamukhi
Prentice island (Stn. 5)	21° 42′ 43.31″ N 88° 17′ 3.62″ E	Situated north of Lothian Island and receives the water of the Saptamukhi River
Canning (Stn. 6)	22° 19′ 03.20″ N 88° 41′ 04.43″ E	Situated in the central part of the Indian Sundarbans and faces the mighty River Matla, a tide-fed river. Due to presence of fish landing stations, passenger jetties and busy market, the area is anthropogenically stressed
Sajnekhali (Stn. 7)	22° 06′ 34.19″ N 88° 48′ 15.78″ E	A Wildlife Sanctuary and a part of Sundarban Tiger Reserve; adjacent to River Bidhya and Gomor. Tourism pressure is extremely high in this station particularly during postmonsoon
Chotomollakhali (Stn. 8)	22° 10′ 40.00″ N 88° 54′ 26.71″ E	Situated in the upper portion of Central Indian Sundarban adjacent to Jhila forest; receives the water of Rangabelia and Korankhali Rivers
Satjelia (Stn. 9)	22° 05′ 17.86″ N 88° 52′ 49.51″ E	Situated adjacent to River Duttar in the upper region of Central Indian Sundarban facing western part of the Jhilla forest
Pakhiralaya (Stn. 10)	22° 07′ 07.23″ N 88° 48′ 29.00″ E	Situated adjacent to River Gomor; opposite to Sajnekhali Wild Life Sanctuary

particular treatment plantation/transplantation of saltmarsh grass, *Porteresia coarctata*, is suggested with the aim to stabilize the substratum followed by plantation of species like *Avicennia marina*, *A. officinalis*, *A. alba*, *Bruguiera gymnorrhiza* etc. depending on the salinity and substratum characteristics. Plantation of low salinity loving *Sonneretia apetala* may also be taken up as the salinity drastically dips during monsoon (8–12 psu) as observed during our visit from 5th to 9th August 2021.

Treatment 3: Vegetative covering on the supralittoral zone

Vegetation is the cheapest way to protect steep slopes of earth embankments. One commonly used plant is vetiver grass (*Chrysopogon zizanioides*). This hardy bunch grows naturally across many of the humid parts of the world. It is extremely resilient and can survive deep inundation of up to 6 m (Howell 2008). Vetiver grass is being

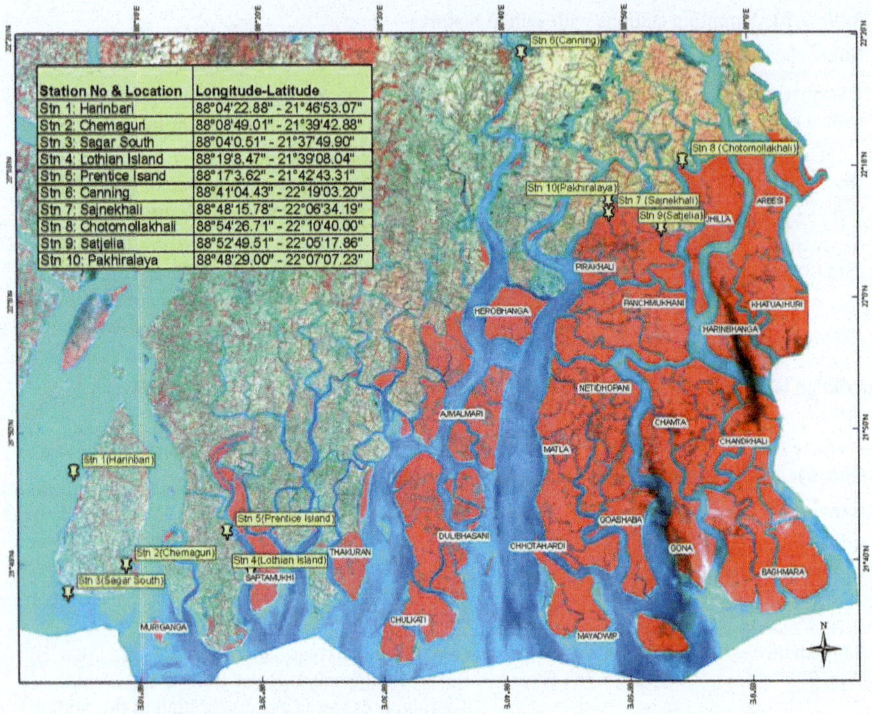

Fig. 5.10 Selected stations in Indian Sundarbans

used as an efficient bio-technology for slope protection in many countries, especially for its attributes: longer life, strong and long finely structured root system, and high tolerance of extreme climate conditions. This technique has recently been employed successfully in certain areas of side-slope protection against rain-cut and wind-induced erosion (Islam et al. 2013). Vetiver shows successful erosion control in fresh water, brackish rivers, and canals. Studies show that vetiver application is about eight times cheaper than masonry wall protection and about five times cheaper than the revetment stone-slope protection system. Thus, vetiver grass plantation for slope protection could be a sustainable, green and cheaper bio-engineering solution (Islam et al. 2013).

Considering the slope of the embankment in the western side of Yuba Abash at Bali Island (22° 5′ 22.11″ N; 88° 45′ 37.67″ E), which is usually 45°–60°, plantation/transplantation of mangrove associate species like *Suaeda maritima* and *Salicornia brachiata* need to be carried out after carpeting a geo-jute matrix on the slope.

A diagrammatic view of stabilizing the supralittoral zone with mangrove associate vegetation and vetiver is highlighted in Fig. 5.24.

In addition to vetiver, *Suaeda* or *Salicornea* sp. are proposed to be planted in high saline—sea facing embankments mainly because of their xerophytic nature

Table 5.11 Water salinity (in psu) in 2021

Sl. No.	Name	Premonsoon	Monsoon	Postmonsoon	Average	Sector
1	Harinbari (Stn. 1)	11.98	6.49	9.54	9.34	Western sector water salinity = 19.24 psu
2	Chemaguri (Stn. 2)	24.22	15.38	20.94	20.18	
3	Sagar South (Stn. 3)	28.19	17.88	23.91	23.33	
4	Lothian island (Stn. 4)	25.74	16.68	22.42	21.61	
5	Prentice island (Stn. 5)	25.81	17.05	22.39	21.75	
Average salinity		23.19	14.70	19.84		
6	Canning (Stn. 6)	17.54	11.63	16.51	15.23	Central sector water salinity = 21.69 psu
7	Sajnekhali (Stn. 7)	28.90	20.31	24.42	24.54	
8	Chotomollakhali (Stn. 8)	27.75	21.79	24.03	24.52	
9	Satjelia (Stn. 9)	25.95	19.10	22.75	22.60	
10	Pakhiralaya (Stn. 10)	24.99	18.57	21.13	21.56	
Average salinity		25.03	18.28	21.77		

Fig. 5.11 *Hertiera fomes* grows luxuriantly within a salinity range between 2.0 and 7.0 psu; the species is presently getting extinct from the Indian part of Sundarban due to salinity rise

Fig. 5.12 *Bruguiera gymnorrhiza* grows luxuriantly within a salinity range between 6.0 and 12.0 psu

Fig. 5.13 *Sonneratia apetala* prefers a salinity range between 5.0 and 12.0 psu

Fig. 5.14 *Nypa fruticans* prefers a salinity range between 4.0 and 15.0 psu

Fig. 5.15 *Avicennia marina* prefers a salinity range between 12.0 and 30.0 psu

Fig. 5.16 *Avicennia alba* prefers a salinity range between 10.0 and 28.0 psu

Fig. 5.17 *Avicennia officinalis* prefers a salinity range between 10.0 and 30.0 psu

Fig. 5.18 *Excoecaria agallocha* with a wide range of salinity tolerance can grow within a salinity range between 5.0 and 30.0 psu

Fig. 5.19 Vertical and horizontal cracks of the embankments

Fig. 5.20 *Phoenix paludosa* along the high saline tidal creeks prefers a salinity range between 16.0 and 32.0 psu

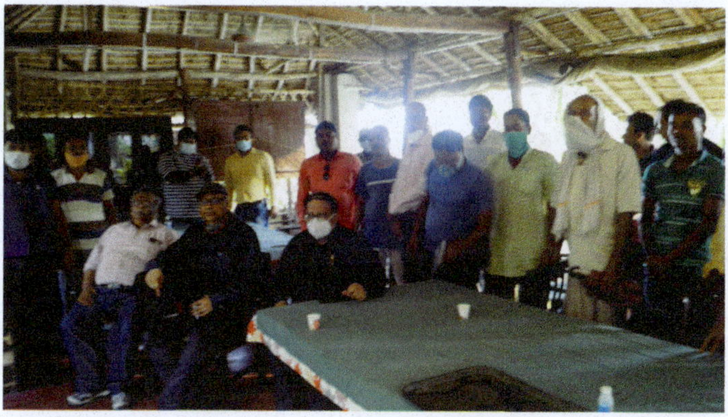

Fig. 5.21 Meeting with the stakeholders at Bali Island to get their views

Fig. 5.22 Damage of embankment at Bali II as observed during the visit of 5th August, 2021

Fig. 5.23 Tetrapod structures (concrete) for wave attenuation

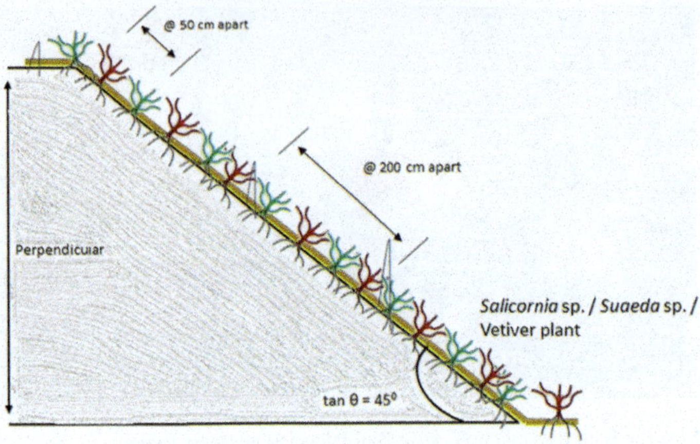

Fig. 5.24 Diagrammatic representation of slope vegetation

(Fig. 5.24) on geo-jute matrix. Different types of geo-jute (JGT) based on mesh opening size and unit mass are available. Absorption capacity of the selected JGTs may be determined using the procedure developed by Khan and Sakib (2012).

The coupling of hard and soft treatments to stabilize the embankments on the western side of the Yuba Abash at Bali Island is represented in Fig. 5.25.

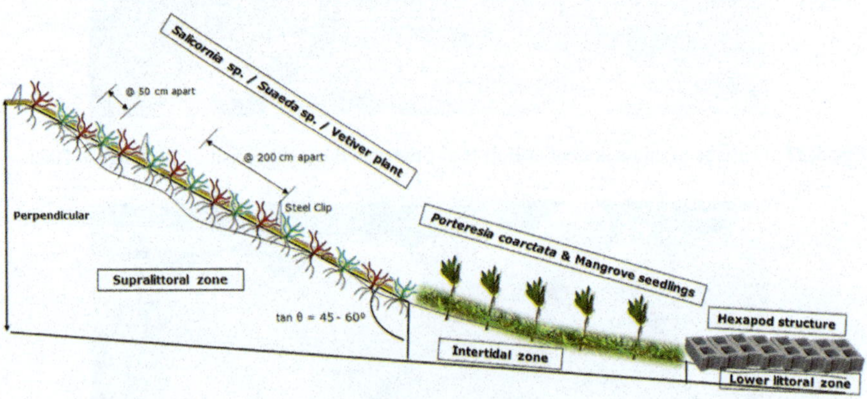

Fig. 5.25 Proposed diagrammatic representation of embankment protection on the western side of Yuba Abash at Bali Island (coupling of hard and soft treatments)

5.1.2 *Mitigation of CO_2 Emission Through Carbon Sequestration*

One of the important objectives of sustainable blue economy is to upgrade the environment in terms of quality so that the biodiversity is not adversely impacted due to pollution. In the framework of Indian Sundarbans, emission of CO_2 is one of the greatest threats that originate from fishing vessels, passenger vessels, trawlers, brick kilns, decomposition of litters and detritus, tourism units and shrimp farms (where generators are frequently used to run the aerators; Fig. 5.26).

All these sources have hiked up the CO_2 level in the near surface atmosphere of Sundarban (Fig. 5.27).

To combat this rising trend of CO_2, mangroves are the best natural tools as these halophytes are potential sink of carbon (Mitra et al. 2010b, 2011d, 2014, 2016; Raha et al. 2013; Bhattacharjee et al. 2013b; Jana et al. 2013; Roy Chowdhury et al. 2014; Bhattacharyya et al. 2015; Ghosh et al. 2015; Guha et al. 2015; Mitra and Gati 2015; Agarwal et al. 2016; Banerjee et al. 2016; Chakraborty et al. 2016; Agarwal et al. 2017; Pramanick et al. 2020; Pal et al. 2020; Ahmed et al. 2021). We carried out an extensive analysis in five different locations distributed in the western and central Indian Sundarbans during April, 2021 to evaluate the carbon sequestration potential of three dominant mangrove species namely *Avicennia marina*, *Sonneratia apetala* and *Excoecaria agallocha*. The results of our field based research are highlighted here.

For convenience, the results of the present study may be broadly divided into Biotic and Abiotic components.

1. **Biotic Components**

Fig. 5.26 Aerators in the shrimp farms are run by generators

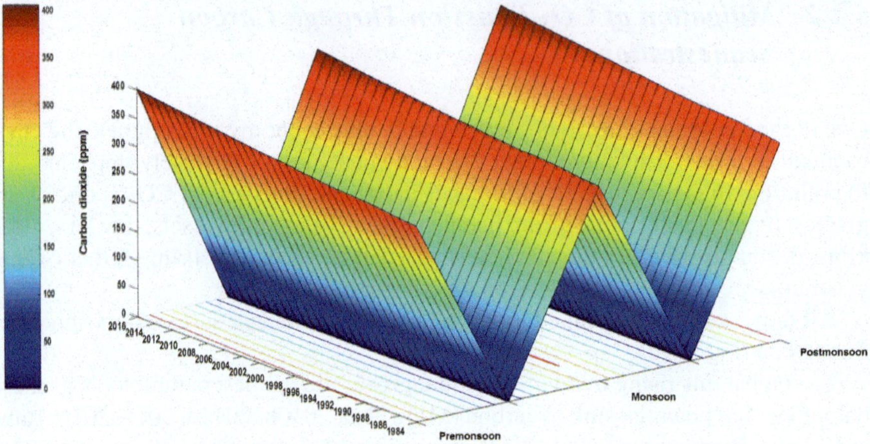

Fig. 5.27 Near-surface carbon dioxide level in the atmosphere of Indian Sundarbans (1984–2016)

The biotic components encompass species-wise AGB, AGC, BGB, sequestered carbon and CO_2-equivalent (based on AGC).

[1.A] In Lothian, the average AGB of *A. marina* was 569.154 tha^{-1}. In case of *S. apetala*, the average AGB was 161.435 tha^{-1}. In case of *E. agallocha*, the average AGB was 77.282 tha^{-1}.

In Lothian, the average AGC of *A. marina* was 260.158 tha^{-1}. In case of *S. apetala*, the average AGC was 74.931 tha^{-1}. In case of *E. agallocha*, the average AGC was 35.318 tha^{-1}.

In Lothian, the average BGB of *A. marina* was 147.980 tha^{-1}. In case of *S. apetala*, the average BGB was 41.973 tha^{-1}. In case of *E. agallocha*, the average BGB was 20.093 tha^{-1}.

In Lothian, the average sequestered carbon of *A. marina* was 21.680 tha^{-1} y^{-1}. In case of *S. apetala*, the average sequestered carbon was 6.244 tha^{-1} y^{-1}. In case of *E. agallocha*, the average sequestered carbon was 2.943 tha^{-1} y^{-1}.

In Lothian, the average CO_2-equivalent of *A. marina* was 954.781 tha^{-1}. In case of *S. apetala*, the average CO_2-equivalent was 274.995 tha^{-1}. In case of *E. agallocha*, the average CO_2-equivalent was 129.618 tha^{-1}.

[1.B] In Prentice, the average AGB of *A. marina* was 454.471 tha^{-1}. In case of *S. apetala*, the average AGB was 147.765 tha^{-1}. In case of *E. agallocha*, the average AGB was 64.705 tha^{-1}.

In Prentice, the average AGC of *A. marina* was 208.259 tha^{-1}. In case of *S. apetala*, the average AGC was 68.579 tha^{-1}. In case of *E. agallocha*, the average AGC was 28.440 tha^{-1}.

In Prentice, the average BGB of *A. marina* was 118.162 tha^{-1}. In case of *S. apetala*, the average BGB was 38.419 tha^{-1}. In case of *E. agallocha*, the average BGB was 16.823 tha^{-1}.

In Prentice, the average sequestered carbon of *A. marina* was 17.355 $tha^{-1}\ y^{-1}$. In case of *S. apetala*, the average sequestered carbon was 5.715 $tha^{-1}\ y^{-1}$. In case of *E. agallocha*, the average sequestered carbon was 2.370 $tha^{-1}\ y^{-1}$.

In Prentice, the average CO_2-equivalent of *A. marina* was 764.310 tha^{-1}. In case of *S. apetala*, the average CO_2-equivalent was 251.685 tha^{-1}. In case of *E. agallocha*, the average CO_2-equivalent was 104.375 tha^{-1}.

[1.C] In Bali Island, the average AGB of *A. marina* was 364.590 tha^{-1}. In case of *S. apetala*, the average AGB was 137.644 tha^{-1}. In case of *E. agallocha*, the average AGB was 61.126 tha^{-1}.

In Bali Island, the average AGC of *A. marina* was 166.122 tha^{-1}. In case of *S. apetala*, the average AGC was 63.933 tha^{-1}. In case of *E. agallocha*, the average AGC was 27.499 tha^{-1}.

In Bali Island, the average BGB of *A. marina* was 94.794 tha^{-1}. In case of *S. apetala*, the average BGB was 35.787 tha^{-1}. In case of *E. agallocha*, the average BGB was 15.893 tha^{-1}.

In Bali Island, the average sequestered carbon of *A. marina* was 13.843 $tha^{-1}\ y^{-1}$. In case of *S. apetala*, the average sequestered carbon was 5.328 $tha^{-1}\ y^{-1}$. In case of *E. agallocha*, the average sequestered carbon was 2.292 $tha^{-1}\ y^{-1}$.

In Bali Island, the average CO_2-equivalent of *A. marina* was 609.667 tha^{-1}. In case of *S. apetala*, the average CO_2-equivalent was 234.634 tha^{-1}. In case of *E. agallocha*, the average CO_2-equivalent was 100.921 tha^{-1}.

[1.D] In Sagar Island, the average AGB of *A. marina* was 350.072 tha^{-1}. In case of *S. apetala*, the average AGB was 135.910 tha^{-1}. In case of *E. agallocha*, the average AGB was 58.818 tha^{-1}.

In Sagar Island, the average AGC of *A. marina* was 148.916 tha^{-1}. In case of *S. apetala*, the average AGC was 62.463 tha^{-1}. In case of *E. agallocha*, the average AGC was 26.403 tha^{-1}.

In Sagar Island, the average BGB of *A. marina* was 91.019 tha^{-1}. In case of *S. apetala*, the average BGB was 35.336 tha^{-1}. In case of *E. agallocha*, the average BGB was 15.293 tha^{-1}.

In Sagar Island, the average sequestered carbon of *A. marina* was 12.410 $tha^{-1}\ y^{-1}$. In case of *S. apetala*, the average sequestered carbon was 5.205 $tha^{-1}\ y^{-1}$. In case of *E. agallocha*, the average sequestered carbon was 2.200 $tha^{-1}\ y^{-1}$.

In Sagar Island, the average CO_2-equivalent of *A. marina* was 546.523 tha^{-1}. In case of *S. apetala*, the average CO_2-equivalent was 229.239 tha^{-1}. In case of *E. agallocha*, the average CO_2-equivalent was 96.900 tha^{-1}.

[1.E] In Kakdwip, the average AGB of *A. marina* was 320.136 tha^{-1}. In case of *S. apetala*, the average AGB was 132.238 tha^{-1}. In case of *E. agallocha*, the average AGB was 56.821 tha^{-1}.

In Kakdwip, the average AGC of *A. marina* was 143.113 tha^{-1}. In case of *S. apetala*, the average AGC was 60.448 tha^{-1}. In case of *E. agallocha*, the average AGC was 25.755 tha^{-1}.

In Kakdwip, the average BGB of *A. marina* was 83.235 tha^{-1}. In case of *S. apetala*, the average BGB was 34.382 tha^{-1}. In case of *E. agallocha*, the average BGB was 14.773 tha^{-1}.

In Kakdwip, the average sequestered carbon of *A. marina* was 11.926 tha^{-1} y^{-1}. In case of *S. apetala*, the average sequestered carbon was 5.037 tha^{-1} y^{-1}. In case of *E. agallocha*, the average sequestered carbon was 2.146 tha^{-1} y^{-1}.

In Kakdwip, the average CO_2-equivalent of *A. marina* was 525.224 tha^{-1}. In case of *S. apetala*, the average CO_2-equivalent was 221.845 tha^{-1}. In case of *E. agallocha*, the average CO_2-equivalent was 94.520 tha^{-1}.

The core findings that emerged from our field-based monitoring are highlighted in separate boxes for three candidate species (Boxes 1–3).

Box 1 Pattern of biotic components for *A. marina*

Station-wise pattern/ spatial pattern	1. We estimated the plot-wise AGB of *A. marina* in all the five stations, which exhibited the order Lothian (569.154 tha^{-1}) > Prentice (454.471 tha^{-1}) > Bali Island (364.590 tha^{-1}) > Sagar Island (350.072 tha^{-1}) > Kakdwip (320.136 tha^{-1}) (Fig. 5.28) 2. We estimated the plot-wise AGC of *A. marina* in all the five stations, which exhibited the order Lothian (260.158 tha^{-1}) > Prentice (208.259 tha^{-1}) > Bali Island (166.122 tha^{-1}) > Sagar Island (148.916 tha^{-1}) > Kakdwip (143.113 tha^{-1}) (Fig. 5.29) 3. For BGB shoot: root empirical ratio was followed and the mean values of each station considering all the plots exhibited the order Lothian (147.980 tha^{-1}) > Prentice (118.162 tha^{-1}) > Bali Island (94.794 tha^{-1}) > Sagar Island (91.019 tha^{-1}) > Kakdwip (83.235 tha^{-1}) (Fig. 5.30) 4. We estimated the plot-wise sequestered carbon in AGB of *A. marina* in all the five stations, which exhibited the order Lothian (21.680 $tha^{-1} y^{-1}$) > Prentice (17.355 $tha^{-1} y^{-1}$) > Bali Island (13.843 $tha^{-1}y^{-1}$) > Sagar Island (12.410 $tha^{-1} y^{-1}$) > Kakdwip (11.926 $tha^{-1} y^{-1}$) (Fig. 5.31) 5. We estimated the plot-wise CO_2-equivalent of *A. marina* in all the five states and the mean values of each station were considered for CO_2-equivalent, which exhibited the order Lothian (954.781 tha^{-1}) > Prentice (764.310 tha^{-1}) > Bali Island (609.667 tha^{-1}) > Sagar Island (546.523 tha^{-1}) > Kakdwip (525.224 tha^{-1}) (Fig. 5.32)

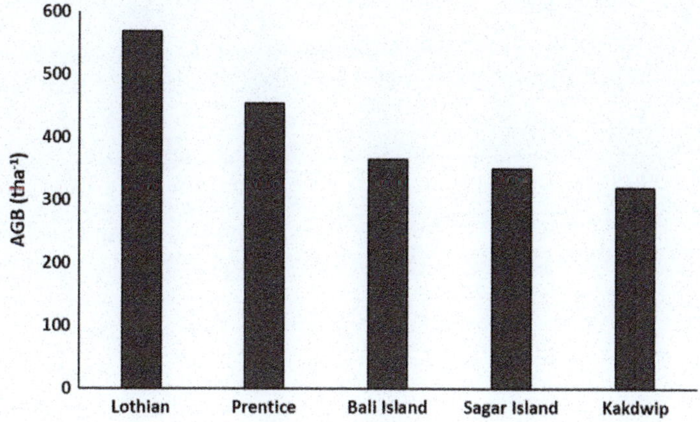

Fig. 5.28 AGB (tha^{-1}) of *Avicennia marina* in five stations

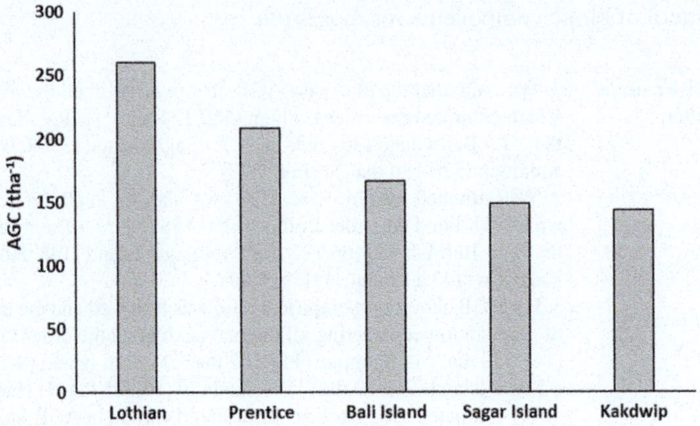

Fig. 5.29 AGC (tha^{-1}) of *Avicennia marina* in five stations

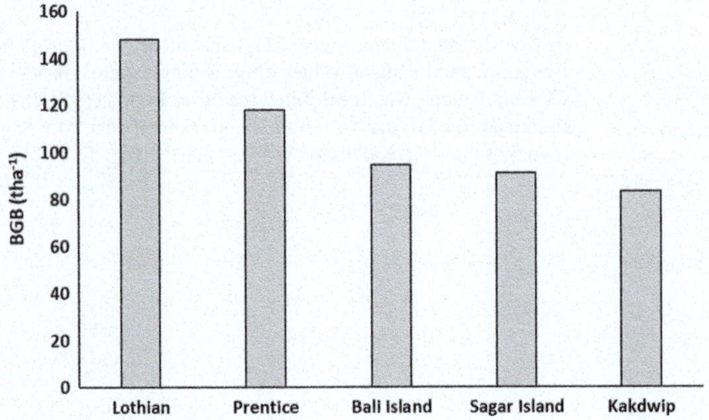

Fig. 5.30 BGB (tha^{-1}) of *Avicennia marina* in five stations

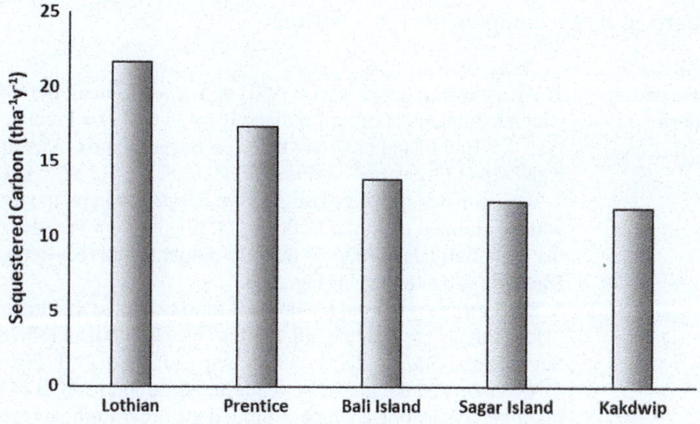

Fig. 5.31 Sequestered Carbon (tha^{-1}y^{-1}) of *Avicennia marina* in five stations

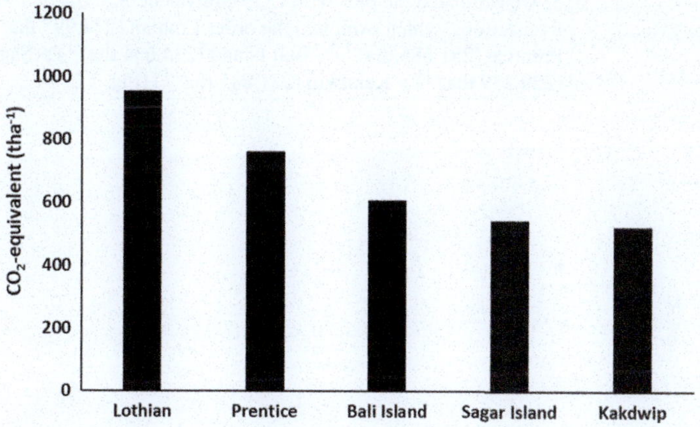

Fig. 5.32 CO_2-equivalent (tha^{-1}) of *Avicennia marina* in five stations

Box 2 Pattern of biotic components for *S. apetala*

Station-wise pattern/ spatial pattern	1. We estimated the plot-wise AGB of *S. apetala* in all the five stations, which exhibited the order Lothian (161.435 tha^{-1}) > Prentice (147.765 tha^{-1}) > Bali Island (137.644 tha^{-1}) > Sagar Island (135.910 tha^{-1}) > Kakdwip (132.238 tha^{-1}) (Fig. 5.33) 2. We estimated the plot-wise AGC of *S. apetala* in all the five stations, which exhibited the order Lothian (74.391 tha^{-1}) > Prentice (68.579 tha^{-1}) > Bali Island (63.933 tha^{-1}) > Sagar Island (62.463 tha^{-1}) > Kakdwip (60.448 tha^{-1}) (Fig. 5.34) 3. BGB estimated through empirical ratio exhibited the order Lothian (41.973 tha^{-1}) > Prentice (38.419 tha^{-1}) > Bali Island (35.787 tha^{-1}) > Sagar Island (35.336 tha^{-1}) > Kakdwip (34.382 tha^{-1}) (Fig. 5.35) 4. We estimated the plot-wise sequestered carbon in AGB of *S. apetala* in all the five stations, which exhibited the order Lothian (6.244 $tha^{-1}y^{-1}$) > Prentice (5.715 $tha^{-1}y^{-1}$) > Bali Island (5.328 $tha^{-1}y^{-1}$) > Sagar Island (5.205 $tha^{-1}y^{-1}$) > Kakdwip (5.037 $tha^{-1}y^{-1}$) (Fig. 5.36) 5. We estimated the plot-wise CO_2-equivalent of *S. apetala* in all the five stations, which exhibited the order Lothian (274.995 tha^{-1}) > Prentice (251.685 tha^{-1}) > Bali Island (234.634 tha^{-1}) > Sagar Island (229.239 tha^{-1}) > Kakdwip (221.845 tha^{-1}) (Fig. 5.37)

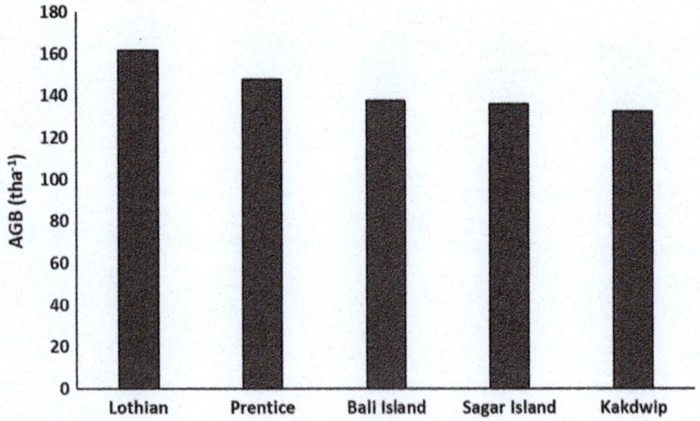

Fig. 5.33 AGB (tha^{-1}) of *Sonneratia apetala* in five stations

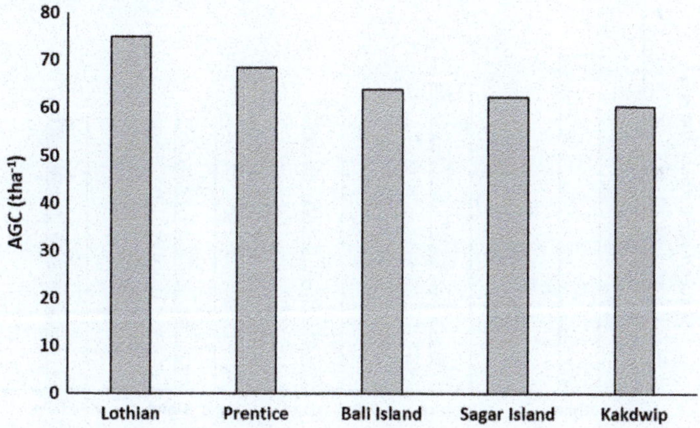

Fig. 5.34 AGC (tha^{-1}) of *Sonneratia apetala* in five stations

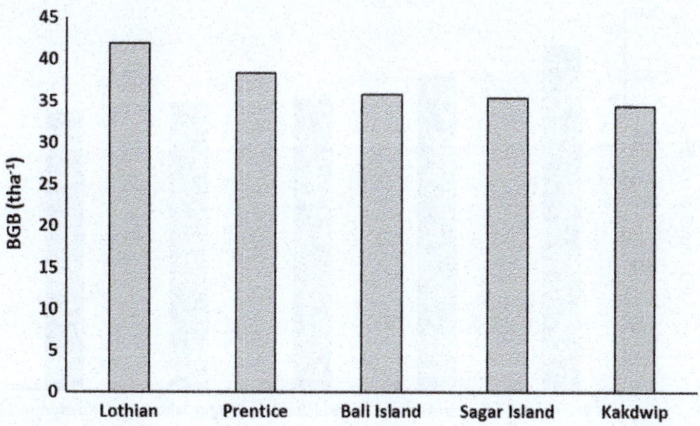

Fig. 5.35 BGB (tha^{-1}) of *Sonneratia apetala* in five stations

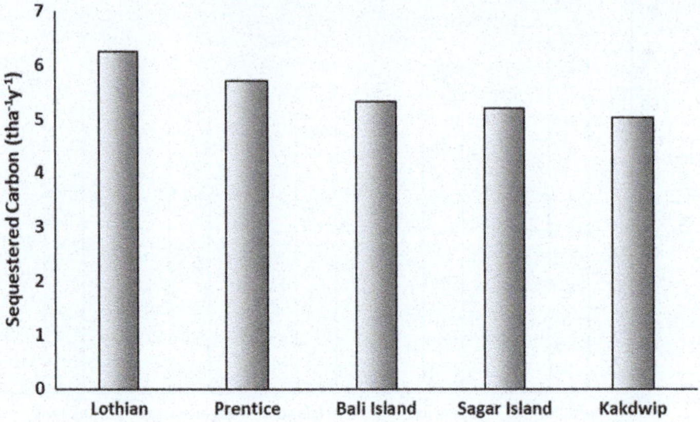

Fig. 5.36 Sequestered Carbon ($tha^{-1}y^{-1}$) of *Sonneratia apetala* in five stations

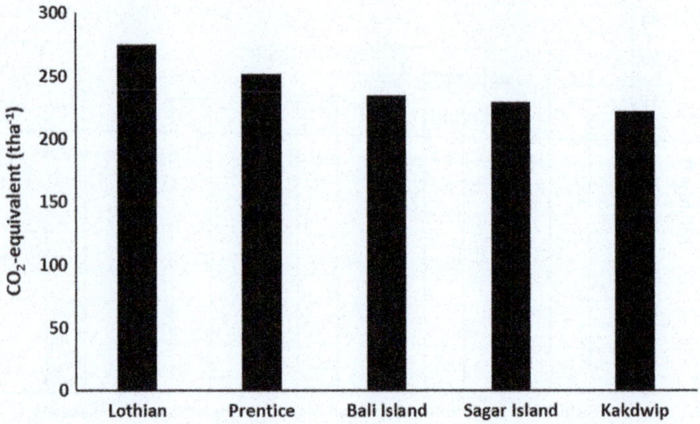

Fig. 5.37 CO_2-equivalent (tha^{-1}) of *Sonneratia apetala* in five stations

Box 3 Pattern of biotic components for *E. agallocha*

Station-wise pattern/ spatial-pattern	1. We estimated the plot-wise AGB of *E. agallocha* in all the five stations and the mean values of each station were considered to evaluate the above ground biomass pattern, which exhibited the order Lothian (77.282 tha^{-1}) > Prentice (64.705 tha^{-1}) > Bali Island (61.126 tha^{-1}) > Sagar Island (58.818 tha^{-1}) > Kakdwip (56.821 tha^{-1}) (Fig. 5.38) 2. We estimated the plot-wise AGC of *E. agallocha* in all the five stations and the mean values of each station were considered to evaluate the growth/biomass pattern, which exhibited the order Lothian (35.318 tha^{-1}) > Prentice (28.440 tha^{-1}) > Bali Island (27.499 tha^{-1}) > Sagar Island (26.403 tha^{-1}) > Kakdwip (25.755 tha^{-1}) (Fig. 5.39) 3. For BGB, we followed the shoot: root empirical ratio, and the mean values of each station considering all the plots exhibited the order Lothian (20.093 tha^{-1}) > Prentice (16.823 tha^{-1}) > Bali Island (15.893 tha^{-1}) > Sagar Island (15.293 tha^{-1}) > Kakdwip (14.773 tha^{-1}) (Fig. 5.40) 4. We estimated the plot-wise sequestered carbon in AGB of *E. agallocha* in all the five stations and the mean values of each station were considered to carbon sequestration pattern, which exhibited the order Lothian (2.943 tha^{-1} y^{-1}) > Prentice (2.370 tha^{-1} y^{-1}) > Bali Island (2.292 tha^{-1} y^{-1}) > Sagar Island (2.200 tha^{-1} y^{-1}) > Kakdwip (2.146 tha^{-1} y^{-1}) (Fig. 5.41) 5. We estimated the plot-wise CO_2-equivalent of *E. agallocha* in all the five stations, which exhibited the order Lothian (129.618 tha^{-1}) > Prentice (104.375 tha^{-1}) > Bali Island (100.921 tha^{-1}) > Sagar Island (96.900 tha^{-1}) > Kakdwip (94.520 tha^{-1}) (Fig. 5.42)

For convenience, station-wise comparative accounts for the three candidate species are highlighted in Tables 5.12, 5.13 and 5.14.

2. **Abiotic Components**

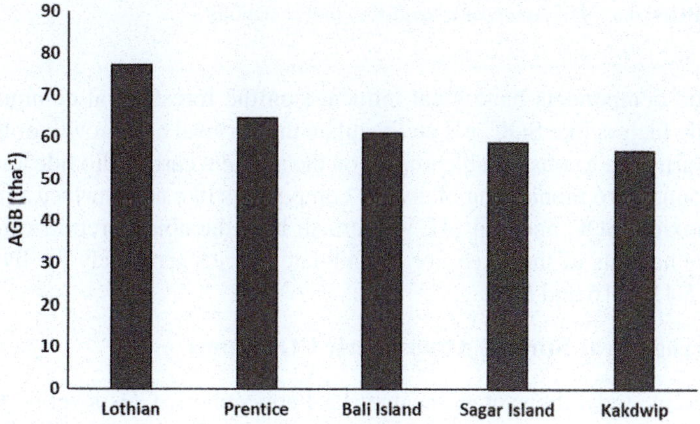

Fig. 5.38 AGB (tha^{-1}) of *Excoecaria agallocha* in five stations

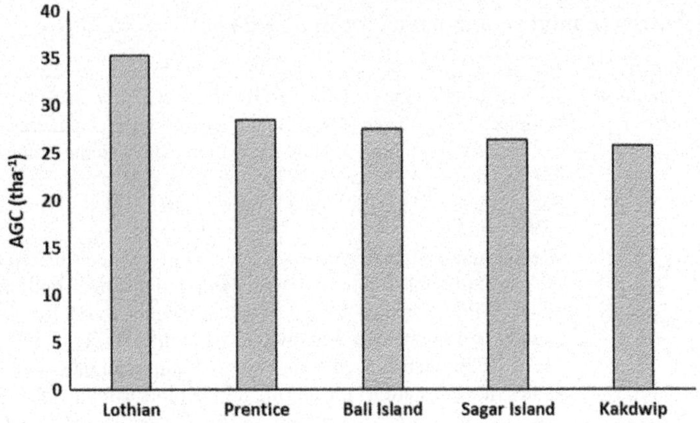

Fig. 5.39 AGC (tha^{-1}) of *Excoecaria agallocha* in five stations

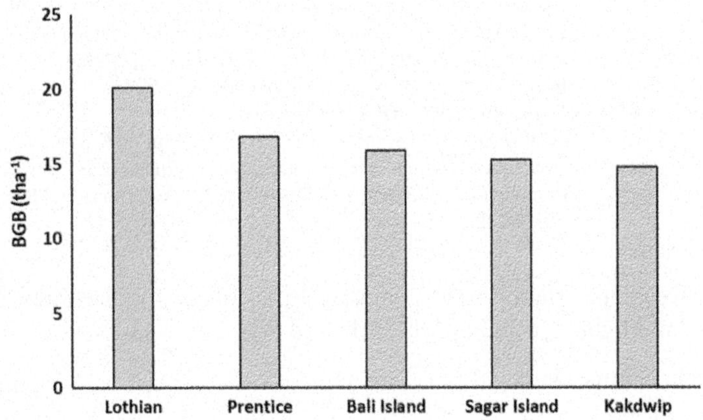

Fig. 5.40 BGB (tha^{-1}) of *Excoecaria agallocha* in five stations

The abiotic components have great influence on the forest floral community e.g., the edaphic factors like SOC, pH etc. regulate the survival and growth of the plants, which in turn pose a considerable impact on the ambient carbon dioxide level. Based on our ground-zero monitoring of abiotic components like near surface atmospheric carbon dioxide, SOC and soil pH, we furnish here the abiotic report cards in and around the habitats of the three floral candidate species separately for five stations in Tables 5.15, 5.16 and 5.17.

2.1 Average Near Surface Atmospheric CO_2 (ppm)

In the present study, the average near surface atmospheric CO_2 level in ***A. marina* habitat** varied as per the order Kakdwip (404 ppm) > Sagar Island (401 ppm) > Bali Island (385 ppm) > Prentice (360 ppm) > Lothian (343 ppm); the average

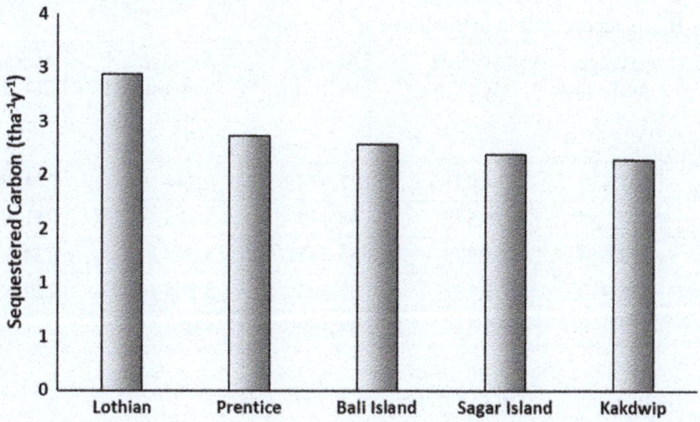

Fig. 5.41 Sequestered Carbon ($tha^{-1}\ y^{-1}$) of *Excoecaria agallocha* in five stations

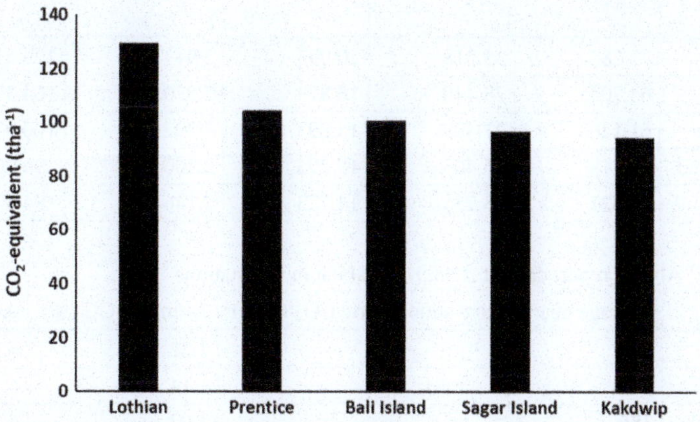

Fig. 5.42 CO_2-equivalent (tha^{-1}) of *Excoecaria agallocha* in five stations

Table 5.12 Biotic report card in five stations of *A. marina*

Stations	Average AGB (tha^{-1})	Average AGC (tha^{-1})	Average BGB (tha^{-1})	Average Sequestered Carbon ($tha^{-1}\ y^{-1}$)	Average CO_2-equivalent (tha^{-1})
Lothian	569.154	260.158	147.980	21.680	954.781
Prentice	454.471	208.259	118.162	17.355	764.310
Bali Island	364.590	166.122	94.794	13.843	609.667
Sagar Island	350.072	148.916	91.019	12.410	546.523
Kakdwip	320.136	143.113	83.235	11.926	525.224

Table 5.13 Biotic report card in five stations of *S. apetala*

Stations	Average AGB (tha^{-1})	Average AGC (tha^{-1})	Average BGB (tha^{-1})	Average Sequestered Carbon (tha^{-1} y^{-1})	Average CO_2-equivalent (tha^{-1})
Lothian	161.435	74.931	41.973	6.244	274.995
Prentice	147.765	68.579	38.419	5.715	251.685
Bali Island	137.644	63.933	35.787	5.328	234.634
Sagar Island	135.910	62.463	35.336	5.205	229.239
Kakdwip	132.238	60.448	34.382	5.037	221.845

Table 5.14 Biotic report card in five stations of *E. agallocha*

Stations	Average AGB (tha^{-1})	Average AGC (tha^{-1})	Average BGB (tha^{-1})	Average Sequestered Carbon (tha^{-1} y^{-1})	Average CO_2-equivalent (tha^{-1})
Lothian	77.282	35.318	20.093	2.943	129.618
Prentice	64.705	28.440	16.823	2.370	104.375
Bali Island	61.126	27.499	15.893	2.292	100.921
Sagar Island	58.818	26.403	15.293	2.200	96.900
Kakdwip	56.821	25.755	14.773	2.146	94.520

Table 5.15 Abiotic report card of *A. marina* habitat in five stations

Stations	Average near surface atmospheric CO_2 (ppm)	Average SOC (%)	Average pH
Lothian	343	1.22	6.7
Prentice	360	1.14	6.73
Bali Island	385	1.02	6.84
Sagar Island	401	0.93	6.88
Kakdwip	404	0.87	6.90

Table 5.16 Abiotic report card of *S. apetala* habitat in five stations

Stations	Average near surface atmospheric CO_2 (ppm)	Average SOC (%)	Average pH
Lothian	325	1.19	6.68
Prentice	341	1.12	6.71
Bali Island	358	1.00	6.75
Sagar Island	372	0.91	6.79
Kakdwip	395	0.84	6.80

Table 5.17 Abiotic report card of *S. apetala* tree habitat in five selected forest stations

Stations	Average near surface atmospheric CO_2 (ppm)	Average SOC (%)	Average pH
Lothian	336	1.17	6.66
Prentice	355	1.11	6.69
Bali Island	379	0.98	6.69
Sagar Island	393	0.86	6.75
Kakdwip	398	0.81	6.78

near surface atmospheric CO_2 level in ***S. apetala* habitat** varied as per the order Kakdwip (395 ppm) > Sagar Island (372 ppm) > Bali Island (358 ppm) > Prentice (341 ppm) > Lothian (325 ppm) and the average near surface atmospheric CO_2 level in ***E. agallocha* habitat** varied as per the order Kakdwip (398 ppm) > Sagar Island (393 ppm) > Bali Island (379 ppm) > Prentice (355 ppm) > Lothian (336 ppm) (*Vide* Tables 5.15, 5.16 and 5.17; Figs. 5.43, 5.44 and 5.45).

2.2 Soil Organic Carbon (SOC)

The average soil organic carbon in ***A. marina* habitat** varied as per the order Lothian (1.22%) > Prentice (1.14%) > Bali Island (1.02%) > Sagar Island (0.93%) > Kakdwip (0.87%); the average soil organic carbon in ***S. apetala* habitat** varied as per the order Lothian (1.19%) > Prentice (1.12%) > Bali Island (1.00%) > Sagar Island (0.91%) > Kakdwip (0.84%) and the average soil organic carbon in ***E. agallocha* habitat** varied as per the order Lothian (1.17%) > Prentice (1.11%) > Bali Island (0.98%) > Sagar Island (0.86%) > Kakdwip (0.81%) (*Vide* Tables 5.15, 5.16 and 5.17; Figs. 5.46, 5.47 and 5.48).

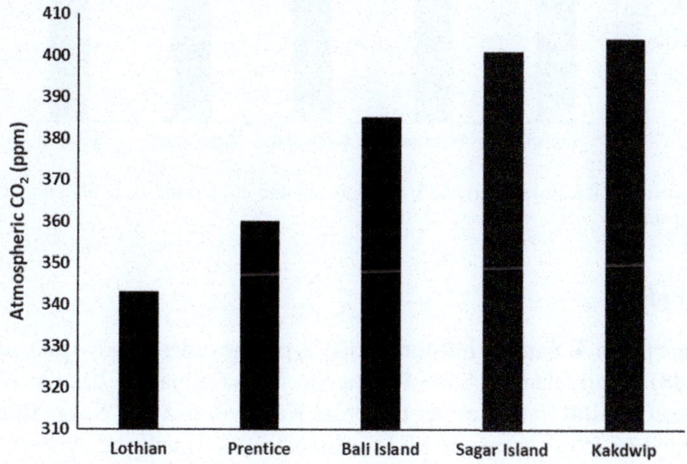

Fig. 5.43 Near surface atmospheric CO_2 (in ppm) in and around the *A. marina* habitat in the five selected stations

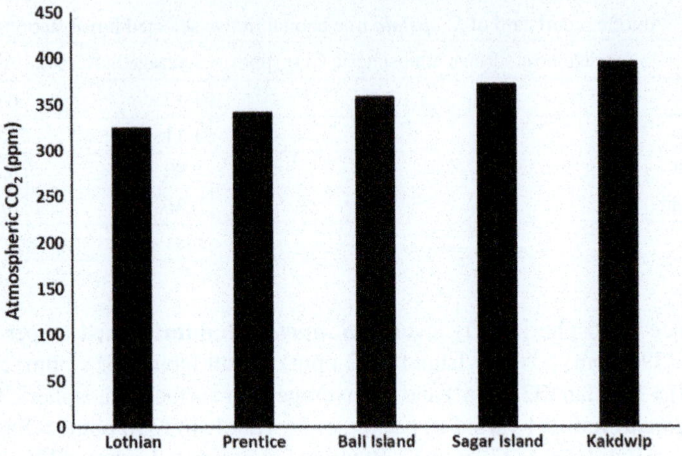

Fig. 5.44 Near surface atmospheric CO_2 (in ppm) in and around the *S. apetala* habitat in the five selected stations

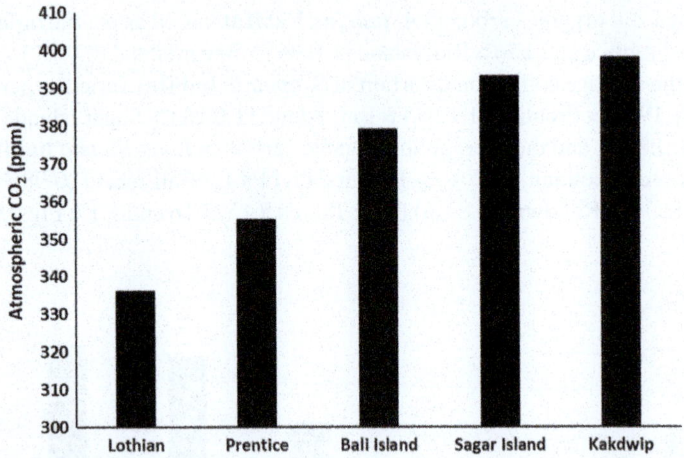

Fig. 5.45 Near surface atmospheric CO_2 (in ppm) in and around the *E. agallocha* habitat in the five selected stations

2.3 Soil pH

The average pH in ***A. marina*** **habitat** varied as per the order Kakdwip (6.90) > Sagar Island (6.88) > Bali Island (6.84) > Prentice (6.73) > Lothian (6.70); the average pH in ***S. apetala*** **habitat** varied as per the order Kakdwip (6.80) > Sagar Island (6.79) > Bali Island (6.75) > Prentice (6.71) > Lothian (6.68) and the average pH in ***E. agallocha*** **habitat** varied as per the order Kakdwip (6.78) > Sagar Island (6.75) > Bali Island (6.69) = Prentice (6.69) > Lothian (6.66) (*Vide* Tables 5.15, 5.16 and 5.17; Figs. 5.49, 5.50 and 5.51).

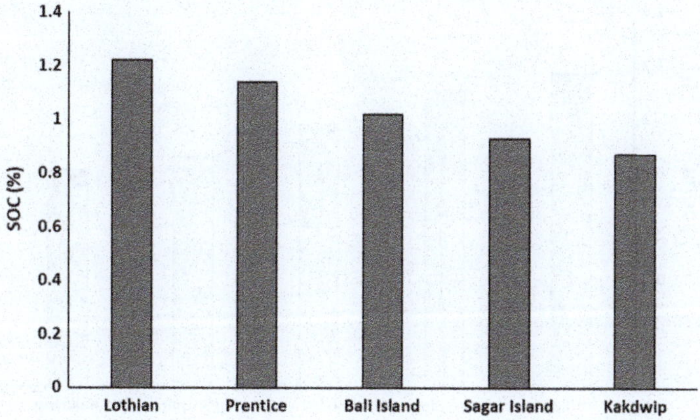

Fig. 5.46 Soil organic carbon (in %) in the *A. marina* habitat of five selected stations

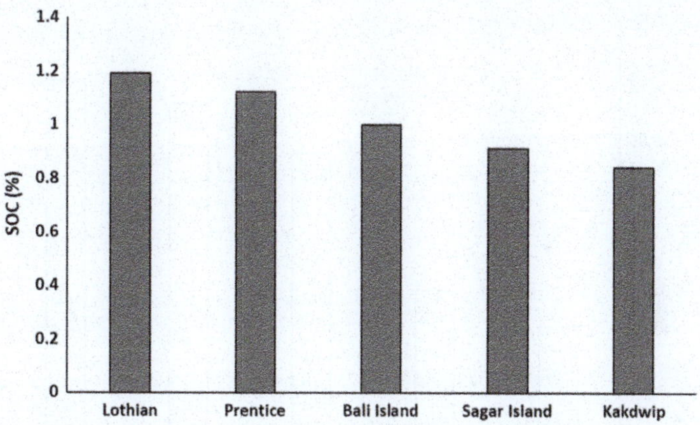

Fig. 5.47 Soil organic carbon (in %) in the *S. apetala* habitat of five selected stations

5.1.3 Carbon Neutral Alternative Livelihood

Livelihood is the basic foundation of economic and food securities in any region. In the framework of Indian Sundarbans, several types of livelihoods are opted, e.g., agriculture, deep sea fishing, shrimp farming, tourism, fish landing related activities, poultry rearing, cattle rearing, apiculture, etc. Almost all these livelihoods are the sources of GHGs. The brick kilns, which are constructed from the dug-out soil of shrimp farms emit considerable amount of carbon dioxide in the atmosphere. The decomposed fishes in the fish landing stations, the degrading mangrove litter and detritus, and the fishing vessels and trawlers also add carbon dioxide to the ambient air. Thus, it is observed that, although there are no major industries in this region, the average CO_2 level in the ambient air is around 390 ppm.

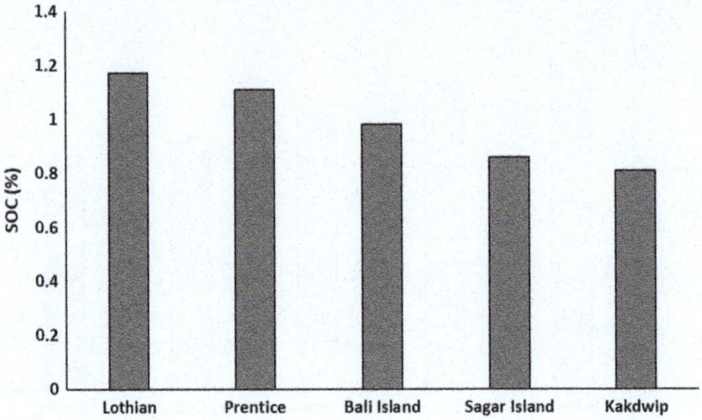

Fig. 5.48 Soil organic carbon (in %) in the selected *E. agallocha* of five selected stations

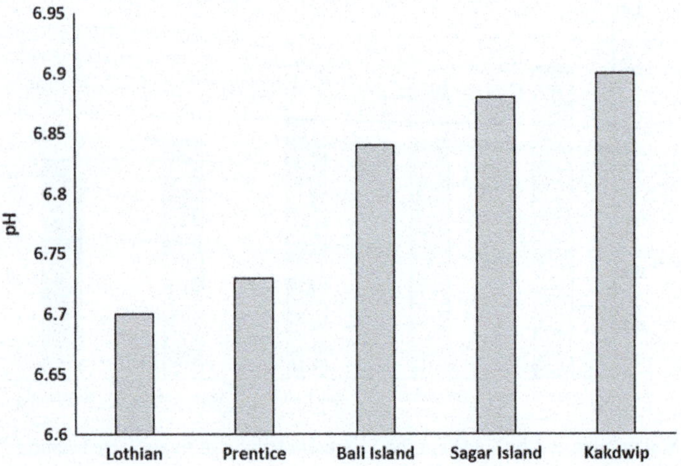

Fig. 5.49 Soil pH in the *A. marina* habitat of five selected stations

Changes of strategies in conventional livelihoods are extremely important, as 'livelihood-as-usual' strategies cannot resolve the threats and challenges associated with the economic security of the residents of Indian Sundarbans. A strategic innovation may thus offer prospects for the kind of changes and choices to make, in order to move away from conventional thinking and acting. These innovations could be embedded in strategic delta livelihood plans, and thereby offer a window of opportunity for initiating and stimulating innovation and associated markets.

On this background, it is essential to reach 'out of the box' and think the viability of few carbon neutral livelihoods befitted for the region (Table 5.18).

The livelihood stated in Table 5.18, may further be extended to mangrove nursery development, as back up of the mass scale mangrove afforestation programme that

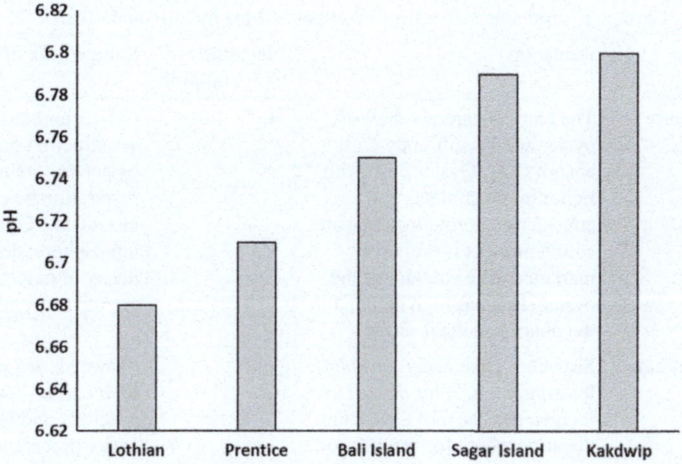

Fig. 5.50 Soil pH in the *S. apetala* habitat of five selected stations

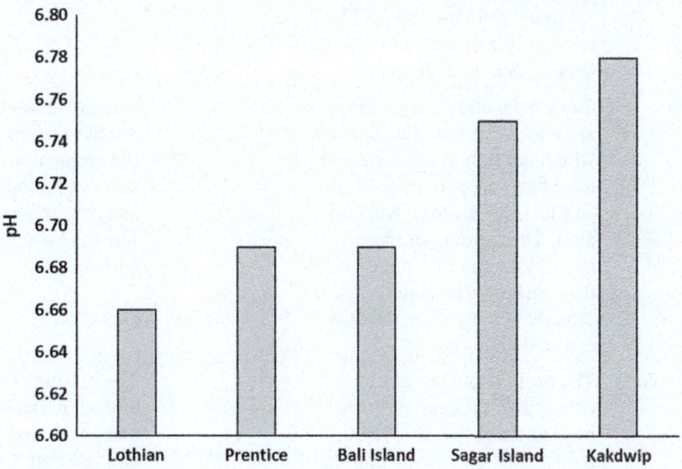

Fig. 5.51 Soil pH in the *E. agallocha* habitat of five selected stations

are taken up every year by the State Forest Department, NGOs and several corporate bodies as a part of CSR. Such type of livelihoods would not only boost up the biodiversity in the region, but at the same time would act as a major sink of carbon dioxide at the local scale (Fig. 5.52).

It is to be stated in this context that policies, programmes and subsidies/incentives are important components to sustain the livelihoods in any region, but in the frame work of Indian Sundarbans, this can have unintended an undesired adverse impact that can hamper the sustainability of livelihoods. These adverse impacts include (a) increased migration in to the area for getting incentives and (b) decreased integration

Table 5.18 Carbon–neutral alternative livelihood befitted for Indian Sundarbans

Alternative livelihood	Remarks	Magnitude of CO_2 emission	Source/Sink of CO_2
Oyster culture	The hard calcareous shell of oyster and its soft body parts absorb CO_2. Oyster flesh with rich mineral contents and considerable protein can be an edible product if properly marketed. The calcium of the oyster shell has high demand in the pharmaceutical sector	+	CO_2 is emitted during the construction process of oyster culture units; respiration by oysters also adds CO_2 to the atmosphere; death and decay of cultured oysters are the sources of CO_2
Seaweed culture	Seaweeds (like *Enteromorpha intestinalis, E. compressa, Ulva lactuca* etc.) absorb CO_2 from the atmosphere for the purpose photosynthesis. Seaweeds are rich in vitamins, minerals and antioxidants and can be consumed in raw form or used for making food products like cookies, ice-cream etc. (Pramanick et al. 2016)	–	Seaweeds are known for their carbon sequestration potential and hence the entire process is practically carbon neutral
Halophyte farming	*Suaeda* sp. and *Salicornia* sp. are mangrove associate species, which are rich in minerals and have been used to make local snacks (like Samosa, Kachuri etc.). The farming of these halophytes can minimize the near surface atmospheric carbon dioxide at local scale (Biswas et al. 2019)	–	Mangrove associate species absorb CO_2 from the ambient media to carry out photosynthesis, and thus the halophyte farm can act as a major carbon sink
Seaweed based fish feed	The protein and mineral contents of the seaweeds are used to make fish feed that reduces the organic load in the pond	++	The boiling of ingredients during dough preparation and packaging process can generate CO_2
Mangrove (*Sonneratia* spp.) fruit-based food products	The fruits of *Sonneratia* are rich in vitamin C (Pramanick et al. 2014) and hence can be used to prepare jam, jelly etc	+	The boiling of fruit pulp during jelly preparation can leave slight carbon foot print

Note '–' means insignificant emission, '+' means minimum emission and '++' moderate emission

Fig. 5.52 Raising mangrove seedling in nursery is a carbon neutral alternative livelihood

of island dwellers in areas having better job opportunities and less environmental risk. A sound supply chain management backed up with total quality assurance can only make the mangrove-centric livelihoods successful in the region.

5.2 Policy Based Management

Protecting the environment of the planet Earth is the prime goal of mankind and is a crucial point to sustain the resource base of blue economy. The domain of protection of natural resources encompasses two major verticals namely conservation and preservation. Conservation is the sustainable use and management of natural resources which includes flora, fauna, air, water, earth deposits etc. Conservation of natural resources primarily focuses on the needs and interests of human beings, for example, biological, economic, cultural and aesthetic values these resources possess. The mangrove ecosystem, for example, contains a wide spectrum of flora and fauna, which provides timber, fuel wood, honey, wax, fishes, medicinal ingredients etc. The seed of conservation germinates for developing a better future. Preservation, on the contrary is an attempt to maintain the existing condition of the environment of a particular ecosystem or habitat that is mostly in wilderness condition. The essence of preservation spreads from the fact that mankind is encroaching the natural habitats at such a rate that many untouched landscapes are now getting sacrificed for industrial development, urban development, farming, tourism, aquaculture etc. Strong lobbies of preservationists support protection of natural reservoir with priority and give less importance to the ecosystem services of natural resources (flora, fauna etc.) to mankind.

A comparative account of conservation and preservation is given in Table 5.19.

The vertical of conservation emphasizes on alternative livelihood schemes through which human encroachment can be reduced and natural reservoir be preserved. We

Table 5.19 Comparative account preservation and conservation

Issue	Preservation	Conservation
Objectives	Landscapes without humans	Flora, fauna, microbes and members of food webs
Justifications	Aesthetic	Intellectual interest; present and future utility preferably for the benefit of mankind
Target	National parks; sanctuary etc.	Hot spots of biodiversity and even natural resources
Obstacles	Encroachment, over-exploitation etc.	Encroachment, over-exploitation, shrinkage of livelihood, lack of proper management of man-natural resource conflict etc.

have strong inclination to recommend the term conservation as the ecosystem services of flora, fauna and other organisms need to be explored and rationally utilized to mitigate the needs of the growing population (that is considered by us as the root cause of all evils in this planet). The generation of alternative livelihood schemes also falls under this category.

Whether the term is preservation or conservation is not the primary question, but protecting the natural resources from the natural and anthropogenic threats is the main goal of both. In context to marine and estuarine ecosystems, it is essential to develop a concrete plan for ensuring protection to the flora and fauna thriving in the system. For achieving this primary goal, programmes and planning with activities must be developed with thrust area and priority (Table 5.20).

5.2.1 Policies and Acts to Regulate Activities in Coastal Zone

The multiple activities in the coastal zone often pose adverse impacts on the fragile ecotone for which CRZ NOTIFICATION 2018 is one of the most important policies followed in India. We present the notification here as stated by the concerned ministry.

Whereas by the notification of the Government of India in the Ministry of Environment and Forests number S.O.19 (E), dated the 6th January, 2011 (hereinafter referred to as the Coastal Regulation Zone Notification, 2011), the Central Government declared certain coastal stretches as Coastal Regulation Zone (hereinafter referred to as the CRZ); under Section 3 of Environment (Protection) Act, 1986;

And whereas, the Ministry of Environment, Forest and Climate Change (MoEF&CC) has received representations from various coastal States and Union Territories (UTs), besides other stakeholders, regarding certain provisions in the CRZ Notification, 2011 related to management and conservation of marine and coastal ecosystems, development in coastal areas, eco-tourism, livelihood options and sustainable development of coastal communities etc.;

Table 5.20 Planning, components, domain of investigation with priority for conserving marine and estuarine ecosystems

S. No.	Planning	Components	Domain of investigation	Priority
1	Integrated landscape development	1. Biodiversity assessment 2. Identification and scaling of threats 3. Identification of degraded and fragile ecosystems 4. Zonation of the operational area into core, buffer and tourism/urbanization/industrial (activity) zones 5. Identification of eco-restoration plans befitted to the landscape	Ecosystem	A
2	Assessment of the impact of climate change on the marine and estuarine ecosystems	1. Assessment and prediction of the impact of climate change on flora, fauna and energy flow through different tiers of trophic levels 2. Assessment and prediction of impact of climate change on the community structure of flora and fauna 3. Identification of keystone species 4. Assessment and prediction of climate change on the socio-economic and demographic profile of coastal communities 5. Preparation of a long-term conservation plan with special emphasis on disaster management	Ecosystem	A
3	Ecosystem services	1. Identification of knowledge, innovations and practices of indigenous and local communities 2. Assessment of impacts of developmental activities on ecosystem goods and services 3. Evaluation of negative and positive impacts using tools of market and non-market valuation 4. Economic analysis of conservation-development scenarios using multiple criteria assessment and cost benefit analysis 5. Ranking of scenarios and identification of options for conservation and sustainable development	Ecosystem	A

(continued)

And whereas, various State Governments/UTs and stakeholders have requested the MoEF&CC to address the concerns related to Coastal Environment and sustainable development w.r.t. the CRZ Notification, 2011;

Whereas MoEF&CC had constituted a committee under the chairmanship of Dr. Shailesh Nayak to examine various issues and concerns of coastal states/Union

Table 5.20 (continued)

S. No.	Planning	Components	Domain of investigation	Priority
4	Biodiversity census	1. Assessment and inventorization of flora and fauna 2. Assessment and scaling of community health of flora and fauna 3. Identification of threatened species 4. Short-term and long-term conservation plan involving the local population (coastal population, island dwellers etc.)	Ecosystem	A
5	Assessment of anthropogenic threats	1. Listing the industries around the landscape and finding out the effluents release 2. Analysis of samples of contaminated soil and water 3. Laboratory analysis of affected plants/animals to determine the concentration of toxicants (bioaccumulation) 4. Identification of indicator organisms	Ecosystem	A
6	Assessment of coastal geomorphology and shoreline configuration	1. Collection of available data on various aspects of coastal geomorphology and analysis in GIS domain to record the changes 2. Identification of drivers which might have changed the geomorphology and develop ecological modeling for expected future changes 3. Quantification and determination of importance of sediment and freshwater flow	Ecosystem	A
7	Assessment of dilution factor in the estuarine and coastal zones	1. Collection of data on freshwater flow, barrage discharge, land run-off and sediment influx 2. Recording and documentation of tidal amplitude 3. Regular monitoring of salinity of sample station (under investigation) and that of source water (particularly for estuarine ecosystem)	Ecosystem	A
8	Assessment of intertidal mudflats/sandy shore/rocky shore and their ecological significance	1. Assessment of biodiversity in the intertidal mudflats/sandy shore/rocky shore along with their adaptations 2. Assessment of avifaunal diversity and migratory species 3. Texture and composition of the intertidal zone	Habitat	A

(continued)

Table 5.20 (continued)

S. No.	Planning	Components	Domain of investigation	Priority
9	Assessment of mangroves/sea grass/saltmarsh grass/coral reefs	1. Identification and marking of monitoring plots for biomass study on regular basis 2. Identification of inter-linkage between the habitats e.g., mangroves and saltmarsh grasses or coral reefs and mangroves etc. (if any)	Habitat	A
10	Coastal lagoon ecology and biodiversity	1. Developmental history and origin of lagoon (through satellite imagery) 2. Flora, fauna and keystone species assessment 3. Primary and secondary productivity of the lagoon 4. Livelihood associated with the lagoon water and other resources	Habitat	A
11	Threatened animals	1. Identification and assessment of status of critical habitats of threatened species 2. Long-term planning for ecorestoration	Habitat	A
12	Monitoring of primary production	1. Preparation of phytoplankton species inventory 2. Tidal, diurnal and seasonal variations in primary productivity 3. Monitoring phytopigment level on regular basis during high tide and low tide	Species	A
13	Monitoring of secondary production	1. Documentation of zooplankton 2. Documentation of nekton (considering both osteichthyes and chondrichthyes) 3. Inventorization of benthic organisms 4. Application of DNA bar coding as tool for biodiversity assessment	Species	A
14	Invasive species	1. Monitoring of ballast discharge 2. Monitoring ecology of invasive species 3. Monitoring the interaction between endemic and alien species 4. Long-term planning to manage invasive species particularly which intrude through ballast discharge	Species	B

(continued)

Territories and various stakeholders, relating to the CRZ Notification 2011 and to recommend appropriate changes in the said Notification;

Whereas the report submitted by Dr. Shailesh Nayak Committee has been examined in the Ministry and consultations have been held with various stakeholders in this regard;

Now, therefore in exercise of the powers conferred by sub-section (1) and clause (v) of sub-section (2) of section 3 of the Environment (Protection) Act, 1986 and in

Table 5.20 (continued)

S. No.	Planning	Components	Domain of investigation	Priority
15	Natural calamities	1. Collection of secondary data on magnitude and occurrence of natural calamities 2. Long-term planning for disaster management 3. Monitoring of resources (flora, fauna, ground water, intertidal zone etc.) before and after the natural calamities through satellite imagery and ground truthing 4. Development of bio-shield to reduce the impact of natural calamities 5. Development of reclamation models for the affected sites 6. Development of species recovery plans particularly for the threatened species	Ecosystem and production sectors	B
16	Livelihood	1. Identification of alternative livelihood schemes befitted to the area of investigation 2. Identification of proper beneficiaries associated with the livelihood schemes 3. Monitoring the viability of schemes through cost benefit analysis and market demand 4. Establishment of local and regional marketing linkage 5. Attempt to develop international market through quality production and subsequent certification	Socio-economic	A
17	Tourism	1. Assessment of existing tourism units 2. Monitoring the effluents released from tourism units and prediction of the level of ingredients of these effluents (viz. nitrate, phosphate etc.) through statistical tools/programmes 3. Assessment of tourist carrying capacity and footfalls 4. Development of eco-tourism involving the local people 5. Training the local people on flora, fauna and other endemic species on regular basis so that they can provide service as eco-guide 6. Development of eco-friendly waste management system (bioremediation)	Socio-economic	B

(continued)

Table 5.20 (continued)

S. No.	Planning	Components	Domain of investigation	Priority
18	Good management practice	1. Collection of successful case studies from district level/state level/country level/international level 2. Group activity through formation of homogeneous group 3. Capacity building in the sphere of aquaculture/agriculture/horticulture etc. 4. Value addition in the field of aquaculture/agriculture/horticulture etc.	Socio-economic	C
19	Production and development sectors	1. Rapid and long-term EIA studies for any developmental activities in the coastal and estuarine ecosystems 2. Impact studies and habitat degradation 3. Monitoring ecosystem services and deletion of natural resources	Socio-economic and developmental activities	B
20	Policy level research	1. Development of a national policy on the coastal and marine environment 2. The efficacy of international, national and state policies and legal instruments in resource management in the coastal and marine environment needs to be assessed 3. Mechanism of networking various governmental, non-governmental and communities based institutions in the participatory management of coastal and marine resources 4. Identification of various components that need to be part of a national or state level policy for restoration and rehabilitation of communities affected by conservation initiatives, natural disasters, etc.	Policy research	C

* *Note* A = High Priority, B = Moderate Priority and C = Low Priority

supersession of Coastal Regulation Zone Notification 2011, the Central Government, with a view to conserve and protect the unique environment of coastal stretches and marine areas, besides livelihood security to the fisher communities and other local communities in the coastal areas and to promote sustainable development based on scientific principles taking into account the dangers of natural hazards, sea level rise due to global warming, does hereby, declare the coastal stretches of the country and the water area up to its territorial water limit, excluding the islands of Andaman and Nicobar and Lakshadweep and the marine areas surrounding these islands, as Coastal Regulation Zone as under:

(i) The land area from High Tide Line (hereinafter referred to as the HTL) to 500 m on the landward side along the sea front.

For the purposes of this Notification, the HTL means the line on the land upto which the highest water line reaches during the spring tide, as demarcated by the National Centre for Sustainable Coastal Management (NCSCM) in accordance with the laid down procedures and made available to various coastal States/UTs.

(ii) CRZ shall apply to the land area between HTL to 50 m or width of the creek whichever is less on the landward side along the tidal influenced water bodies that are connected to the sea and the distance upto which development along such tidal influenced water bodies is to be regulated shall be governed by the distance upto which the tidal effects are experienced which shall be determined based on salinity concentration of 5 parts per thousand (ppt or psu) measured during the driest period of the year and distance up to which tidal effects are experienced shall be clearly identified and demarcated accordingly in the Coastal Zone Management Plans (hereinafter referred to as the CZMPs).

This CRZ limit of 50 m or width of the creek whichever is less, shall however be subject to revision and final approval of the respective CZMPs as per this Notification, framed with due consultative process/public hearing etc. and environmental safeguards enlisted therein. Till such time the CZMP to this notification is approved, the limit of 100 m or width of the creek whichever is less, shall continue to apply.

Explanation: For the purposes of this sub-paragraph the expression tidal influenced water bodies mean the water bodies influenced by tidal effects from sea in the bays, estuaries, rivers, creeks, backwaters, lagoons and ponds etc. that are connected to the sea.

(iii) The intertidal zone i.e., land area between the HTL and the Low Tide Line (hereinafter referred to as the LTL).

(iv) The water and the bed area between the LTL to the territorial water limit (12 Nm) in case of sea and the water and the bed area between LTL at the bank to the LTL on the opposite side of the bank, of tidal influenced water bodies.

2. Classification of the CRZ—For the purpose of conserving and protecting the coastal areas and marine waters, the CRZ area shall be classified as follows, namely:

2.1. CRZ-I areas are environmentally most critical and shall be further classified as under:

2.1.1. CRZ-I A:

(a) CRZ-I A shall constitute the following ecologically sensitive areas and the geo-morphological features which play a role in the maintaining the integrity of the coast viz.:

   (i) Mangroves. In case mangrove area is more than 1000 sq. m, a buffer of 50 m along the mangroves shall be provided and such area shall also constitute CRZ-I A.

   (ii) Corals and coral reefs;

(iii) Sand Dunes;
(iv) Biologically active Mudflats;
(v) National parks, marine parks, sanctuaries, reserve forests, wildlife habitats and other protected areas under the provisions of Wild Life (Protection) Act, the Forest (Conservation) Act or Environment (Protection) Act; including Biosphere Reserves;
(vi) Salt Marshes;
(vii) Turtle nesting grounds;
(viii) Horseshoe crab habitats;
(ix) Sea grass beds;
(x) Nesting grounds of birds;
(xi) Areas or structures of archaeological importance and heritage sites.

(b) A detailed environment management plan shall be formulated by the States/UTs for such ESAs in respective territories, as mapped out by NCSCM, based on guidelines as contained in Annexure-I and integrated in the CZMPs.

2.1.2. CRZ-I B:

The intertidal zone i.e., the area between Low Tide Line and High Tide Line shall constitute the CRZ-I B.

2.2. CRZ-II

CRZ-II shall constitute the developed land areas up to or close to the shoreline, within the existing municipal limits or in other existing legally designated urban areas, which are substantially built-up with a ratio of built up plots to that of total plots being more than 50% and have been provided with drainage and approach roads and other infrastructural facilities, such as water supply and sewerage mains etc.

2.3. CRZ-III

Land areas that are relatively undisturbed (viz rural areas etc.) and those do not fall under CRZ-II, shall constitute CRZ-III. CRZ-III shall be further classified into following categories:

2.3.1. CRZ-III A

Such densely populated CRZ-III areas, where the population density is more than 2161 per sq km as per 2011 census base, shall be designated as CRZ-III A. In CRZ-III A, area up to 50 m from the HTL on the landward side shall be earmarked as the No Development Zone (NDZ), provided the CZMPs as per this Notification, framed with due consultative process, have been approved, failing which, a NDZ of 200 m shall continue to apply.

2.3.2. CRZ-III B

All other CRZ-III areas with population density of less than 2161 per sq km, as per 2011 census base, shall be designated as CRZ-III B. In CRZ-III B, the area up to 200 m from the HTL on the landward side shall be earmarked as the No Development Zone (NDZ).

2.3.3. Land area up to 50 m from the HTL, or width of the creek whichever is less, along the tidal influenced water bodies, shall also be earmarked as the NDZ in CRZ III areas.

Note: The NDZ shall not be applicable in such areas falling within notified Port limits.

2.4. CRZ-IV:

The CRZ-IV shall constitute the water area and shall be further classified as under:

2.4.1. CRZ-IV A

The water area and the sea bed area between the Low Tide Line up to twelve (12) nautical miles on the seaward side shall constitute CRZ-IV A.

2.4.2. CRZ-IV B

CRZ-IV B areas shall include the water area and the bed area between LTL at the bank of the tidal influenced water body to the LTL on the opposite side of the bank, extending from the mouth of the water body at the sea up to the influence of tide, i.e., salinity of five parts per thousand (ppt) during the driest season of the year.

3. Areas requiring special consideration in the CRZ

Following coastal areas shall be accorded special consideration for the purpose of protecting the critical coastal environment and difficulties faced by local communities:

3.1. Critically Vulnerable Coastal Areas (CVCA)

Sunderban region of West Bengal and other ecologically sensitive areas identified as under Environment (Protection) Act, 1986 such as Gulf of Khambat and Gulf of Kutch in Gujarat, Malvan, Achra-Ratnagiri in Maharashtra, Karwar and Coondapur in Karnataka, Vembanad in Kerala, Gulf of Mannar in Tamil Nadu, Bhaitarkanika in Odisha, Coringa, East Godavari and Krishna in Andhra Pradesh shall be treated as Critical Vulnerable Coastal Areas (CVCA) and managed with the involvement of coastal communities including fisher folk who depend on coastal resources for their sustainable livelihood.

3.2. CRZ for inland Backwater islands and islands along the mainland coast.

3.3. CRZ falling within municipal limits of Greater Mumbai.

4. Prohibited activities within CRZ

The following activities shall be prohibited, in general, within the entire CRZ. Exceptions to these and other permissible/regulated activities in specific CRZ categories viz. CRZ-I, II, III and IV, shall however be governed by the provisions under para 5 of this Notification.

(i) Setting up of new industries and expansion of existing industries, operations or processes.
(ii) Manufacture or handling of oil, storage or disposal of hazardous substances as specified in the notification of Ministry of Environment, Forest & Climate Change.
(iii) Setting up of new fish processing units.
(iv) Land reclamation, bunding or disturbing the natural course of seawater.
(v) Discharge of untreated waste and effluents from industries, cities or towns and other human settlements.
(vi) Dumping of city or town wastes including construction debris, industrial solid wastes, fly ash for the purpose of land filling.
(vii) Port and harbour projects in high eroding stretches of the coast.
(viii) Mining of sand, rocks and other sub-strata materials.
(ix) Dressing or altering active sand dunes.
(x) In order to safeguard the aquatic system and marine life, disposal of plastic into the coastal waters shall be prohibited. Adequate measures for management and disposal of plastic materials shall be undertaken in the CRZ.

5. Regulation of permissible activities in the CRZ 5.1 CRZ-I

5.1.1. CRZ-I A

These areas are ecologically most sensitive and generally no activities shall be permitted to be carried out in the CRZ-I A areas, with following exceptions:

(i) Eco-tourism activities such as mangrove walks, tree huts, nature trails, etc., in identified stretches areas subject to such eco-tourism plan featuring in the approved CZMPs as per this Notification, framed with due consultative process/public hearing etc. and further subject to environmental safeguards and precautions related to the Ecologically Sensitive Areas, as enlisted in the CZMPs.
(ii) In the mangrove buffer, only such activities shall be permitted like laying of pipelines, transmission lines, conveyance systems/mechanisms and construction of road on stilts etc. that are required for public utilities.
(iii) Construction of roads and roads on stilts, by way of reclamation in CRZ-I areas, shall be permitted only in exceptional cases for defense, strategic purposes and public utilities, subject to a detailed marine/terrestrial environment impact assessment, to be recommended by the Coastal Zone Management Authority and approved by the Ministry of Environment, Forest and Climate Change; and in case construction of such roads passes through mangrove areas or is likely to damage the mangroves, a minimum three times the mangrove area affected/destroyed/cut during the construction process shall be taken up for compensatory plantation of mangroves.

5.1.2. CRZ-I B—The inter tidal areas

Activities shall be regulated/permissible in the CRZ-I B areas as under:

(i) Land reclamation and bunding etc. shall be permitted only for activities such as;

(a) Foreshore facilities like ports, harbours, Jetties, wharves, quays, slipway, bridges and sea links etc.

(b) Projects for defence, strategic and security purpose;

(c) Road on stilts, provided that such roads shall not be authorized for permitting development on the landward side of such roads, till the existing High Tide Line.

Provided further that the use of reclaimed land may be permitted only for public utilities such as mass rapid or multimodal transit system, construction and installation of all necessary associated public utilities and infrastructure to operate such transit or transport system including those for electrical or electronic signaling system, transit stopover of permitted designs; except for any industrial operation, repair and maintenance.

(d) Measures for control of erosion.

(e) Maintenance and clearing of waterways, channels and ports.

(f) Measures to prevent sand bars, installation of tidal regulators, laying of storm water drains or for structure for prevention of salinity ingress and freshwater recharge.

(ii) Activities related to waterfront or directly needing foreshore facilities such as ports and harbours, jetties, quays, wharves, erosion control measures, breakwaters, pipelines, lighthouses, navigational safety facilities, coastal police stations and the like.

(iii) Power by non-conventional energy sources and associated facilities.

(iv) Transfer of hazardous substances from ships to Ports, terminals and refineries and vice versa.

(v) Facilities for receipt and storage of petroleum products and liquefied natural gas as specified in Annexure-II, subject to implementation of safety regulations including guidelines issued by the Oil Industry Safety Directorate in the Ministry of Petroleum and Natural Gas and guidelines issued by MoEF&CC, provided that such facilities are for receipt and storage of fertilizers and raw materials required for fertilizers, like ammonia, phosphoric acid, sulphur, sulphuric acid, nitric acid etc.

(vi) Storage of non-hazardous cargo i.e., edible oil, fertilizers and food grains in notified Ports.

(vii) Hatchery and natural fish drying.

(viii) Existing fish processing units may utilize 25% additional plinth area for modernization purposes (only for additional equipments and pollution control measures) subject to the following:
  (a) FSI of such reconstruction not exceeding the permissible FSI permissible as per prevalent town and country planning regulations.
  (b) Additional plinth area is constructed only to the landward side.
  (c) Approval of the concerned SPCB/PCC.

(ix) Treatment facilities for waste and effluents and conveyance of treated effluents.

(x) Storm water drains.

(xi) Projects classified as Strategic, Defense related projects and Projects of Department of Atomic Energy.

(xii) Manual mining of atomic mineral(s) notified under Part-B of First Schedule of Mining and Minerals (Development) Act, 1957 occurring as such or in association with one or other minerals in the inter-tidal zone by such agencies as authorized by Department of Atomic Energy, as per mining plan approved by the Department of Atomic Energy.

  Provided that the manual mining operations are carried out only by deploying persons using baskets and hand spades for collection of ore or mineral within the intertidal zone and as per approved mining plan, without deploying or using drilling and blasting or Heavy Earth Moving Machinery in the intertidal zone.

(xiii) Exploration and extraction of oil and natural gas and all associated activities and facilities thereto;

(xiv) Foreshore requiring facilities for transport of raw materials, facilities for intake of cooling water and outfall for discharge of treated wastewater or cooling water from thermal power plants, in conformity with the environmental standards notified by MoEF&CC and relevant directions of Central Pollution Control Board (CPCB)/SPCB/PCC.

(xv) Pipelines, conveying systems including transmission lines.

(xvi) Weather radar for monitoring of cyclones movement and prediction and associated facilities.

(xvii) Salt harvesting and associated facilities.

(xviii) Desalination plants and associated facilities.

5.2. CRZ-II

(i) Activities as permitted in CRZ-I B, shall also be permissible in CRZ-II, as applicable.

(ii) Construction of buildings for residential purposes, schools, hospitals, institutions, offices, public places etc. shall be permitted only on the landward side of the existing road, or on the landward side of existing authorized fixed structures; provided that no permission for construction of buildings shall be given on landward side of any new roads which are constructed on the seaward side of an existing road.

(iii) Buildings permitted as in (ii) above, shall be subject to the local town and country planning regulations as applicable from time to time, and the norms for the Floor Space Index or Floor Area Ratio prevailing as on the date of this Notification. It shall be the responsibility of the concerned Town Planning Authority to ensure that the Solid Wastes are handled as per respective SWM rules and no untreated sewage is discharged on to the coast or coastal waters.

(iv) Reconstruction of authorized buildings shall be permitted, without change in present land use, subject to the local town and country planning regulations as applicable from time to time, and the norms for the Floor Space Index or Floor Area Ratio, prevailing as on the date of this Notification. It shall be the responsibility of the concerned Town Planning Authority to ensure that the Solid Wastes are handled as per respective SWM rules and no untreated sewage is discharged on to the coast or coastal waters.

(v) Development of vacant plots in designated areas for construction of beach resorts/hotels subject to the conditions/guidelines at Annexure-III.

(vi) Temporary tourism facilities shall be permissible in the beaches. Such temporary facilities shall only include shacks, toilets/washrooms, change rooms, shower panels; walk ways constructed using interlocking paver blocks etc., drinking water facilities, seating arrangements etc. Such facilities shall however be permitted only subject to the tourism plan featuring in the approved CZMPs as per this Notification, framed with due consultative process/public hearing etc. and further subject to environmental safeguards enlisted in the CZMPs.

5.3. CRZ-III

(i) Activities as permitted in CRZ-I B, shall also be permissible in CRZ-III, as applicable.

(ii) Regulation of activities in the NDZ:

Following shall be permissible/regulated in the NDZ:

(a) No construction shall be permitted within NDZ in CRZ III, except for repairs or reconstruction of existing authorized structure not exceeding existing Floor Space Index, existing plinth area and existing density; for permissible activities under the notification including facilities essential for activities and construction/reconstruction of dwelling units of traditional coastal communities including fisher folk, incorporating necessary disaster management provisions and proper sanitation arrangements.

(b) Agriculture, horticulture, gardens, pastures, parks, playfields and forestry.

(c) Construction of dispensaries, schools, public rain shelter, community toilets, bridges, roads, provision of facilities for water supply, drainage, sewerage, crematoria, cemeteries and electric sub-station which are required for the local inhabitants may be permitted on a case to case basis by CZMA.

(d) Construction of units or auxiliary thereto for domestic sewage, treatment and disposal with the prior approval of the concerned Pollution Control Board or Committee.

(e) Facilities required for local fishing communities such as fish drying yards, auction halls, net mending yards, traditional boat building yards, ice plant, ice crushing units, fish curing facilities and the like;

(f) Wherever there is a national or state highway passing through the NDZ of CRZ-III areas, temporary tourism facilities such as toilets, change rooms, drinking water facility and temporary shacks can be taken up on the seaward side of the road.

On landward side of such roads in the NDZ, Resorts/hotels and associated tourism facilities shall be permitted. Such facilities shall, however, be permitted only subject to the incorporation of tourism plan in the approved CZMPs as per this Notification and the conditions/guidelines at Annexure-III, as applicable.

(g) Temporary tourism facilities shall be permissible in the NDZ of CRZ-III areas. Such temporary facilities shall only include shacks, toilets/washrooms, change rooms, shower panels; walk ways constructed using interlocking paver blocks etc., drinking water facilities, seating arrangements etc. Such facilities shall however be permitted only subject to the tourism plan featuring in the approved CZMPs as per this Notification.

(iii) For CRZ-III Areas beyond the NDZ, activities shall be permissible/regulated as under:

(a) Development of vacant plots in designated areas for construction of beach resorts/hotels subject to the conditions/guidelines at Annexure-III;

(b) Construction or reconstruction of dwelling units so long it is within the ambit of traditional rights and customary uses such as existing fishing villages etc. Building permission for such construction or reconstruction will be subject to local town and country planning rules with overall height of construction not exceeding 9 m. with two floors (ground + one floor);

(c) The local communities including fishermen can be permitted to facilitate tourism through 'home stay' without changing the plinth area/design or facade of the existing houses.
(d) Construction of public rain shelters, community toilets, water supply drainage, sewerage, roads and bridges etc.
(e) Limestone Mining:
Selective mining of limestone minerals may be permitted in specific identified areas under the mining plans, which are adequately above the height of HTL, based on the recommendations of reputed National Institutes in the mining field such as CSIR, Central Mining Research Institute etc. provided that the extraction of minerals is carried out not below a height of 1 m above HTL and an adequate barrier is created so as to safeguard against saline water incursion and subject to appropriate safeguards related to pollution of coastal waters and prevention of coastal erosion.

(iv) Drawing of groundwater and construction related thereto shall be prohibited within 200 m of HTL except for the use of local communities in areas inhabited by them. In the areas between 200 and 500 m of the HTL, groundwater withdrawal can be permitted only through manual means from ordinary wells for drinking, horticulture, agriculture and fisheries etc. where no other source of water is available. Restrictions for such drawal may be imposed by the designated Authority by State Government/UT Administration in the areas affected by sea water intrusion.

5.4. CRZ-IV

(i) Traditional fishing and allied activities undertaken by local communities.
(ii) Land reclamation and bunding etc. to be permitted only for activities such as;
(a) Foreshore facilities like ports, harbours, Jetties, wharves, quays, slipway, bridges and sea links etc.
(b) Projects for defence, strategic and security purpose.
(c) Measures for control of erosion.
(d) Maintenance and clearing of waterways, channels and ports.
(e) Measures to prevent sand bars, installation of tidal regulators, laying of storm water drains or for structure for prevention of salinity ingress and freshwater recharge.
(iii) Activities related to waterfront or directly needing foreshore facilities such as ports and harbours, jetties, quays, wharves, erosion control measures, breakwaters, pipelines, navigational safety facilities, and the like.
(iv) Power by non-conventional energy sources and associated facilities.

(v) Transfer of hazardous substances from ships to Ports.
(vi) Storage of non-hazardous cargo like edible oil, fertilizers and food grains in notified Ports.
(vii) Facilities for discharging treated effluents into the water course.
(viii) Projects classified as Strategic and Defence related projects.
(ix) Projects of Department of Atomic Energy.
(x) Exploration and extraction of oil and natural gas and all associated activities and facilities thereto;
(xi) Foreshore requiring facilities for transport of raw materials, facilities for intake of cooling water and outfall for discharge of treated wastewater or cooling water from thermal power plants. Foreshore requiring facilities for transport of raw materials, facilities for intake of cooling water and outfall for discharge of treated wastewater or cooling water from thermal power plants, in conformity with the environmental standards notified by MoEF&CC and relevant directions of CPCB/SPCB/PCC.
(xii) Pipelines, conveying systems including transmission lines.
(xiii) Weather radar for monitoring of cyclone prediction, movement and associated facilities.
(xiv) Construction of memorials/monuments and allied facilities by the concerned State Government in CRZ-IV (A) areas, in exceptional cases, with adequate environmental safeguards, subject to the following, namely:-
    (a) The concerned State Government shall submit justification for locating the project in CRZ-IV A area along with details of alternate sites considered and weightage matrix on various parameters including environmental parameters, to State CZMA who will examine the project and make recommendation to the Central Government (MoEF&CC) for grant of Terms of Reference (ToRs) for preparation of an environmental impact assessment report by the State Government;
    (b) On grant of ToRs by the Central Government, the concerned State Government shall submit the draft Environmental Impact Assessment report (EIA) with Environmental Management Plan (EMP), draft Risk Assessment Report with Disaster Management Plan (DMP) including on-site and off-site emergency plan and evacuation plan during emergency, to the State Pollution Control Board for conduct of public hearing for the proposed project in accordance with the procedure laid down under the Environment Impact Assessment notification;
    (c) The concerned State Government shall, after addressing the relevant issues raised by the public during the public hearing referred to in sub-item (b), submit the final EIA, EMP, Risk Assessment and DMP, to the State CZMA for their examination and recommendation to MoEF&CC;

(d) The Central Government may, if it considers necessary so to do, dispense with the requirement of public hearing referred to in sub-item (b), if it is satisfied that the project will not involve rehabilitation and resettlement of the public or the project site is located away from human habitation.

6. Coastal Zone Management Plans (CZMPs)

(i) All coastal States/UTs shall revise/update their respective coastal zone management plan (CZMP) framed under CRZ Notification, 2011, as per provisions of this Notification and submit to MoEF&CC for approval at the earliest. All the project activities attracting the provisions of this Notification shall be required to be appraised as per the updated CZMPs to this Notification. Until and unless the CZMPs are so revised/updated, provisions of this Notification shall not come in force and the CZMPs as per provisions of CRZ Notification, 2011 shall continue to be followed for appraisal and CRZ clearance to such projects;

(ii) The CZMPs may be prepared/updated by the coastal State Government or Union territory by engaging reputed and experienced scientific institution(s) or the agencies including the National Centre for Sustainable Coastal Management (hereinafter referred to as the NCSCM) of MoEF&CC and in consultation with the concerned stakeholders;

(iii) The coastal States and Union Territory shall prepare draft CZMPs in 1:25,000 scale map identifying and classifying the CRZ areas within the respective territories in accordance with the guidelines given in Annexure-IV of the notification, which involve public consultation;

All developmental activities listed in this notification shall be regulated by the State Government, Union Territory Administration, the local authority or the concerned CZMA within the framework of such approved CZMPs as the case may be in accordance with provisions of this notification;

(iv) The draft CZMPs shall be submitted by the State Government or Union territory to the concerned CZMA for appraisal, including appropriate consultations, and recommendations in accordance with the procedure(s) laid down in the Environment (Protection) Act, 1986;

(v) MoEF&CC shall thereafter consider and approve the respective CZMPs of concerned State Governments/UTs.

(vi) The CZMPs shall not normally be revised before a period of five years after which, the concerned State Government or the Union territory may consider undertaking a revision.

7. CRZ clearance for permissible/regulated activities—Delegations

(i) All permitted/regulated project activities attracting the provisions of this notification shall be required to obtain CRZ clearance prior to their commencement.

(ii) All development activities/projects in CRZ-I and CRZ-IV areas, which are regulated/permissible as per this notification, shall be dealt with by MOEF&CC for CRZ clearance, based on the recommendation of the concerned CZMA.

(iii) For all other permissible/regulated activities as per this Notification, which fall purely in CRZ-II/CRZ-III areas, the CRZ clearance shall be considered by the concerned CZMAs. Such projects in CRZ-II and III, which also happen to be traversing through CRZ-I and/or CRZ-IV areas, CRZ clearance shall, however be considered only by MoEF&CC, based on recommendations of the concerned CZMA.

(iv) Projects/activities which attract the provisions of this Notification as also the provisions of EIA Notification 2006, shall be dealt with for a composite Environmental and CRZ clearance under EIA Notification 2006 by the concerned approving Authority, based on recommendations of concerned CZMA, as per delegations i.e., State Environmental Impact Assessment Authority (hereinafter referred to as the SEIAA) or MoEF&CC for category "B" and category "A" projects respectively.

(v) In case of building/construction projects with built-up area less than the threshold limit stipulated for attracting the provisions of the EIA Notification, these shall be approved by the concerned local State or Union Territory Planning Authorities in accordance with this notification, after obtaining recommendations of concerned CZMA.

(vi) Only for self dwelling units up to a total built up area of 300 m^2, approval shall be accorded by the concerned local Authority, without the requirement of recommendations of concerned CZMA. Such authorities shall, however, examine the proposal from the perspective of the CRZ Notification before according approval.

8. Procedure for CRZ Clearance for Permissible/regulated Activities

(i) The project proponents shall apply with the following documents to the concerned State or the Union territory Coastal Zone Management Authority for seeking prior clearance under CRZ Notification:

(a) Project summary details as per Annexure-V of the notification.

(b) Rapid EIA Report including marine and terrestrial component, as applicable, except for building construction projects/housing schemes.

(c) Comprehensive EIA with cumulative studies for projects, if located in low and medium eroding stretches, as per the CZMP to this notification.

(d) Risk Assessment Report and Disaster Management Plan.

(e) CRZ map in 1:4000 scale, drawn up by any of the agencies identified by the MoEF&CC vide its Office order number J-17011/8/92-IA-III dated 14th March 2014 using the demarcation of the High Tide Line or LTL, as carried out by NCSCM.

- (f) Project layout superimposed on the above map duly indicating the project boundaries and the CRZ category of the project location.
- (g) The CRZ map normally covering 7 km radius around the project site also indicating the CRZ-I, II, III and IV areas including other notified ecologically sensitive areas.
- (h) "Consent to establish" or NOC from the concerned State Pollution Control Boards or Union Territory Pollution Control Committees for the projects involving discharge of industrial effluents.

(ii) The concerned CZMA shall examine the documents as in (i) above, in accordance with the approved CZMP and in compliance with CRZ notification and make recommendations within a period of sixty days from date of receipt of complete application:-

- (a) To MoEF&CC or SEIAA for category "A" and category "B" projects respectively, as the case may be, for the projects also attracting EIA Notification, 2006; as per delegations in the EIA Notification 2006. For such Category "B" projects located in CRZ-I or CRZ-IV areas, final recommendation for CRZ clearance shall however be made only by MoEFCC to the concerned SEIAA to enable it accord a composite EC and CRZ clearance to the proposal.
- (b) To MoEF&CC for the projects/activities not covered in the EIA notification, 2006, but attracting CRZ Notification and located in CRZ-I or CRZ-IV areas.
- (c) To SEIAA for the projects/activities not covered in the EIA notification, 2006, but attracting CRZ Notification and located in CRZ-II or CRZ-III areas.
- (d) To the concerned State or Union territory planning authorities, in case of construction projects with built-up area less than the threshold limit stipulated for attracting the provisions of the EIA Notification 2006.

(iii) MoEF&CC, SEIAA or the concerned State/Union territory planning authority shall consider such projects for clearance based on the recommendations of the concerned CZMA, within a period of sixty days.

(iv) In case the CZMAs are not in operation due to their reconstitution or any other reasons, then it shall be responsibility of the Department of Environment in the State Government or Union territory Administrations, who are the custodian of the CZMPs of respective States or Union Territories, to provide comments and recommend the proposals in terms of the provisions of the said notification.

(v) The clearance accorded to the projects under this notification shall be valid for a period of seven years, provided that the construction activities are completed and the operations commence within seven years from the date of issue of such clearance.

The validity may be further extended for a maximum period of three years, provided an application is made to the concerned authority by the applicant within the validity period, along with recommendation for extension of validity of the clearance by the concerned State/Union Territory Coastal Zone Management Authority.

(vi) Post clearance monitoring:-
   (a) It shall be mandatory for the project proponent to submit half-yearly compliance reports in respect of the stipulated terms and conditions of the environmental clearance in hard and soft copies to the regulatory authority(s) concerned, on 1st June and 31st December of each calendar year and all such compliance reports submitted by the project proponent shall be published in public domain and its copies shall be given to any person on application to the concerned CZMA.
   (b) The compliance report shall also be displayed on the website of the concerned regulatory authority.

(vii) To maintain transparency in the working of the CZMAs, it shall be the responsibility of the CZMA to create a dedicated website and post the agenda, minutes, decisions taken, clearance letters, violations, action taken on the violations and court matters including the Orders of the Hon'ble Court as also the approved CZMPs of the respective State Government or Union Territory.

9. Enforcement of the CRZ Notification:

(i) For the purpose of implementation and enforcement of the provisions this notification and compliance with conditions stipulated there under, the powers either original or delegated are available under Environment (Protection) Act, 1986 with the MoEF&CC, State Government or the Union territory Administration, NCZMA and SCZMAs;

(ii) The composition, tenure and mandate of NCZMA and State Government or the Union territory CZMAs have already been notified by MoEF&CC in terms of Orders of Hon'ble Supreme Court in Writ Petition 664 of 1993;

(iii) The State Government or the Union territory CZMAs shall primarily be responsible for enforcing and monitoring of this notification and to assist in this task, the State Government and the Union territory shall constitute district level Committees under the Chairmanship of the District Magistrate concerned comprising at least three representatives of local traditional coastal communities including from fisherfolk. The State Government may consider further delegation of the enforcement of this notification to the level of respective District Magistrates.

(iv) The dwelling units of the traditional coastal communities including fisher folk, tribals as were permissible under the provisions of the CRZ notification, 2011, but which have not obtained formal approval from concerned authorities under the aforesaid notification shall be considered by the respective Union territory CZMAs and the dwelling units shall be regularized subject to the following condition, namely:-

(a) These are not used for any commercial activity.
(b) These are not sold or transferred to non-traditional coastal community.

10. Areas requiring special consideration:

10.1. Critically Vulnerable Coastal Areas (CVCAs):

(i) For all the CVCAs enlisted in para 3.3 of this notification, Integrated Management Plans (IMPs) shall be prepared, which shall, inter alia, keep in view the conservation and management of mangroves, needs of local communities such as, dispensaries, schools, public rain shelter, community toilets, bridges, roads, jetties, water supply, drainage, sewerage and the impact of sea level rise and other natural disasters. The IMPs will be prepared in line with the guidelines for preparation of Coastal Zone Management Plans;
(ii) Till such time the IMPs are approved and notified, construction of dispensaries, schools, public rain/cyclone shelters, community toilets, bridges, roads, jetties, water supply, drainage, sewerage which are required for traditional inhabitants shall be permitted on a case to case basis, by the CZMA with due regards to the views of coastal communities including fisher folk.

10.2. CRZ for inland Backwater islands and islands along the mainland coast:

(i) All the inland islands in the coastal backwaters and islands along the mainland coast shall also be covered under this CRZ Notification.
(ii) In view of the unique coastal systems of backwater and backwater islands along with space limitations in such coastal stretches, CRZ of 20 m from the HTL on the landward side shall uniformly apply to such islands and activities shall be regulated as under:
    (a) Existing dwelling units of local communities may be repaired or reconstructed within 20 m from the HTL of these islands. However, no new construction shall be permitted.
    (b) Foreshore facilities such as fishing jetty, fish drying yards, net mending yard, fishing processing by traditional methods, boat building yards, ice plant, boat repairs and the like, may be taken up in CRZ limits subject to due environmental safeguards.

(iii) Integrated Island Management Plans (IIMPs), as applicable to smaller islands in Lakshadweep and Andaman & Nicobar, as per Island Protection Zone Notification, shall be formulated by respective States/UTs for all such islands and submitted to MoEF&CC for approval at the earliest. Until and unless the IIMPs are framed, provisions of this Notification shall not come in force and the CZMPs as per provisions of CRZ Notification 2011 shall continue to be followed.

10.3. CRZ areas falling within municipal limits of the Greater Mumbai:

(i) In order to protect and preserve the "green lung" of the Greater Mumbai area, all open spaces, parks, gardens, playgrounds indicated in development plans within CRZ-II shall be categorized as No Development Zone. A Floor Space Index up to 15% shall be allowed only for construction of civic amenities, stadium and gymnasium meant for recreational or sports related activities and the residential or commercial use of such open spaces shall not be permissible.

(ii) Construction of sewage treatment plants in CRZ-I area for the purpose of treating the sewage from the municipal area shall be taken only by the municipal authorities in exceptional circumstances, where no alternate site is available to set up such facilities, subject to recommendations of the CZMA and approval by the Central Government and in case the construction of such plant is inevitable in a mangrove area, a minimum three times the mangrove area affected/destroyed/cut during the construction process shall be taken up for compensatory plantation of mangroves.

Annexure-I

Conservation, Protection and Management Framework for ESAs

The coastal and marine Ecologically Sensitive Areas (ESAs) and the geomorphological features play a vital role in maintaining the functions of the coast. Mangroves, beaches, coral reefs etc., aid in controlling coastal erosion, shoreline change, saltwater intrusion and also serve as natural defence against coastal hazards such as storm surges, cyclones and tsunamis. The ESAs maintain the biological integrity of the coast by providing direct and indirect ecosystem services to the coastal livelihood. In addition, several invaluable archaeological and heritage sites are also located along the coast. Hence conservation and protection of the above areas/features/sites become necessary.

1. General measures

(i) All ESAs shall be identified and boundary delineated by NCSCM using satellite data.

(ii) The State/UT Governments through the authorized agencies shall prepare CZMP as per the guidelines contained in the Notification highlighting the conservation and protection of the ESAs.
(iii) Those activities permissible under this notification shall be included in the CZMPs.

Specific conditions shall be adopted for the conservation, protection and management of each of the ESAs as under:-

1.1. Mangroves:

(i) Mangroves declared as forest under Forest Conservation Act, 1980:
Notwithstanding anything contained in this notification, such mangroves declared by the concerned State Governments/UT Administrations/Central Government as forest land under the F(C) Act, 1980 shall attract the provisions of the F(C) Act, 1980 only.
(ii) Mangroves not declared under Forest Conservation Act, 1980.
(a) Mangroves in Government land shall be protected based on a detailed plan to be prepared by the concerned State/UT Governments. In case the mangrove area is more than 1000 m^2, a buffer of 50 m along the periphery of mangrove area shall be provided. This buffer zone of 50 m may be utilized for public facilities for developing parks, research facilities related to mangrove biodiversity, facilities for conservation and the like.
(b) Mangroves in private land will not require a buffer zone.

1.2. Corals and coral reefs and associated biodiversity:

(i) Destruction of coral and coral reefs and the surroundings is a prohibited activity.
(ii) All coral and coral reefs shall be protected except for those small quantities required for research purposes.
(iii) Coral and coral reefs transplantation activities shall be through recognized research institutions wherever required for regeneration after obtaining necessary approvals under Wildlife (Protection) Act 1972.
(iv) The dead and/or destroyed coral areas shall be taken up for rejuvenation and rehabilitation. The conservation and protection of corals and coral reefs shall be taken up as follows:
(a) Active and live coral and coral reefs identified and delineated shall be declared and notified as ESA under E (P) Act 1986.
(b) It shall be ensured that no activities that are detrimental to the health of corals, coral reefs and its associated biodiversity such as mining, effluent and sewage discharge, dredging, ballast water discharge, ship washings, fishing other than traditional non-destructive fisheries, construction activities and the like are taken up in and around the coral areas.

1.3. The National Parks, marine parks, Sanctuaries, reserve forests, wildlife habitats and other protected areas declared under the provisions of Wild Life (Protection) Act, 1972 (53 of 1972), the F(C)Act 1980 (69 of 1980) or E(P)Act 1986 (29 of 1986); including Biosphere Reserves would be conserved and protected as follows:

(i) Conservation and protection of the above listed areas shall be as per the provisions.
(i) of the respective Acts/notifications/guidelines.
(ii) Efforts shall be made to increase the forest area in the coastal region in order to prevent loss of life and property from increased storms, tides and floods.
(iii) The concerned State Governments/UTs shall provide for adequate funds for such measures to undertake shelter belt plantation or bio-shields with planting material suitable to the location.

1.4. Salt marshes:
The conservation and protection of salt marshes shall be as follows:

(i) The salt marsh areas shall be conserved and protected and efforts shall be made to promote the endemic biodiversity in the salt marshes.
(ii) Only those activities required for overhead conveying/transmission of cables and underground laying of transmission line cables and so on, shall be permissible.
(iii) Traditional fishing is permissible in salt marshes.
(iv) Temporary tourism facilities around the salt marsh areas could be considered subject to adhering to strict norms laid down in the guidelines.
(v) Certain salt marshes which have less biodiversity, identified by NCSCM and demarcated in CZMP can be considered for salt pan activities.

1.5. Turtle nesting grounds shall be protected and conserved as follows:

(i) Turtle nesting grounds identified by the concerned States/UTs shall be protected as per Wildlife (Protection) Act of 1972.
(ii) No activities shall be permitted in and around the turtle nesting ground including those causing light and sound pollution except for those required for conservation and protection of these sites.
(iii) Strict management plans for protecting the turtle nesting grounds shall be undertaken and implemented by the concerned State/UT Authorities.

1.6. Horseshoe crabs habitats shall be protected and conserved as follows:

(i) The habitat identified shall be taken up for conservation and protection.
(ii) No activities shall be taken up in and around these habitats which affect the horse shoe crab ecosystem.

1.7. Sea grass beds shall be protected and conserved as follows:

(i) Identified sea grass beds shall be conserved and protected.
(ii) No developmental activities that have adverse effect on the sea grass bed shall be undertaken.
(iii) Efforts shall be made to propagate sea grass beds along the coastal waters where ever possible by States/UTs as it acts as a carbon sink.

1.8. Nesting grounds of birds shall be protected and conserved as follows:

(i) The nesting ground of birds including their local migratory route shall be protected. No developmental activities which have adverse impact on the nesting grounds and the migratory routes shall be undertaken including construction of wind mills, transmission lines and the like in the locality.
(ii) Efforts shall be made to increase the forest cover and mangrove cover including enriching the biodiversity of salt marsh and other coastal water bodies so as to provide for suitable habitat for the avifauna.

1.9. Geo-morphologically Important Zones shall be protected and managed as follows:

(i) Sand dunes identified shall be conserved and protected as follows:
  (a) Sand dunes identified shall be notified under E(P)Act 1986;
  (b) No developmental activities be permissible except for providing eco-friendly temporary tourism facilities on stilts such as walk-ways, tents and the like;
  (c) Mining of sand from sand dunes is a prohibited activity except for the removal of rare earth minerals with proper replenishment using the tailings or other suitable sand.
  (d) No activities on the sand dunes shall be taken up that would lead to erosion/destruction of sand dunes;
  (e) Afforestation, if any, on the sand dunes shall be done only with native flora.
  (f) The States/UTs shall prepare management plans for the demarcated sand dunes.
(ii) Sandy beaches:
  (a) Mining of beach sand is prohibited except for the removal of rare earth minerals with proper replenishment using the tailings or other suitable sand.
  (b) When the permissible developmental activities are taken up on the beaches if loss of beach in the neighbourhood is predicted, necessary beach nourishment to compensate for the losses shall be undertaken by the project authorities and its long term maintenance shall be ensured by them.
  (c) The States/UTs shall prepare management plans for the demarcated beaches.

(iii) Biologically active Mudflats:
   (a) Biologically active mudflats will be identified by NCSCM in association with State Govt./UT Administration.
   (b) The States/UTs shall prepare management plans for such demarcated biologically active mudflats.

1.10. Areas or structures of archaeological importance and heritage value sites:

   (i) State Archaeological agencies shall be responsible for conservation and protection of all archaeological structures and heritage sites identified by Archaeological Survey of India, as per the provisions of the respective Acts/notifications/guidelines.
   (ii) No activities that are detrimental to the identified areas or structures of archaeological and heritage value shall be permitted.
   (iii) It shall be ensured that these structures or areas are preserved and activities undertaken without changing the façade/plinth of such structures. Such structures could be considered for use in accordance with the relevant norms after undertaking careful designing of the interiors without changing the exterior architectural design of the structure.

Annexure-II

List of Petroleum and Chemical Products Permitted for Storage in CRZ, Except CRZ-I A

(i) Crude oil;
(ii) Liquefied Petroleum Gas;
(iii) Motor spirit;
(iv) Kerosene;
(v) Aviation fuel;
(vi) High speed diesel;
(vii) Lubricating oil;
(viii) Butane;
(ix) Propane;
(x) Compressed Natural Gas;
(xi) Naphtha;
(xii) Furnace oil;
(xiii) Low Sulphur Heavy Stock;
(xiv) Liquefied Natural Gas;
(xv) Fertilizers and raw materials for manufacture of fertilizers
(xvi) Acetic acid
(xvii) Mono ethylene glycol.

Annexure-III

Guidelines for Development of Beach Resorts/Hotels in the Designated CRZ Areas

1. CRZ-II

Construction of beach resorts/hotels in designated areas of CRZ-II for occupation of tourist or visitors shall be subject to the following conditions, namely:-

(i) Construction shall be permitted only to the landward side of an existing road or existing authorized fixed structures.
(ii) Live fencing and barbed wire fencing with vegetative cover may be allowed around private properties subject to the condition that such fencing shall in no way hamper public access to the beach;
(iii) No flattening of sand dunes shall be carried out;
(iv) No permanent structures for sports facilities shall be permitted except construction of goal posts, net posts and lamp posts;
(v) Construction of basement may be allowed subject to the condition that no objection certification is obtained from the State Ground Water Authority to the effect that such construction will not adversely affect fee flow of groundwater in that area;
(vi) The State Ground Water Authority shall take into consideration the guidelines issued by Central Government before granting such no objection certificate;
(vii) The quality of treated effluents, solid wastes, emissions and noise levels and the like, from the project area must conform to the standards laid down by the competent authorities including the Central or State Pollution Control Board and under the Environment (Protection) Act, 1986;
(viii) Necessary arrangements for the treatment of the effluents and solid wastes must be made and it must be ensured that the untreated effluents and solid wastes are not discharged into the water or on the beach; and no effluent or solid waste shall be discharged on the beach;
(ix) If the project involves diversion of forestland for non-forest purposes, clearance as required under the Forest (Conservation) Act, 1980 shall be obtained and the requirements of other Central and State laws as applicable to the project shall be met with; and approval of the State or Union territory Tourism Department shall be obtained.

2. CRZ-III

Construction of beach resorts/hotels in designated areas of CRZ-III for occupation of tourists or visitors shall be subject to the following conditions, namely:-

(i) Live fencing and barbed wire fencing with vegetative cover may be allowed around private properties subject to the condition that such fencing shall in no way hamper public access to the beach;
(ii) No flattening of sand dunes shall be carried out;
(iii) No permanent structures for sports facilities shall be permitted except construction of goal posts, net posts and lamp posts;

(iv) Construction of basement may be allowed subject to the condition that no objection certification is obtained from the State Ground Water Authority to the effect that such construction will not adversely affect fee flow of groundwater in that area;

(v) The State Ground Water Authority shall take into consideration the guidelines issued by Central Government before granting such no objection certificate;

(vi) Though no construction is allowed in the no development zone for the purposes of calculation of Floor Space Index, the area of entire plot including the portion which falls within the no development zone shall be taken into account;

(vii) The total covered area on all floors shall not exceed 33 per cent of the plot size i.e., the Floor Space Index shall not exceed 0.33 and the open area shall be suitably landscaped with appropriate vegetal cover;

(viii) The construction shall be consistent with the surrounding landscape and local architectural style;

(ix) The overall height of construction up to the highest ridge of the roof, shall not exceed 9 m and the construction shall not be more than two floors (ground floor plus one upper floor);

(x) Groundwater shall not be tapped within 200 m of the High Tide Line; within the 200 m 500 m zone it can be tapped only with the concurrence of the Central or State Ground Water Board;

(xi) Extraction of sand, leveling or digging of sandy stretches, except for structural foundation of building or swimming pool, shall not be permitted within 500 m of the High Tide Line;

(xii) The quality of treated effluents, solid wastes, emissions and noise levels and the like, from the project area must conform to the standards laid down by the competent authorities including the Central or State Pollution Control Board and under the Environment (Protection) Act, 1986;

(xiii) Necessary arrangements for the treatment of the effluents and solid wastes must be made and it must be ensured that the untreated effluents and solid wastes are not discharged into the water or on the beach; and no effluent or solid waste shall be discharged on the beach;

(xiv) To allow public access to the beach, at least a gap of 20 m width shall be provided between any two hotels or beach resorts; and in no case shall gaps be less than 500 m apart; and

(xv) If the project involves diversion of forestland for non-forest purposes, clearance as required under the Forest (Conservation) Act, 1980 shall be obtained and the requirements of other Central and State laws as applicable to the project shall be met with; and approval of the State or Union territory Tourism Department shall be obtained.

Note: Construction of beach resorts or hotels shall not be permitted in Ecologically sensitive areas (such as marine parks, mangroves, coral reefs, breeding and spawning grounds of fish, wildlife habitats and such other area as may be notified by the Central or State Government Union territories).

Annexure-IV

Guidelines for Preparation of Coastal Zone Management Plans

1. Demarcation of High Tide Line and Low Tide Line
   Demarcation of High Tide Line (HTL) and Low Tide Line (LTL) as carried out by NCSCM for the entire coastline of the country, has been made available to the Coastal States/Union Territories and only such demarcation of HTL and LTL shall be applicable for all purposes under this Notification.

2. A 'Hazard line' has been demarcated by the Survey of India (SOI) taking into account the extent of the flooding on the land area due to water level fluctuations, sea level rise and shoreline changes(erosion/accretion) occurring over a period of time. The hazard line mapped by SOI has been shared with the coastal States/UTs through NCSCM. The hazard line shall be used as a tool for disaster management plan for the coastal environment, including planning of adaptive and mitigation measures. With a view to reduce the vulnerability of the coastal communities and ensuring sustainable livelihood, while drawing the CZMPs, the land use planning for the area between the Hazard line and HTL shall take into account such impacts of climate change and shoreline changes.

3. Preparation of CZM Maps
   (i) Base Maps of 1:25,000 scale shall be acquired from the Survey of India (SOI) and wherever 1:25,000 maps are not available, 1:50,000 maps shall be enlarged to 1:25,000 for the purpose of base map preparation and these maps will be of the standard specification given below:

Unit:	7.5 min × 7.5 min
Numbering:	Survey of India Sheet Numbering System
Horizontal Datum:	Everest or WGS 84
Vertical Datum:	Mean Sea Level (MSL)
Topography:	Topography in the SOI maps will be updated using latest satellite imageries or aerial photographs

   (ii) Coastal Zone Management (CZM) Maps of scale 1:25,000 shall be got prepared by any of the agencies identified by the MoEF&CC vide its Office order number J-17011/8/92-IA-III dated 14th March 2014 using the demarcation of the High Tide Line or LTL, as carried out by NCSCM.
   (iii) Various regulatory lines viz. at a distance of 20 m, 50 m, 200 m and 500 m from HTL respectively, as applicable in various CRZ categories, and the Hazard line shall be demarcated and transferred to the CZM Maps.
   (iv) HTL, LTL and CRZ boundaries, as applicable, shall also be demarcated in the CZM maps along the banks of tidal influenced inland water bodies.

(v) Classification of different coastal zones shall be done as per the CRZ notification and Standard national or international colour codes shall be used.

4. Local level CZM Maps

(i) Local level CZM Maps are for the use of local bodies and other agencies to facilitate implementation of the Coastal Zone Management Plans.
(ii) Cadastral (village) maps in 1:3960 or the nearest scale, as available with revenue authorities shall be used as the base maps.
(iii) HTL, LTL, other CRZ regulatory lines and the Hazard line shall be demarcated in the cadastral maps and classifications shall be transferred into local level CZM maps.

5. Classification of CRZ areas

(i) The CZM Maps shall clearly demarcate the land use plan of the area and map out the Ecologically Sensitive Areas (ESAs) or the CRZ-IA areas as per mapping made available by NCSCM to coastal State/UTs. All such ESAs shall be appropriately demarcated with colour codes.
(ii) Buffer zone along mangrove areas of more than 1000 m^2 shall be stipulated with a different colour distinguishing from the mangrove area. The buffer zone shall also be classified as CRZ-I area.
(iii) In the CRZ areas, the fishing villages, common properties of the fishermen communities, fishing jetties, ice plants, fish drying platforms or areas infrastructure facilities of fishing and local communities such as dispensaries, roads, schools, and the like, shall be indicated on the cadastral scale maps. States shall prepare detailed plans for long term housing needs of coastal fisher communities in view of expansion and other needs, provisions of basic services including sanitation, safety, and disaster preparedness.
(iv) The water areas of CRZ-IV shall be demarcated and clearly demarcated if the water body is sea, lagoon, backwater, creek, bay, and estuary and for such classification of the water bodies the terminology used by Naval Hydrographic Office shall be relied upon.
(v) The fishing Zones in the water bodies and the fish breeding areas shall be clearly marked.
(vi) In CVCAs, the land use maps shall be superimposed on the Coastal Zone Management Plan clearly demarcating the CRZ-I, II, III, IV.
(vii) The existing authorized developments on the seaward side shall be clearly demarcated.
(viii) The features like cyclone shelters, rain shelters, helipads and other infrastructure including road network may be clearly indicated on the CZM Maps for the purpose of rescue and relief operations during cyclones, storms, tsunami and the like.

(ix) Construction of buildings or other activities shall be permitted under the CZMPs provided adequate arrangements are made for proper management and disposal of solid and liquid wastes in accordance with the environmental standards, rules and statutes etc. Under no circumstances, untreated effluents shall be disposed off in the coastal waters.

6. Public consultations on the CZMP.

   (i) The draft CZMPs prepared shall be given wide publicity and suggestions and objections received in accordance with the Environment (Protection) Act, 1986. Public hearing on the draft CZMPs shall be held at district level by the concerned CZMAs.
   (ii) Based on the suggestions and objections received the CZMPs shall be revised and approval of MoEF&CC shall be obtained.
   (iii) The approved CZMP shall be put up on the website of MoEF&CC, concerned website of the State, Union Territory CZMA and hard copy made available in the Panchayat Office, District Collector Office and the like.

7. Revision of Coastal Zone Management Plans

   (i) Whenever there is a doubt the concerned State or Union territory Coastal Zone Management Authority shall refer the matter to the National Centre for Sustainable Coastal Management who shall verify the CZMP based on latest satellite imagery and ground truthing.
   (ii) If required the rectified map would be submitted to MoEF&CC for consideration.

Annexure-V

Project Information Details

1. PROJECT DETAILS

   A. Project Name
   B. Survey No./Village/Co-ordinates
   C. District
   D. State
   E. Whether the proposal is for (Select relevant field)
      (i) Fresh Clearance under CRZ
      (ii) Amendment to an already issued CRZ clearance
      (iii) Extension of validity of an already issued CRZ clearance
   F. Name of the Applicant
   G. Address of the Applicant
   H. Contact details (Telephone nos. and e-mail address)
   I. Cost of the project (Rs in crores)

2. BENEFITS OF THE PROJECT

A. Details of Project Benefits
B. Employment Likely to be Generated (Yes/No)
   (i) Total Manpower Requirement
   (ii) Temporary Employment (Numbers)
   (iii) Temporary Employment—During Construction (Numbers)
   (iv) Temporary Employment—During Operation (Numbers)

3. DESCRIPTION OF THE PROJECT UNDER CONSIDERATION (Select the Category of the Project):

A. Resort/Buildings/civic amenities
   (i) Total area/Built-up area (in sqm.)
   (ii) Height of structure
   (iii) FSI ratio
   (iv) Name of concerned town planning authority/Panchayat etc.
   (v) Details of provision of car parking area

B. Coastal Roads/Roads on Stilt
   (i) Area of land reclamation
   (ii) Estimated quantity of muck/earth for reclamation
   (iii) Traffic carrying capacity
   (iv) Dimensions of road

C. Pipelines from thermal power blow down
   (i) Length of pipeline
   (ii) Length traversing CRZ area
   (iii) Depth of excavation
   (iv) Width of excavation
   (v) Length of pipeline from seashore to deep sea
   (vi) Depth of outfall point from surface of sea water
   (vii) Temperature of effluent above ambient at disposal point

D. Marine Disposal of Treated Effluent through pipelines
   (i) Location of intake/outfall
   (ii) Depth of outfall point
   (iii) Length of pipeline
   (iv) Length traversing CRZ area
   (v) Depth of excavation
   (vi) Width of excavation
   (vii) Length of pipeline from shore to deep sea/creek (viii) Depth of outfall point from surface of water
   (viii) Depth of water at disposal point
   (ix) BOD, COD, TSS, oil & grease, heavy metals in the effluent

E. Facility for storage of goods/chemicals

(i) Name of chemical
(ii) End use of the chemical
(iii) No. of tanks for storage
(iv) Capacity of tanks

F. Offshore structures

(i) Exploration or development
(ii) Depth of sea bed
(iii) No. of rigs
(iv) No. of platform
(v) Details of group gathering stations

G. Desalination Plant

(i) Capacity of desalination
(ii) Total brine generation
(iii) Temperature of effluent above ambient at disposal point
(iv) Ambient salinity
(v) Disposal point

H. Mining of rare earth/atomic minerals

(i) Capacity of mining
(ii) Type of mineral to be extracted
(iii) End use of the mineral

I. Sewage Treatment Plants

(i) Capacity
(ii) Total area of construction
(iii) Compliance of effluent parameters as laid down by cpcb/spcb/other authorised agency
(iv) Whether discharge is in sea water/creek? If yes

- Distance of marine outfall point from shore/from the tidal river bank
- Depth of outfall point from sea water/river water surface
- Depth of seabed/riverbed at outfall point

J. Lighthouse

(i) Total ground area of foundation/platform
(ii) Height of the structure

K. Wind Mills

(i) Capacity (MW)
(ii) Height of the windmill
(iii) Diameter of the windmill
(iv) Length of blade
(v) Speed of rotation

(vi) Transmission lines (overhead or underground)

L. Others

(i) Please specify with salient features
(ii) Upload relevant Documents (upload PDF only)

4. PROJECT LOCATION AS PER CRZ CLASSIFICATION (If project site falls in different/multiple CRZ categories the same may also be elaborated)

5. CLAUSE OF CRZ NOTIFICATION UNDER WHICH PROJECT IS A PERMISSIBLE/REGULATED ACTIVITY

6. MANDATORY FIELDS FOR PROJECT ASSESSMENT

A. CRZ map in 1:4000 scale indicating HTL, LTL demarcation and distance of the nearest project boundary (in meters) from HTL to be stated
   (i) Upload Map (kml file)
B. Project layout superimposed on CRZ Map 1:4000 scale with classification of project location including other notified ESAs prepared
   (i) Upload Map (kml file)
C. CRZ map 1:25,000 scale covering 7 km radius around Project site
   (i) Upload Map (kml file)

7. PROJECT LOCATED IN (Select Type)

(i) Non eroding Coast
(ii) Low and Medium eroding coast
(iii) High eroding Coast

8. DETAILS OF FOREST/MANGROVES LAND INVOLVED (YES/NO) IF YES

(i) Detail of area diverted
(ii) Forest clearance to be submitted (Upload document)
(iii) No. of trees to be cut under the project
(iv) Compensatory afforestation plan to be submitted (Upload document)

9. DISTANCE OF PROPOSED PROJECT FROM ESA/MARINE PARK/WILD LIFE SANCTUARY

(i) Within 10 km radius from the project site (Yes/No) If

- Permission from NBWL to be submitted (Upload document)

YES

10. NOC OR CONSENT TO ESTABLISH FROM STATE/UT POLLUTION CONTROL BOARDS OBTAINED (YES/NO)
    If YES

(i) Copy of NOC to be provided (Upload document)
(ii) Conditions imposed to be stated (Upload document)

11. EIA studies (relevant fields to be filled)

A. Terrestrial studies:
(i) Summary Details of EIA (Terrestrial) Studies
(ii) Upload Recommendation made in EIAs (Upload document)
(iii) State period of Study

B. Marine Studies
(i) Summary Details of EIA (Marine) Studies
(ii) Upload Recommendation made in EIAs (Upload document)
(iii) State period of Study

12. DISASTER MANAGEMENT PLAN/NATIONAL OIL SPILL DISASTER CONTINGENCY PLAN (if applicable)

13. PROJECT INVOLVING DISCHARGE OF LIQUID EFFLUENTS:

(i) Capacity of STP
(ii) Quantity of effluent generated
(iii) Quantity of effluent treated
(iv) Method of treatment and disposal

14. PROJECT INVOLVING DISCHARGE OF SOLID WASTE:

(i) Type of solid waste
(ii) Quantity of solid waste generated
(iii) Method of disposal
(iv) Mode of transport

15. WATER REQUIREMENT (KLD)

(i) Quantity of water required
(ii) Source of water
(iii) If Ground water (Upload a copy of approval from CGWA or authorised body)
(iv) If other Source (Upload a copy of permission from competent authority)
(v) Mode of transport
(vi) Commitment of water supply (Upload document)

16. DETAILS OF WATER TREATMENT AND RECYCLING (If any) (Multiple Entries Allowed)

Type/Source	Quantity of Waste Water Generated (Kilos Litre per Day)	Treatment Capacity (Kilos Litre per Day)	Treatment Method	Mode of Disposal	Quantity of Discharged Water (Kilos Litre per Day)	Quantity of Treatment Water used in Recycling/Reuse (Kilo Litre per Day)

17. DETAILS OF RAINWATER HARVESTING
    - (i) No. of Storage tanks
    - (ii) Total capacity of tanks
    - (iii) No. of Recharge Pits
    - (iv) Capacity of pits

18. ENERGY REQUIREMENT AND SOURCES
    - (i) Total Power Requirements (kW h)
    - (ii) Upload Copy of Agreement (upload pdf only)
    - (iii) Stand By Arrangement (Details)

19. ENERGY EFFICIENCY/SAVING MEASURES
    - (i) Source/Mode
    - (ii) Details of savings

20. RECOMMENDATION OF STATE COASTAL ZONE MANAGEMENT AUTHORITY
    - (i) Upload Copy of CZMA recommendations (Upload pdf only)
    - (ii) Compliance status of the Conditions Imposed

21. WHETHER PROPOSAL ATTRACTS EIA NOTIFICATION, 2006. (Yes/No) If YES,
    - (i) the category thereof
    - (ii) Status of proposal for EC (as applicable)

22. SOCIAL AND ENVIRONMENTAL ISSUES AND MITIGATIONS MEASURES SUGGESTED INCLUDING BUT NOT LIMITED TO R&R, WATER, AIR, HAZARDOUS WASTES, ECOLOGICAL ASPECTS, ETC. (Brief Details to be Provided)

23. DETAILS OF COURT CASES Whether there is any Court Cases pending against the project and/or land in which the project is proposed to be set up? (Yes/No)

    If Yes,

    Pending or Disposed (Select relevant)

Fig. 5.53 Tourism units on the Mandarmani Beach violated the CRZ norms

(i) Name of the Court (Supreme Court, High Court, NGT)
(ii) Case Details
(iii) Orders/Directions of the court, if any and its relevance with the proposed project (Upload document)

CRZ has been formulated to protect the ecologically fragile coastal ecosystems, but in many cases the regulation has been violated which caused a threat to the biotic and abiotic resources in the coastal region. A very prominent example in this context is the hotels/tourism units constructed in Mandarmani region of coastal West Bengal. These units have been constructed violating the CRZ norms (Fig. 5.53), but there is also a light of hope as many of these tourism units like Arya Beach Resort, Golden Retreat, Suhana Resort have been demolished. Legal actions have been taken on many more and notices have already been served to shut down the establishments.

It is also a matter of concern that there is no restriction in regulating the tourist pressure in many of the beaches of the lower Gangetic delta (Fig. 5.54). This may increase the short term gain in the beach tourism sector, but excessive foot falls can negatively impact the biotic community of the beaches as many of the tourists are not aware of the fragile biotic community of the coasts. To make this sector sustainable eco-tourism needs to be implemented involving the local people who are much aware of the beach flora and fauna along with their habitats. Sensitive part of the beach with high population density/diversity of biotic community may be converted to beach hot spot by providing extra protection through formation of Beach Protection Committee (BPC).

Fig. 5.54 Tourist pressure in the beaches of the lower Gangetic delta

5.2.2 Institutional Support to Upgrade the Livelihood of Sundarban People

The upgradation and sustainability of the livelihood of 4.5 million populations of Sundarbans are direct functions of the reservoir of endemic flora and fauna. To conserve the rich genetic resources of the deltaic complex four important pillars need to be strengthened as listed here.

1. Vulnerability reduction—Vulnerability of island dwellers using measures that protect residents from expected future natural hazards (preferably cyclones of varied magnitude in this region that trigger tidal surges, mighty wave actions etc.) to reduce loss of lives and assets. Effective cyclone prediction, rapid evacuation from the disaster zone and building of sufficient disaster shelter and health care units coupled with development of natural shield (like mangroves and associated halophytes) can reduce the degree of vulnerability.
2. Poverty reduction—This is accomplished by consolidating sustainable livelihood clusters and job training programs, and by providing improved access to health and education services. Alternative livelihood based on management of endemic resources need to be implemented through capacity building.
3. Biodiversity conservation—Conservation of the mangrove forest resources involving the participation of the local population is a roadmap to socio-economic development and
4. Institutional change—These measures complement all initiatives by supporting their effective implementation.

Inter-relationships exist among the four pillars. For example, development of livelihood clusters and training programs in the buffer zone of Sundarbans will offer new employment opportunities to those currently dependent on illegal or dangerous forest-dependent livelihoods, and the subsequent reduction of pressure on the forest ecosystem will help conserve biodiversity. Another example of inter-relationships can be visualized in context to realigning embankments to ensure their future ability to protect the populations residing behind them. This relocation of embankments will provide new areas for mangrove restoration and growth, thereby enhancing forest conservation efforts and providing bio-shields to decrease adverse effects of cyclonic storms. Taken together, these pillars provide a strategy for promoting growth in the stable zone, while building the resilience of populations in the transition zone. By improving conditions in the transition zone, residents will be empowered in the long term to integrate themselves into economic growth nodes in or beyond the stable zone. The intended long-term result will be a net decrease in the poverty level of the population at risk in the transition zone, with a concomitant reduction in pressure on the Sundarbans Reserve Forest/core zone and an increase in forest coverage. To ensure this occurs effectively, changes in institutional arrangements will be required, and those changes are a critical element in the overall strategy. We carried out a total in-depth survey on the sustainability of different verticals of blue economy considering the relevant institutions and conclude that with a strong institutional support the fishery, mangrove based tourism and biodiversity sectors are highly sustainable in this region of lower Gangetic delta. Shrimp culture can only be sustainable if the environmental issues are nullified with the implementation of bio-treatment ponds (Table 5.21), the technology of which is already standardized.

5.3 Take Home Messages

[A] Blue economy stands on the pillars of wide variety of biotic and abiotic resources of the oceans, seas, bays and estuaries, which could determine the course of development and sustainability of this vertical. The spirit of blue economy which aims at optimum and sustainable use of aquatic resources for growth and development is extremely crucial for the coastal nations especially for the small island developing states. Indian Sundarbans in the lower Gangetic delta region is a cluster of some 105 islands, which is enriched with genetic diversities scattered in the mangrove forest, underlying saline soil, estuarine water and intertidal mudflats. Apart from the biotic resources, the tidal range, wind velocity, solar radiation, extensive beaches are important abiotic resources of the region that can provide momentum to the regional blue economy. However, several natural and anthropogenic threats often add sticks to the spokes of blue economy causing retardation of the progress in the sectors of fisheries and aquaculture, shipping and transportation of goods, estuary/sea based recreation and tourism, marine biodiversity conservation and non-conventional energy sector etc.

[B] We carried out a SWOT analysis during 2021 in three sectors of Indian Sundarbans (viz. western, central and eastern sectors), which are markedly different in terms of salinity and geomorphological features that impact the biotic resources of the region. To obtain data for SWOT analysis, stratified random sampling technique

Table 5.21 Major actors (sector-wise) for sustainable blue economy in Indian Sundarbans

Sector	**Major Actors**	**Magnitude of sustainability**	**Remark**
Fishery	• State Fishery Department • State Forest Department • Local NGOs • Local Fisherman Association • Academic and Research Institutes like Central Institute of Brackishwater Aquaculture (CIBA), Central Institute of Fisheries Education (CIFE) etc.		Abundance of fish resources; strong institutional back up, but during fish catch conservation related laws are hardly followed
Aquaculture	• State Fishery Department • State Forest Department • Local NGOs • CRZ Authority (State Level Committee) • Industries related to export of shrimps, crabs etc.		Unorganized sector; institutional support required in terms of subsidy; bio-treatment plant needs to be installed using bivalves (preferably oysters), seaweeds and marsh grass (*Porteresia coarctata*) that can absorb nutrient load and organic substances
Fish drying	• State Fishery Department • State Forest Department • Local NGOs • Local Fisherman Association		Unorganized sector; institutional support required; strong supply chain needs to be established with poultry and fish feed manufacturing units
Mangrove based tourism	• State Tourism Department • State Fishery Department • State Forest Department • Local NGOs • Universities & Academic institutions • State Environment Department • State Pollution Control Board • Hotel owners Association		Partially organized sector; Awareness on eco-tourism needs to be generated; Strict laws and policies need to be implemented to prohibit tourism related pollutants like plastics, leftover food items, drink cans etc. from throwing in the estuaries or mudflats; Combined Effluent Treatment Plants (CETP) need to be established integrating all the local hotels (preferably in the western Indian Sundarbans)

(continued)

Table 5.21 (continued)

Shipping & Marine transport	• Kolkata Port Trust • Coast Guard • Ministry of Water ways • State Fishery Department • State Environment Department • State Pollution Control Board		Draft to be maintained through dredging at frequent intervals; Ballast discharge problems are acute and hardly any high technology based institutional support exists
Ship breaking	• Kolkata Port Trust • Coast Guard • Ministry of Water ways • State Fishery Department • State Environment Department • State Pollution Control Board		Ecofriendly technology needs to be adopted, which is lacking at present; occupational health issues need to be addressed
Biodiversity & Carbon sequestration	• State Tourism Department • State Fishery Department • State Forest Department • State Biodiversity Board • Local NGOs • Universities & Academic institutions		Awareness on every tiers of the society needs to be generated; carbon credit needs to implemented on the basis of carbon sequestration data bank of endemic halophytes keeping local people and island dwellers in the loop; Local People Participation (LPP) model is an effective step to initiate and propagate alternative livelihoods and lessen the illegal intrusion in to the core zone

Index:

Highly sustainable	Moderately sustainable	Doubtful in present state	Highly doubtful in terms of technology

was employed and data was collected from 988 respondents comprising of 375 fisherman, 264 vessel owners, 170 hotel/tourism unit owners and 179 researchers using questionnaires. The results of the analysis are highlighted in Tables 5.1, 5.2, 5.3 and 5.4.

[C] The embankments are comparable with the life support system to the people dwelling in the lower Gangetic region preferably in Sundarbans. We observed from a series of field study that the nature of embankment failure in the Indian part of Sundarban is very complicated and is a function of physical, chemical, geological and biological factors coupled with human intervention in the entire failure loop. The frequent cyclonic depressions in the lower Gangetic delta region damage the embankments. We designed a biogeoengineering treatment to protect the western part of Yuba Abash in Bali Island (specifically Bali II) located in the central Indian

Sundarbans (22° 05′ 44.1″ N and 88° 46′ 38.1″ E) through three treatments, the results of which are still under investigations.

[D] We carried out a total in-depth survey on the sustainability of different verticals of blue economy considering the relevant institutions and conclude that with a strong institutional support the fishery, mangrove based tourism and biodiversity sectors are highly sustainable in the lower Gangetic delta. Shrimp culture can only be sustainable if the environmental issues are nullified with the implementation of bio-treatment ponds, the technology of which is already standardized. Mangrove-centric carbon neutral livelihoods coupled with unique carbon sequestration potential of mangroves, if properly focused with strong supply chain management can provide momentum to sustainability of blue economy in Indian Sundarbans.

References

Agarwal S, Zaman S, Biswas S, Pal N, Pramanick P, Mitra A (2016) Spatial variation of mangrove seedling carbon with respect to salinity: a case study with *Bruguiera gymnorrhiza* seedling. Int J Adv Res Biol Sci 3(8):7–12

Agarwal S, Islam S, Zaman S, Mitra A (2017) Carbon stocks in the mangrove vegetation of Lothian Wild Life Sanctuary of Indian Sundarbans: implication for climate change mitigation. Global J Eng Sci Res 4(10):1–9

Ahmed S, Zaman S, Mitra A (2021) Carbon sequestration in a common true mangrove flora, Excoecaria agallocha of Indian Sundarbans. In: Mitra A, Calma MM, Chakrabarty SP (eds) Natural resources and their ecosystem services, vol II, Chapter 7. HSRA Publication, pp. 138–143. ISBN 978-93-5506-064-8

Banerjee K, Vyas P, Chowdhury R, Mallik A, Mitra A (2010) The affects of salinity on the mangrove growth in the lower Gangetic delta. J Indian Ocean Stud 18(3):389–397

Banerjee K, Sengupta K, Raha AK, Mitra A (2013) Salinity based allometric equations for biomass estimation of Sundarban mangroves. Biomass Bioenergy (elsevier) 56:382–391

Banerjee K, Amin G, Fazli P, Pramanick P, Zaman S, Mitra A (2016) *Excoecaria agallocha*: a potential Mangrove species in context to carbon storage in high saline zone. J Environ Sci Comp Sci Eng Tech 5(2):115–120

Banerjee K, Gatti RC, Mitra A (2017) Climate change-induced salinity variation impacts on a stenoecious mangrove species in the Indian Sundarbans. Ambio (springer) 46:492–499

Bhattacharjee AK, Zaman S, Bhattacharyya SB, Pramanick P, Raha AK, Mitra A (2013a) How mangroves respond to hypersaline condition? Preparedness for predicted sea level rise. Int J Sci Res 2(6):360–364

Bhattacharjee AK, Zaman S, Raha AK, Mitra A (2013b) Impact of salinity on above ground biomass and stored carbon in a common mangrove *Excoecaria agallocha* of Indian Sundarbans. Am J Biopharmacol Biochem Life Sci 2(2):1–11

Bhattacharyya S, Mitra A, Raha AK (2015) Stored carbon in above ground biomass of dominant mangrove floral species in Sagar Island of Indian Sundarbans. J Chem Biol Phys Sci 5(4):4664–4672

Biswas S, Pal N, Zaman S, Mitra A (2019) Samosa and kachuri from mangrove associate species: an innovative utilization of coastal flora. Int J Res Appl Sci Eng Technol 7(1):73–79

Chakraborty A, Datta U, Zaman S, Pal N, Pramanick P, Mitra A, Mitra A (2016) Spatio-temporal variation of stored carbon in *Sonneratia apetala*: a case study from Indian Sundarbans. Int J Innov Res Sci Eng Tech 5(3):2975–2980

Ghosh R, Mitra A (2015) Effect of salinity on nutritional value of saltmarsh grass (*Porteresia coarctata*) from Gangetic delta, northeast coast of India. Indian J Geo-Mar Sci 44(7):1043–1052
Ghosh R, Trivedi S, Pramanick P, Zaman S, Mitra A (2015) Seagrass: a store house of Carbon. J Energy Environ Carbon Credit 5(2):23–29
Guha A, Zaman S, Mitra A, Rudra T, Mitra A (2015) Variation of stored carbon in dominant seaweed species of Indian Sundarbans. J Chem Biol Phys Sci 5(1):979–989
Howell J (2008) Study of road embankment erosion and protection. https://assets.publishing.service.gov.uk/media/57a08bb7ed915d3cfd000ebc/Seacap19-TP6.pdf
Islam MS, Shahriar BAM, Shahin HM (2013) Study on growth of vetiver grass in tropical region for slope protection. Int J Geomate 5(2):729–734
Jana HK, Zaman S, Chakraborty S, Pramanick P, Mondal KC, Mitra A (2013) Spatio-temporal variation of stored carbon in *Porteresia coarctata* along the east and west coasts of India. Int J Eng Manag Sci 4(3):377–381
Khan I, Sakib MN (2012) Determination of absorption capacity of geojute. Undergraduate thesis, Department of Civil Engineering, Bangladesh University of Engineering and Technology, Dhaka, Bangladesh
Mitra A (2013) Sensitivity of mangrove ecosystem to changing climate. Springer, New Delhi, Heidelberg, New York, Dordrecht, London. ISBN-10: 8132215087; ISBN-13: 978-8132215080. ISBN 978-81-322-1509-7 (eBook)
Mitra A (2018) Salinity: a primary growth driver of mangrove forest. Sustain for 1:1–9
Mitra A (2019) Estuarine pollution in the lower Gangetic Delta, vol XVI. Springer International Publishing, p 371. ISBN 978-3-319-93305-4
Mitra A (2020) Mangrove forests in India exploring ecosystem services, vol XV. Springer, Chem, p 361. eBook ISBN 978-3-030-20595-9. https://doi.org/10.1007/978-3-030-20595-9
Mitra A, Gati RC (2015) Carbon census in the mangrove ecosystem of lower Gangetic delta. Economol J V:11–27
Mitra A, Zaman S (2014) Carbon sequestration by coastal floral community. The Energy and Resources Institute (TERI) TERI Press, India. ISBN 978-81-7993-551-4
Mitra A, Zaman S (2015) Blue carbon reservoir of the blue planet, vol XII. Springer, p 299. ISBN 978-81-322-2106-7. https://doi.org/10.1007/978-81-322-2107-4
Mitra A, Zaman S (2016) Basics of marine and estuarine ecology. Springer India, pp 1–481. ISBN 978-81-322-2707-6
Mitra A, Zaman A (2020) Environmental science—a ground zero observation on the Indian Subcontinent, vol XIV. Springer, Chem, p 478. eBook ISBN 978-3-030-49131-4. https://doi.org/10.1007/978-3-030-49131-4
Mitra A, Zaman S (2021) Estuarine acidification: exploring the situation of mangrove dominated Indian Sundarbans estuaries, vol XII. Springer, Chem, p 402. https://doi.org/10.1007/978-3-030-84792-0. e-Book ISBN 978-3-030-84792-0
Mitra A, Chowdhury R, Sengupta K, Banerjee K (2010a) Impact of salinity on mangroves of Indian Sundarbans. J Coast Environ 1(1):71–82
Mitra A, Banerjee K, Raha AK (2010b) Mangroves: a unique sink of carbon in the Tropics; Publicity Division, Directorate of Forests, Govt. of West Bengal, pp 1–32
Mitra A, Banerjee K, Sengupta K (2011a) Impact of AILA, a tropical cyclone on salinity, pH and dissolved oxygen of an aquatic sub-system of Indian Sundarbans. Natl Acad Sci Lett Ind 81(Part II):198–205
Mitra A, Halder P, Banerjee K (2011b) Changes of selected hydrological parameters in Hooghly estuary in response to a severe tropical cyclone (Aila). Indian J Mar Sci 40(1):32–36
Mitra A, Mondol K, Banerjee K (2011c) Spatial and tidal variations of physico-chemical parameters in the lower Gangetic delta region, West Bengal, India. J Spat Hydrol 11(1):62–69
Mitra A, Sengupta K, Banerjee K (2011d) Standing biomass and carbon storage of above-ground structures in dominant mangrove trees in the Sundarbans. For Ecol Manag (Elsevier) 261(7):1325–1335. https://doi.org/10.1016/j.foreco.2011d.01.012

Mitra A, Sengupta K, Banerjee K (2012) Spatial and temporal trends in biomass and carbon sequestration potential of *Sonneratia apetala* Buch-Ham in Indian Sundarbans. Proc Natl Acad Sci India Sect B Biol Sci 82(2):317–323

Mitra A, Zaman S, Pramanick P, Bhattacharyya SB, Raha AK (2014) Stored carbon in dominant seaweeds of Indian Sundarbans. Pertanika J Trop Agric Sci 37(2):263–274

Mitra A, Rudra T, Guha A, Ray A, Pramanick P, Pal N, Zaman S (2016) Ecosystem service of Avicennia alba in terms of carbon sequestration. J Environ Sci Comp Sci Eng Tech 5(1):155–160

Pal N, Zaman S, Pramanick P, Mitra A (2017) Impact of aquatic salinity on mangrove seedlings: a case study on *Heritiera fomes* (Common name: Sundari). Biomed J Sci Tech Res 1(4):1–5. https://doi.org/10.26717/BJSTR.2017.01.000348

Pal N, Zaman S, Mitra A (2020) Carbon scrubbing property of *Rhizophora apiculata* inhabiting Indian Sundarbans. In: Mitra A, Calma MM, Chakrabarty SP (eds) Proceedings: natural resources and their ecosystem services, pp 50–54. HSRA Publication. ISBN 978-81-947216-7-3

Pramanick P, Zaman S, Bera D, Raha AK, Mitra A (2014) Mangrove fruit products: a search for alternative livelihood for Island dwellers of Gangetic delta. Int J Pharma Res Scholars 3(1):131–137

Pramanick P, Zaman S, Rudra T, Guha A, Mitra A (2016) Comparative account of heavy metal level in two species of Chlorophyceae inhabiting Indian Sundarbans. Int J Res Chem Environ 6(2):65–68

Pramanick P, Mitra A, Zaman S, Mitra A (2020) Regulatory service of saltmarsh grass in context to carbon storage. In: Mitra A, Calma MM, Chakrabarty SP (eds) Proceedings: natural resources and their ecosystem services. HSRA Publication, pp 42–49. ISBN 978-81-947216-7-3

Raha AK, Bhattacharyya SB, Zaman S, Banerjee K, Sengupta K, Sinha S, Sett S, Chakraborty S, Datta S, Dasgupta S, Roy Chowdhury M, Ghosh R, Mondal K, Pramanick P, Mitra A (2013) Carbon census in dominant mangroves of Indian Sundarbans. The J Energy Environ Sci 127:345–354

Roy Chowdhury M, Zaman S, Jha CS, Sengupta K, Mitra A (2014) Mangrove biomass and stored carbon in relation to soil properties: a case study from Indian Sundarbans. Int J Pharma Res Schol 3(I-2):58–69

Trivedi S, Zaman S, Ray Chaudhuri T, Pramanick P, Fazli P, Amin G, Mitra A (2016) Inter-annual variation of salinity in Indian Sundarbans. Indian J Mar Sci 45(3):410–415

Zaman S, Bhattacharyya SB, Pramanick P, Raha AK, Chakraborty S, Mitra A (2013) Rising water salinity: a threat to mangroves of Indian Sundarbans. In: Anwarul Abedin Md, Habiba U, Shaw R (eds) Water insecurity: a social dilemma. Emerald Group Publishing Limited. ISBN: 9781781908822

The manufacturer's authorised representative in the EU is Springer Nature Customer Service Centre GmbH, Europaplatz 3, 69115 Heidelberg, Germany. If you have any concerns regarding our products, please contact ProductSafety@springernature.com

Printed and bound by CPI Group (UK) Ltd, Croydon, CR0 4YY
28/09/2026
02224078-0007